Springer Theses

Recognizing Outstanding Ph.D. Research

For further volumes:
http://www.springer.com/series/8790

Aims and Scope

The series "Springer Theses" brings together a selection of the very best Ph.D. theses from around the world and across the physical sciences. Nominated and endorsed by two recognized specialists, each published volume has been selected for its scientific excellence and the high impact of its contents for the pertinent field of research. For greater accessibility to non-specialists, the published versions include an extended introduction, as well as a foreword by the student's supervisor explaining the special relevance of the work for the field. As a whole, the series will provide a valuable resource both for newcomers to the research fields described, and for other scientists seeking detailed background information on special questions. Finally, it provides an accredited documentation of the valuable contributions made by today's younger generation of scientists.

Theses are accepted into the series by invited nomination only and must fulfill all of the following criteria

- They must be written in good English.
- The topic should fall within the confines of Chemistry, Physics, Earth Sciences, Engineering and related interdisciplinary fields such as Materials, Nanoscience, Chemical Engineering, Complex Systems and Biophysics.
- The work reported in the thesis must represent a significant scientific advance.
- If the thesis includes previously published material, permission to reproduce this must be gained from the respective copyright holder.
- They must have been examined and passed during the 12 months prior to nomination.
- Each thesis should include a foreword by the supervisor outlining the significance of its content.
- The theses should have a clearly defined structure including an introduction accessible to scientists not expert in that particular field.

Lena Josefine Daumann

Spectroscopic and Mechanistic Studies of Dinuclear Metallohydrolases and Their Biomimetic Complexes

Doctoral Thesis accepted by
University of Queensland, Australia

Author
Dr. Lena Josefine Daumann
School of Chemistry and Molecular Biosciences
University of Queensland
Brisbane, QLD
Australia

Supervisors
Prof. Lawrence Gahan
Assoc. Prof. Gerhard Schenk
School of Chemistry and Molecular Biosciences
University of Queensland
Brisbane, QLD
Australia

ISSN 2190-5053 ISSN 2190-5061 (electronic)
ISBN 978-3-319-06628-8 ISBN 978-3-319-06629-5 (eBook)
DOI 10.1007/978-3-319-06629-5
Springer Cham Heidelberg New York Dordrecht London

Library of Congress Control Number: 2014939390

Printed on acid-free paper

Springer is part of Springer Science+Business Media (www.springer.com)

Parts of this thesis have been published in the following journal articles:

1. L.J. Daumann, G. Schenk, D.L. Ollis, L.R. Gahan, "*Spectroscopic and mechanistic studies of dinuclear metallohydrolases and their biomimetic complexes*" *Dalton Trans.* **2014,** 43, 910–928.
2. L.J. Daumann, J.A. Larrabee, D. Ollis, G. Schenk, L.R. Gahan, *"Immobilization of the enzyme GpdQ on magnetite nanoparticles for organophosphate pesticide bioremediation"* J Inorg Biochem. **2014**, 131, 1–7.
3. L.J. Daumann, L. Marty, G. Schenk, L.R. Gahan, "*Asymmetric Zinc(II) Complexes as Functional and Structural Models for Phosphoesterases*" Dalton Trans. **2013**, 42, 9574–9584.
4. L.J. Daumann, J.A. Larrabee, P. Comba, G. Schenk, L.R. Gahan, "*Dinuclear Cobalt(II) Complexes as Metallo-β-lactamase Mimics*" Eur. J. Inorg. Chem. **2013**, 17, 3082–3089.
5. L.J. Daumann, P. Comba, J. Larrabee, R. Stranger, G. Cavigliasso, G. Schenk, L.R. Gahan "*Synthesis, Magnetic Properties and Phosphoesterase Activity of Binuclear Cobalt(II) Complexes*" Inorg. Chem. **2013**, 52 (4), 2029–2043.
6. L.J. Daumann, B.Y. McCarthy, K.S. Hadler, T.P. Murray, L.R. Gahan, J.A. Larrabee, D.L. Ollis, G. Schenk "*Promiscuity Comes at a Price: Catalytic Versatility vs. Efficiency in Different Metal Ion Derivatives of the Potential Bioremediator GpdQ*" BBA - Proteins and Proteomics **2013**, 1834, 1, 425–432.
7. L.J. Daumann, L.R. Gahan, P. Comba, G. Schenk "*Cadmium(II) Complexes; Mimics of Organophosphate Pesticide Degrading Enzymes and Metallo-β-lactamases*" Inorg. Chem. **2012**, 51, 7669–7681.
8. L.J. Daumann, K.E. Dalle, G. Schenk, R.P. McGeary, P.V. Bernhardt, D.L. Ollis, L.R. Gahan "*The Role of Zn-OR and Zn-OH Nucleophiles and the Influence of p-Substituents in the Reactions of Binuclear Phosphatase Mimetics*" Dalton Trans. **2012**, 41, 1695–1708.

Supervisors' Foreword

This thesis describes the structural and functional studies of the enzyme Glycerophosphodiesterase (GpdQ) from *Enterobacter aerogenes* and examines the properties of small mimics of this enzyme and related binuclear metallohydrolases such as the metallo-ß-lactamases. This thesis has a broad scope in its investigations, the topics covered include structural, biochemical, kinetic and mechanistic studies, chemical synthesis, as well as the use of polymeric and nanoparticle supports. Its main outcome is an enhanced understanding of the active site structural features of the enzyme GpdQ, in particular the contribution of metal ions to catalysis. The results obtained are also applicable to other metalloenzymes. A significant breakthrough has been the development of a GpdQ mutant with improved ability to hydrolyze organophosphate pesticides. Importantly, GpdQ and corresponding biomimetics were successfully immobilized on solid supports for potential applications in the area of bioremediation of organophosphate pesticides. Both the enzyme and biomimetics can be stored on the solid support without loss of activity for long periods of time. Furthermore, a number of Zn(II), Cd(II), and Co(II) complexes were spectroscopically and mechanistically characterized, some of them among the most active biomimetics toward organophosphates reported to date. This thesis presents a comprehensive account and the results obtained have enhanced our understanding of hydrolytic cleavage of important substrates like phosphoesters and β-lactams and the results have opened up opportunities for significant further developments in bioinorganic chemistry.

Brisbane, March 2014

Prof. Lawrence Gahan
Assoc. Prof. Gerhard Schenk

Acknowledgments

First and foremost I would like to thank my advisor, Lawrence Gahan for his dedication and support during my Ph.D. Lawrie, thank you for making me a better scientist, I owe my deepest gratitude to you. I hope that one day I'll be a great mentor just as you are.

Many thanks also to my associate advisor Gary Schenk. Thank you Gary for being always so encouraging and for being enthusiastic to talk about my data (and the world) over coffee.

I am greatly indebted to the past and present members of the research group who have assisted me in countless ways. Thanks to Dr. Kieran Hadler, Dr. Fernanda Ely, Marcelo Pedroso, and Bianca McCarthy for teaching me how to do everything in the protein lab and for being fantastic companions both at work and beyond. Obrigados gathinas! Many thanks to Laurène Marty, Harriet McAtee, and Tim Zerk for being great and enthusiastic summer students and to Kristian Dalle who developed the synthesis of CH_3**HL2** and its Zn(II) complex and solved the crystal structure of this respective complex.

Thanks to Prof. Paul Bernhardt for X-ray crystal data collection and for teaching me all about structure refinement. You became something like a third supervisor during these years.

Merci vilmal Prof. Peter Comba for always welcoming me back into your lab in Heidelberg and letting me be part of your group again. Also, Danke to the Heidelberg crew for their help: Stefan, Michi, Mimo, Migro, Goli, and Simone.

Many thanks to Prof. James Larrabee at Middlebury College. Jim, thank you for teaching me MCD and for your hospitality during my stays. Merci also to Sunhee Choi, you are my role model! Many thanks Kalie, Dan, and Nate for making me feel welcome every time and taking me for lunch.

Many thanks to Barry Wood who operated the XPS instrument and conducted the data fitting and to Graeme Auchterloine who operated the TEM instrument.

Thanks to Prof. Rob Stranger and Dr. Germain Cavigliasso at the Research School of Chemistry, Australian National University for taking the time to do the DFT calculations with me. Thank you also for kindly providing the orbital energy plots. A big "Thank you" also to our collaborator Prof. David Ollis and his coworkers Tracy and Sylvia for training me in PCR, protein purification, and making mutants and providing the 8-3 plasmid.

Thank you Dr. Tri Le for the assistance in setting up the pulse program for the ^{113}Cd NMR and NMR advice in general.

I would also like to acknowledge Prof. Hiroshi Sakiyama from Yamagata University in Japan for his support in using the software Magsaki.

For financial support, I would like to thank the Commonwealth Government for the IPRS and UQCent scholarships and the UQ Graduate School, the Royal Australian Chemical Institute, the German Exchange Service, the SCMB Student Support Fund, the Society of Biological Inorganic Chemistry, and M.G. & R.A. Plowman for the travel grants that enabled me to visit collaborator labs and to present at conferences.

I would like to also express my appreciation to my friends; the "Australian family": Tanja, Paul, George, Fernanda, Marcelo, Dave, Liz, Bianca, Ruth, Ania, Maya, Neve, Helga, Shad, Saud, Rayan, Maram, Yola…and all the other amazing people I have met on this 3-year journey.

Russ, Roni, and Evey, thank you for being wonderful housemates during my 3 years in Chapel Hill. You made me feel at home from day one!

Thank you my ladies in Germany, Steffi, Maria, Katharina and Regina, for being good friends although thousands of kilometers away.

Finally I would like to thank my family; my parents, Uta, Claus, and Kurt and my sister Hannah for their endless support and encouragement when I needed it most. And most importantly thank you Dani, my partner, spouse, best friend, travel companion, and motivator, for coming with me to Australia and making this time here even more special and memorable. Thank you for everything!

Contents

Abbreviations

AAS	Atomic Absorption Spectroscopy
AChE	Acetylcholine esterase
Ala	Alanine
AOM	Angular Overlap Model
APTS	3-aminopropyltriethoxysilane
Asn	Asparagine
Asp	Aspartate
B	Racah Parameter
B3-LYP	Becke, three-parameter, Lee–Yang–Parr
BDNPP	Bis(2,4-dinitrophenyl) phosphate
BM	Bohr Magneton
BP	Becke, Perdew
BPNPP	Bis(4-nitrophenyl) phosphate
BS	Broken Symmetry
C	Racah Parameter
CAPS	*N*-cyclohexyl-3-aminopropanesulfonic acid
CHES	*N*-cyclohexyl-2-aminoethanesulfonic acid
CLEA	Cross Linked Enzyme Aggregates
COSY	Two-dimensional correlation spectroscopy
CV	Cyclic Voltammetry
Cys	Cysteine
D	Axial zero field splitting parameter
DFT	Density Functional Theory
DIPEA	N,N-Diisopropylethylamine
DNA	Deoxyribonucleic acid
DNPP	2,4-dinitrophenyl phosphate
DPP	Diphenyl phosphate
E	Rhombic zero field splitting parameter
E. aerogenes	*Enterobacter aerogenes*
E. coli	*Escherichia coli*
EcMetAp	*E. coli* Methionine Aminopeptidase
EDC	1-ethyl-3-(3-dimethylaminopropyl)carbodiimide
EPR	Electron Paramagnetic Resonance
ESI-MS	Electrospray Injection Mass Spectrometry

EtPNPP	Ethyl 4-nitrophenyl phosphate
eV	Electron volt
FDPP	Pentafluorophenyl diphenylphosphinate
GpdQ	Glycerophosphodiesterase
HEPES	4-(2-hydroxyethyl)-1-piperazineethanesulfonic acid
His	Histidine
HMBC	Heterobinuclear multiple bond connectivity
HPNP	2-hydroxypropyl-4-nitrophenyl phosphate
HSQC	Heterobinuclear single quantum correlation
IR	Infrared
J	Exchange coupling constant
K_{assoc}	Metal association constant
k_{cat}	First order rate constant
K_d	Metal dissociation constant
K_m	Michaelis constant
LMCT	Ligand to Metal Charge Transfer
MCD	Magnetic Circular Dichroism
mdeg	Millidegrees
MES	2-(*N*-morpholino)ethane sulfonic acid
MNP	Magnetite Nano Particles
MR	Merrifield Resin
MβL	Metallo-β-lactamase
nm	Nanometers
NMO	N-methyl morpholino N-oxide
NMR	Nuclear Magnetic Resonance
OP	Organophosphate
OpdA	Organophosphate degrading enzyme A
OPH	Organophosphate Hydrolase
ORTEP	Oak Ridge Thermal Ellipsoid Plot
P. diminuta	*Pseudomonas diminuta*
PAMAM	Poly amido amine dendrimer
PAP	Purple Acid Phosphatase
pH	*Potentia hydrogenii*
Phe	Phenylalanine
pK_a	Protonation constant
PNPP	4-nitrophenyl phosphate
ppm	Parts per million
PTE	Phosphotriesterase
PTFE	Polyterafluoroethylene
PyBOP	Benzotriazol-1-yl-oxytripyrrolidinophosphonium hexafluoro phosphate
SA	Streptavidin
Ser	Serine
SQUID	Superconducting Quantum Interference Device
T	Tesla

TEM	Transmission Electron Microscopy
Thr	Threonine
TRIS	Tris(hydroxymethyl)amino methane
Tyr	Tyrosine
V_{max}	Maximum velocity/rate
V_o	Initial velocity/rate
VTVH	Variable Temperature Variable Field
wt	Wild type
XPS	X-ray photoelectron spectroscopy
ZFS	Zero field splitting

Chapter 1
Introduction

1.1 Dinuclear Metallohydrolases

The family of dinuclear metallohydrolases comprises a large number of enzymes with very different functionalities (Table 1.1).[1] The dual action of two metals can have different advantages over catalysis by just one metal ion [1]. First, the thermodynamic driving force for redox reactions is lowered due to charge-delocalization; second, the activation barrier is also lowered for solvent and enzyme reorganization [1]. In addition substrates can be more effectively oriented and the electrostatic activation of substrates is more easily achieved [1]. Furthermore, two metal ions facilitate the formation of hydrolysis-initiating nucleophiles more readily, and lower lying energy transition states for hydrolysis reactions can be achieved [1].

The two metal binding sites in dinuclear metallohydrolases are often defined as the α- and β-site, respectively [6]. The metal ion content and oxidation states of the dinuclear metallohydrolases are as diverse as their proposed mechanisms, nucleophiles and substrates utilized. Phosphoesterases such as the **G**lycero**p**hos-pho**d**iesterase from *Enterobacter aerogenes* (GpdQ), the **o**rgano**p**hosphate **d**egrading **a**gent (OpdA) from *Agrobacterium radiobacter* and the **p**hos-pho**t**ri**e**sterase (PTE) from *Pseudomonas diminuta* have attracted interest due to their ability to degrade toxic pesticides, whereas research appeal for purple acid phosphatases (PAPs), metallo-β-lactamases (MβL) or ureases is mainly due to their role in human disorders and therapy [31]. The active site structures of dinuclear metallohydrolases exhibit a remarkable similarity, with the metal ions being typically coordinated to 4–6 amino acid residues in addition to terminal or bridging (exchangeable) water/hydroxide ligands. The amino acid residues are mainly histidine, tyrosine, glutamic acid, aspartate, serine, asparagine, and (carboxylated) lysine.

[1] Figures and Tables in this chapter have appeared in [L.J. Daumann et al., "Spectroscopic and mechanistic studies of dinuclear metallohydrolases and their biomimetic complexes" Dalton Trans. **2014**, 43, 910–928.]—Reproduced by permission of The Royal Society of Chemistry.

L. J. Daumann, *Spectroscopic and Mechanistic Studies of Dinuclear Metallohydrolases and Their Biomimetic Complexes*, Springer Theses, DOI: 10.1007/978-3-319-06629-5_1,

Table 1.1 Overview over hydrolytic metalloenzymes

Enzyme	Metal ion composition	Biological function	Research incentive
Glycerophospho-diesterase (GpdQ) [2–10]	Fe(II)-Zn(II), Co(II)-Co(II), Mn(II)-Mn(II), Cd(II)-Cd(II)	Hydrolysis of 3'-5' phosphodiester bond of glycerophosphodiesters	Potential bioremediator
Purple acid phosphatases (PAP) [11–19]	Fe(III)-Fe(II), Fe(III)-Mn(II), Fe(III)-Zn(II)	Bone metabolism (animals) and phosphate uptake (plants)	Target for anti-osteoporotic drugs
Metallo-β-lactamases [20]	Zn(II)-Zn(II)	Hydrolysis of β-lactam substrates (antibiotics)	Study potential inhibitors to fight antibiotic resistance
Rv0805 [21, 22]	Fe(III)-Mn(II)[a]	Cleavage of cAMP/cGMP phosphodiester bonds	Potential bioremediator
Phosphotriesterases (OpdA, OPH) [23–26]	Co(II)-Co(II), Mn(II)-Mn(II), Cd(II)-Cd(II), Fe(II)-Zn(II)	Organophosphate hydrolysis	Potential bioremediator
Ureases [27]	Ni(II)-Ni(II)	Hydrolysis of urea	Treatment of bacterial infections, regulation of nitrogen uptake in plants
Ser/Thr phosphatase [11, 28]	Fe(II)-Fe(II), Fe(II)-Zn(II), Mn(II)-Mn(II)	Gene expression	Target in chemotherapy
Leucine aminopeptidases [29]	Zn(II)-Zn(II)	Amino acid synthesis	Target for drugs against leukemia
Arginase [1, 30]	Mn(II)-Mn(II)	Hydrolysis of Arginine	Involved in vascular diseases

[a] proposed in vivo

1.1.1 Phosphoesterases

Phosphoesterases are ubiquitous in nature. They are involved in signal transduction in bacteria (Rv0805) [21, 22], phospholipid remodeling and synthesis (GpdQ) [2, 7, 8], regulation of phosphate levels in plants, fungi and mammals (PAPs) [28, 31–36], to name just a few. Some phosphoesterases like the organophosphate-degrading agent OpdA from *A. radiobacter* have apparently recently evolved due to the widespread use of organophosphate pesticides [25, 37]. Another well studied enzyme of this type is the PTE from *P. diminuta* and *Flavobacterium* species, which is also known under the name OPH (organophosporus hydrolase) [38, 39]. It is proposed that these enzymes provide bacteria living in soil with essential nutrients like phosphate by hydrolyzing pesticides. These enzymes have attracted attention as they are able to hydrolyze toxic organophosphate nerve agents and pesticides [2, 9, 40, 41]. The following section will give an overview of the problems arising through the increasing use of organophosphates.

1.1.1.1 Organophosphates

It is estimated that thousands of people die or suffer from severe organophosphate (OP) pesticide poisoning *per annum* [42]. In developing countries, especially, the benefits of pesticides have to stand against the risks for farmers handling and applying them. The question of how to eliminate the residues of these compounds from the environment to avoid hazards for aquatic organisms and human health has been raised [42–45]. OPs have also been used as nerve agents, for example, sarin or VX [44]. All of these compounds exert their toxicity by irreversibly inactivating the enzyme acetylcholine esterase (AChE) [45]. It is proposed that the mode of action involves the organophosphate molecule undergoing nucleophilic attack by a serine-OH moiety in the active site of AChE, subsequently blocking the site for the neurotransmitter acetylcholine [44, 45]. With the enzyme inhibited, acetylcholine builds up in the synapse and nerve impulses are continually transmitted. The resulting dysfunction of the parasympathetic nervous system leads, in many cases, to the death of the organism. OPs are differentiated into mono-, di- and triesters. Figure 1.1 shows some examples of organophosphates that have been used as pesticides and nerve agents [44–47]. Most pesticides bear a P=S moiety whereas nerve agents contain the P=O motif. Thiophosphates are less potent AChE inhibitors than the corresponding oxophosphates [44, 45]. This is due to the former being more stable and thus less likely to react with AChE. However, desulfurization in vivo leads to a high acute toxicity of the thiophosphate pesticides [48]. A typical example is the conversion of parathion to paraoxon. This reaction is quite slow causing a delay of the fatal effects therefore providing an opportunity for therapy and detoxification of the organism. Mesomeric and inductive effects of substituents affect the stability and electophilicity further; electron withdrawing groups/atoms such as fluorine enhance the reactivity of the phosphorus and thus the toxicity and AChE inhibitory effects [48].

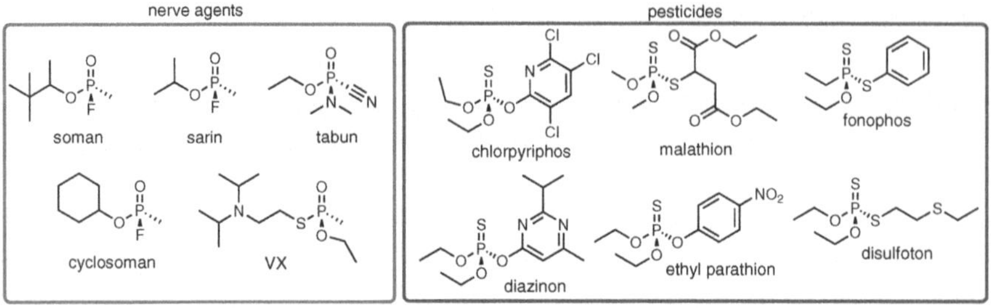

Fig. 1.1 Phosphoesters that have been used as nerve agents or pesticides

The hydrolysis of OPs to less harmful compounds is an important aspect for environmental decontamination. The phosphate ester bond is prone to cleavage; this occurs mostly by hydrolysis but photolytic degradation by sunlight is also possible [49]. Biological pathways for decontamination such as bioremediation have attracted recent attention [9]. In the context of this thesis the term 'bioremediation' is used when plants, microorganisms or enzymes derived from them are used to uptake or break down environmental contaminants e.g. heavy metals, chemicals from oil spills or pesticides [43, 50–52].

1.1.1.2 The Phosphotriesterase OpdA

OpdA is one of the most efficient and fastest enzymes known to degrade phosphate triesters, working at near diffusion limited rates towards favored substrates [25]. It has been commercially introduced to Australia by CSIRO and Orica Watercare under the name Landguard™ OP-A, as bioremediator in agriculture (treatment of sheep dip, water run-off and soil decontamination) [23, 53]. Landguard™ OP-A is capable of degrading common pesticides like chlorpyriphos or diazinon and is to date the only enzyme-based product for pesticide bioremediation in Australia. Therapeutic use of OpdA against OP poisoning has also been recently demonstrated in a number of studies [54–56]. For example, rats that had been poisoned with a lethal dose of the highly toxic pesticide dichlorvos showed 100 % survival after one single treatment with OpdA [54]. The pharmacokinetics of this enzyme have also been studied in the African green monkey to develop new therapeutic approaches for OP poisoning [55]. OpdA has a similar active site to OPH with a more buried 5-coordinate α-site in the resting state, and a more solvent exposed 6-coordinate β-site [24–26, 57, 58]. The metals are bridged by a carboxylated lysine residue in addition to a μ-hydroxide. The active site of OpdA is shown in Fig. 1.2 along with the overall structure.

Depending on the metal ion substitution in OpdA, the enzyme displays a high mechanistic flexibility for OP degradation [26]. For the Zn(II) and Cd(II) derivatives one kinetically relevant protonation equilibrium was observed in ethyl paraoxon hydrolysis at a low pK_a range of around 4–5, which is typical for a

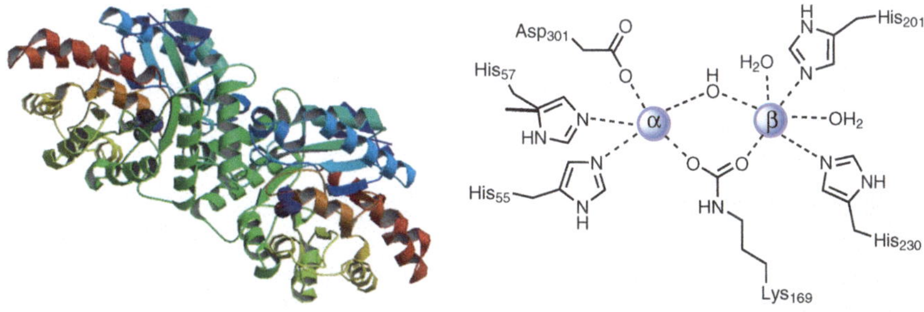

Fig. 1.2 Overall structure of OpdA (*left*) and active site (*right*). PDB number 2D2G [59, 60]

μ-hydroxide [26]. For the Co(II) derivative however, a terminal hydroxide is the proposed nucleophile [26]. This mechanistic flexibility might be important for a rapid development of OpdA towards environmental changes. The native metal ion content in vivo is currently unknown for this enzyme. Although OpdA is an extremely efficient enzyme it has two drawbacks. The pH optimum is at a very limited pH range (so the pH has to be adjusted in many cases before water treatment to use the full potential of the enzyme), and it is only capable of degrading triester substrates. This has led to the investigation of other more promiscuous enzymes like the glycerophosphodiesterase GpdQ from *E. aerogenes.* [2].

1.1.1.3 The Potential Bioremediator GpdQ

GpdQ, a phosphodiesterase, was first purified by Gerlt and co-workers and is believed to be the first phosphoesterase capable of hydrolysing stable alkyl diesters like dimethyl phosphate and diethyl phosphate [2, 8]. Its biological function is the cleavage of 3'-5'-phosphodiesterbonds in glycerol-3-phosphoethanolamine [61]. Similar to OpdA, GpdQ is a member of the diverse metallophosphoesterase family. While OpdA is predominantly a phosphotriesterase, GpdQ is a highly stable enzyme able to degrade di- and monoester substrates, making it more versatile and thus an ideal candidate for bioremediation [2, 7, 8, 62]. Since it is able to degrade a variety of OPs and products of the degradation of the OP nerve agent VX, it has promise in its application not only as bioremediator but also as defense agent against chemical weapons [2]. It also functions over a large pH range; activity is observed at pH values as high as 11 and as low as 3 [4, 5]. The overall structure is a hexamer which consist of a trimer of identical dimeric subunits. Why this enzyme exists as a hexamer is currently unknown. The active site in the resting state is mononuclear with only the 6-coordinate α-site being occupied by a metal [6]. This site is composed of Asp8, His10, Asp 50 and His197. The β-site, vacant in the resting state, contains the amino acid ligands Asp50, Asn80, His156 and His195 [3, 5, 62]. The hexamer and the active site are shown in Fig. 1.3.

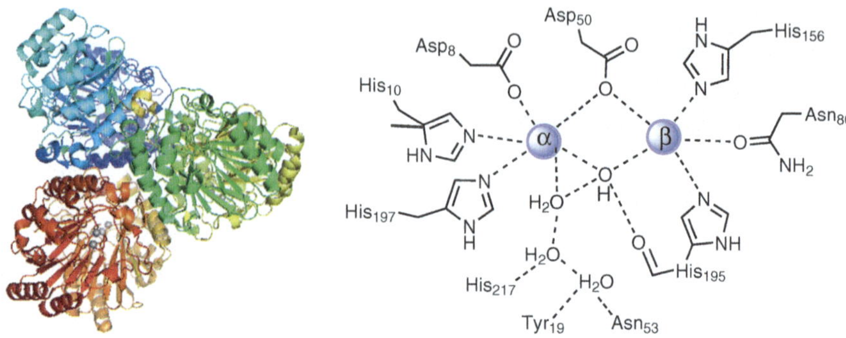

Fig. 1.3 Hexamer of GpdQ (*left*) and active site structure (*right*). PDB number 3D03 [6, 59]

A more detailed introduction on what is known about the mechanism and structure of this enzyme is given in Chap. 3. The metal ion content in vivo is currently unknown, but it was shown that the reactivity of apo-GpdQ can be readily reconstituted with Co(II), Mn(II), Fe(II), Zn(II) and even Cd(II) [3–5, 10, 62]. As Co(II) serves as an excellent spectroscopic probe, the Co(II)Co(II) derivative is the most well studied metal derivative [3, 5]. It was initially proposed, on the basis of spectroscopic and crystallographic results, that the native metal ion composition was Fe(II)Zn(II) but that study lacked a thorough investigation of the catalytic mechanism of this metal derivative of GpdQ [62].

1.1.2 Metallo-β-lactamases

The second class of dinuclear metallohydrolases of particular interest in this thesis are the metallo-β-lactamases (MβLs). β-Lactams are the most widely used therapeutic agents against bacterial infections [63, 64]. MβLs are one strategy responsible for antibiotic resistance in bacteria as they cleave the lactam bond rendering antibiotics inactive [20, 64]. The MβLs exhibit a broad substrate profile and are capable of cleaving carbapenems, cephalosporins and penicillins [63–65]. There are two main groups of β-lactamases (serine-β-lactamases and MβLs). While the class of metal-free serine-β-lactamases can be inhibited in their action, the search for inhibitors continues, as there is to date no clinical useful inhibitor for MβLs. Bacteria are constantly evolving their ability to inactivate antibiotic drugs, the 'driving evolutionary force' is the increased use of antibiotics during the past 60 years [64]. For example, in 1943 when penicillin was first used, all Gram-positive bacterial infections were susceptible to treatment with penicillin; three years later, however, resistance among those bacteria had been observed [66]. It is certain that in order to treat bacterial infections successfully in the future the mechanism of action of MβLs has to be fully understood so that clinically useful inhibitors and new therapeutic approaches can be developed. Chapters 5 and 6

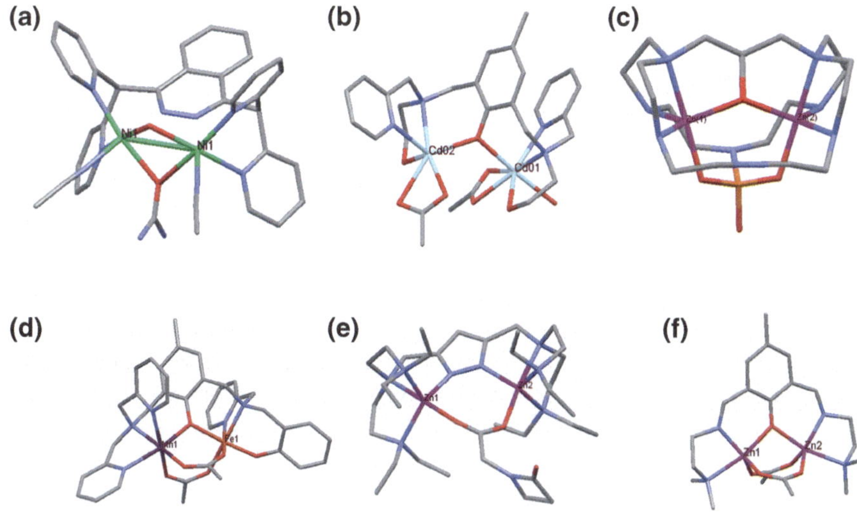

Fig. 1.4 Examples of dinuclear complexes mimicking hydrolytic enzymes. **a** Urease [68] **b** GpdQ [10] **c** Alkaline phosphatase [69] **d** PAP [70] **e** MβL [71] **f** Aminopeptidase [72]

will give a more detailed introduction of what is known about the mechanism of dinuclear MβLs. In addition some of the substrates that are used for kinetic, clinical and spectroscopic purposes will be discussed.

1.2 Biomimetics of Dinuclear Metallohydrolases

The study of biomimetics can be of great benefit for the understanding of enzymatic reactions. The term biomimetic refers, in the context of this work, to a compound that mimics structural, functional and spectroscopic properties of an enzyme [67]. Often only one or two of these aspects are achieved for a model system and they usually display substantially lower activity. There are, however, advantages over the enzyme: model complexes are generally more stable and robust than their enzymatic counterpart, they can be readily crystallized and provide easy accessible structural information on metal ion coordination. Also as these model systems are considerably less complex, kinetic and spectroscopic data interpretation is simplified and—by comparison to data derived for the enzyme—the mechanism of action and structural features can be elucidated and thus related back to the parent metalloenzyme. Also models can be obtained on a larger scale and are often less costly to synthesize, a distinct benefit for potential applications. A few structures of model complexes for dinuclear hydrolytic enzymes are shown in Fig. 1.4. The approaches for ligand and complex design are diverse.

Over the last decades biomimetics have helped to characterize intermediates in enzyme reactions as well as provide structural and mechanistic insights into the native systems. For example, X-ray absorption near-edge spectroscopy data from Cd(II) thiolate and imidazole complexes provided insight into the coordination number and nature of coordinating amino acids in the Cd(II)-containing carbonic anhydrase of a marine diatom, which suggested that the active site of this protein employs a tetrahedral coordination geometry and that Cd(II) is bound to a least one cysteine residue [73]. A dicobalt(II) complex, $[Co_2(\mu\text{-OH})(\mu\text{-Ph}_4\text{DBA})(\text{TMEDA})_2(\text{OTf})]$ (Ph_4DBA is the dinucleating ligand dibenzofuran-4,6-bis(diphenylacetate)), reported by Larrabee et al. turned out to be an excellent spectroscopic model for the 5 and 6-coordinate sites of methionine aminopeptidase and GpdQ [74–77]. The magnetic exchange coupling from this μ-OH bridged complex was shown to be ferromagnetic and the results from this study can be used to distinguish μ-aqua and μ-hydroxo bridging motifs in Co(II)Co(II) enzyme systems [74, 76]. Neves et al. were able to synthesize, and spectroscopically and kinetically characterize a dinuclear Fe(III)Zn(II) complex with a terminal Fe(III)-bound water molecule at a position equivalent to that of the proposed nucleophile in red kidney bean PAP [78]. The phosphatase activity of this complex yielded a bell-shaped strongly pH-dependent rate profile with a pH-optimum at 6.5, and pK_a values of 5.3 and 8.1 showing that this complex is an excellent structural and functional model of the active site of PAPs [78]. Another major contribution by a biomimetic was the observation and characterization of a blue intermediate during hydrolysis of nitrocefin (a commonly used β-lactam substrate) by a dizinc(II) complex reported by Lippard and co-workers [79]. The absorption maximum of the intermediate at 640 nm was similar to that found in the MβL enzyme from *Bacillus fragilis* [79–81]. The authors proposed a structure of the intermediate based on IR, NMR and kinetic measurements [79]. These are just few examples of how model complexes can contribute valuable insight into enzyme catalyzed processes.

Ligand and complex design in biomimetic systems is diverse but a few general concepts are normally followed: (i) the metal ions employed are often the same as in the native systems e.g. Ni(II) in urease models [68]; and (ii) pyridine or pyrazole residues are often used to mimic the histidine residues in the enzymes; phenol, carboxylate, pyrazolate or water molecules serve as mimics for bridging residues like aspartate, lysine or water/hydroxide; and (iii) dinucleating ligands are used to bring the two metal ions into close proximity.

1.3 Research Aims and Outline

In this work a detailed study of the enzyme GpdQ and biomimetics of dinuclear metallohydrolases will be conducted. The specific aims of this thesis are:

- Investigation of the influence of the hexameric structure on the activity of GpdQ and generation of differed mixed metal ion derivatives to investigate implications of different metals on mechanism and to determine preferences in vivo. A variety of mutants will be used to determine influences of the hydrogen bond network around the active site and techniques such as atomic absorption spectroscopy and magnetic circular dichroism will be utilized (Chap. 3).

- Comparison of the kinetic activity of dinuclear Zn(II) complexes as mimics for GpdQ and implications of different potential nucleophiles in the model complex mechanism will be discussed, along with an example on how to fine tune ligands to give more active biomimetics. Techniques such as ^{18}O-labeling studies will be used, in addition to mass spectrometry and crystallography, to understand the nature of substrate binding and hydrolysis (Chap. 4).
- Two Cd(II) complexes will be examined for their ability to hydrolyze OPs and β-lactam substrates. The mechanism of action will be probed using ^{113}Cd-NMR, ^{18}O-labeling, solution IR and UV-Vis studies. The relevance of Cd(II) as biological metal and the observation of a blue intermediate in β-lactam hydrolysis will be discussed (Chap. 5).
- The synthesis and comprehensive studies, including spectroscopic, kinetic and magnetic characterization using MCD, SQUID and DFT-calculations, of five dinuclear Co(II) complexes as highly active mimics for GpdQ (Chap. 6).
- Synthesis of asymmetric Zn(II) and Co(II) model complexes as functional and structural models for GpdQ will be reported (Chap. 7).
- Immobilization of GpdQ on modified magnetite nanoparticles and of biomimetics attached to resins will be presented. Comprehensive characterization of activity and properties of the immobilized systems will be conducted as a starting point for future applications of these systems in bioremediation (Chap. 8).

In the next chapter the experimental details for ligands and complexes employed in this thesis as well as descriptions of experimental techniques will be reported.

References

1. G.C. Dismukes, Chem. Rev. **96**, 2909–2926 (1996). (Washington, DC, U.S.)
2. E. Ghanem, Y. Li, C. Xu, F.M. Raushel, Biochemistry **46**, 9032–9040 (2007)
3. K.S. Hadler, L.R. Gahan, D.L. Ollis, G. Schenk, J. Inorg. Biochem. **104**, 211–213 (2010)
4. K.S. Hadler, N. Mitić, F. Ely, G.R. Hanson, L.R. Gahan, J.A. Larrabee, D.L. Ollis, G. Schenk, J. Am. Chem. Soc. **131**, 11900–11908 (2009)
5. K.S. Hadler, N. Mitić, S.H. Yip, L.R. Gahan, D.L. Ollis, G. Schenk, J.A. Larrabee, Inorg. Chem. **49**, 2727–2734 (2010)
6. K.S. Hadler, E.A. Tanifum, S.H. Yip, N. Mitić, L.W. Guddat, C.J. Jackson, L.R. Gahan, K. Nguyen, P.D. Carr, D.L. Ollis, A.C. Hengge, J.A. Larrabee, G. Schenk, J. Am. Chem. Soc. **130**, 14129–14138 (2008)
7. C.J. Jackson, P.D. Carr, H.K. Kim, J.W. Liu, D.L. Ollis, Acta Crystallogr., Sect. F: Struct. Biol. Cryst. Commun. **62**, 659–661 (2006)
8. C.J. Jackson, P.D. Carr, J.W. Liu, S.J. Watt, J.L. Beck, D.L. Ollis, J. Mol. Biol. **367**, 1047–1062 (2007)

9. C. Scott, G. Pandey, C.J. Hartley, C.J. Jackson, M.J. Cheesman, M.C. Taylor, R. Pandey, J.L. Khurana, M. Teese, C.W. Coppin, K.M. Weir, R.K. Jain, R. Lal, R.J. Russell, J.G. Oakeshott, Indian J. Microbiol. **48**, 65–79 (2008)
10. R.E. Mirams, S.J. Smith, K.S. Hadler, D.L. Ollis, G. Schenk, L.R. Gahan, J. Biol. Inorg. Chem. **13**, 1065–1072 (2008)
11. K.S. Hadler, T. Huber, A.I. Cassady, J. Weber, J. Robinson, A. Burrows, G. Kelly, L.W. Guddat, D.A. Hume, G. Schenk, J.U. Flanagan, BMC Res Notes **1**, 78 (2008)
12. T. Klabunde, N. Strater, B. Krebs, H. Witzel, FEBS Lett. **367**, 56–60 (1995)
13. K. Koizumi, K. Yamaguchi, H. Nakamura, Y. Takano, J. Phys. Chem. A **113**, 5099–5104 (2009)
14. N. Mitić, K.S. Hadler, L.R. Gahan, A.C. Hengge, G. Schenk, J. Am. Chem. Soc. **132**, 7049–7054 (2010)
15. N. Mitić, C.J. Noble, L.R. Gahan, G.R. Hanson, G. Schenk, J. Am. Chem. Soc. **131**, 8173–8179 (2009)
16. M. Olczak, B. Morawiecka, W. Watorek, Acta Biochim. Pol. **50**, 1245–1256 (2003)
17. G. Schenk, L.R. Gahan, L.E. Carrington, N. Mitić, M. Valizadeh, S.E. Hamilton, J. de Jersey, L.W. Guddat, Proc. Natl. Acad. Sci. U.S.A **102**, 273–278 (2005)
18. G. Schenk, L.W. Guddat, Y. Ge, L.E. Carrington, D.A. Hume, S. Hamilton, J. de Jersey, Gene **250**, 117–125 (2000)
19. G. Schenk, R.A. Peralta, S.C. Batista, A.J. Bortoluzzi, B. Szpoganicz, A.K. Dick, P. Herrald, G.R. Hanson, R.K. Szilagyi, M.J. Riley, L.R. Gahan, A. Neves, J. Biol. Inorg. Chem. **13**, 139–155 (2008)
20. M.W. Crowder, J. Spencer, A.J. Vila, Acc. Chem. Res. **39**, 721–728 (2006)
21. A.R. Shenoy, M. Capuder, P. Draskovic, D. Lamba, S.S. Visweswariah, M. Podobnik, J. Mol. Biol. **365**, 211–225 (2007)
22. A.R. Shenoy, N. Sreenath, M. Podobnik, M. Kovacevic, S.S. Visweswariah, Biochemistry **44**, 15695–15704 (2005)
23. R.M. Dawson, S. Pantelidis, H.R. Rose, S.E. Kotsonis, J. Hazard. Mater. **157**, 308–314 (2008)
24. G. Schenk, F. Ely, K.S. Hadler, N. Mitić, L.R. Gahan, D.L. Ollis, N.M. Plugis, M.T. Russo, J.A. Larrabee, J. Biol. Inorg. Chem. **16**, 777–787 (2011)
25. I. Horne, T.D. Sutherland, R.L. Harcourt, R.J. Russell, J.G. Oakeshott, Appl. Environ. Microbiol. **68**, 3371–3376 (2002)
26. F. Ely, K.S. Hadler, L.R. Gahan, L.W. Guddat, D.L. Ollis, G. Schenk, Biochem. J. **432**, 565–573 (2010)
27. S. Ciurli, S. Benini, W.R. Rypniewski, K.S. Wilson, S. Miletti, S. Mangani, Coord. Chem. Rev. **192**, 331–355 (1999)
28. N. Mitić, S.J. Smith, A. Neves, L.W. Guddat, L.R. Gahan, G. Schenk, Chem. Rev. **106**, 3338–3363 (2006). (Washington, DC, U.S.)
29. W.T. Lowther, B.W. Matthews, Chem. Rev. **102**, 4581–4607 (2002). (Washington, DC, U.S.)
30. Z.F. Kanyo, L.R. Scolnick, D.E. Ash, D.W. Christianson, Nature **383**, 554–557 (1996)
31. G. Schenk, N. Mitić, L. Gahan, S. Smith and K. Hadler, in *Eurobic 9: Proceedings of the 9th European Biological Inorganic Chemistry Conference*, (2008), pp. 29–38
32. A. Durmus, C. Eicken, B.H. Sift, A. Kratel, R. Kappl, J. Hüttermann, B. Krebs, Eur. J. Biochem. **260**, 709–716 (1999)
33. G. Schenk, Y. Ge, L.E. Carrington, C.J. Wynne, I.R. Searle, B.J. Carroll, S. Hamilton, J. de Jersey, Arch. Biochem. Biophys. **370**, 183–189 (1999)
34. L.W. Guddat, A.S. McAlpine, D. Hume, S. Hamilton, J. de Jersey, J.L. Martin, Structure **7**, 757–767 (1999)
35. G. Schenk, T.W. Elliott, E. Leung, L.E. Carrington, N. Mitić, L.R. Gahan, L.W. Guddat, BMC Struct. Biol. **8**, 6 (2008)
36. N. Sträter, J. Beate, M. Scholte, B. Krebs, A.P. Duff, D.B. Langley, R. Han, B.A. Averill, H.C. Freeman, J.M. Guss, J. Mol. Biol. **351**, 233–246 (2005)

37. H. Yang, P.D. Carr, S.Y. McLoughlin, J.W. Liu, I. Horne, X. Qiu, C.M.J. Jeffries, R.J. Russell, J.G. Oakeshott, D.L. Ollis, Protein Eng. **16**, 135–145 (2003)
38. F.M. Raushel, Curr. Opin. Microbiol. **5**, 288–295 (2002)
39. N. Sethunathan, T. Yoshida, Can. J. Microbiol. **19**, 873 (1973)
40. A.H. Mansee, W. Chen, A. Mulchandani, J. Ind. Microbiol. Biotechnol. **32**, 554–560 (2005)
41. A.A. DiNovo, D.A. Schofield, J. Appl. Micorbiol. **109**, 548–557 (2010)
42. J. Jeyaratnam, World Health Stat. Q. **43**, 139–144 (1990)
43. I. Horne, M. Selleck, M.R. Williams, K.M. Weir, J.G. Oakeshott, R.J. Russell, T.D. Sutherland, C.W. Coppin, Clin. Exp. Pharmacol. Physiol. **31**, 817–821 (2004)
44. J. Newmark, The neurologist **13**, 20–32 (2007)
45. S.W. Wiener, R.S. Hoffman, J. Intensiv, Care Med. **19**, 22–37 (2004)
46. K. Jaga, C. Dharmani, Pan Am. J. Public Health **14**, 171–185 (2003)
47. L. Gharahbaghian, T. Bey, Int. J. Disaster Med. **1**(2), 103–108 (2003)
48. R.L. Metcalf, *Insect Control* (Wiley-VCH, Weinheim, 2000)
49. T.M. Sakellarides, M.G. Siskos, T.A. Albanis, Int. J. Environ. Anal. Chem. **83**, 33–50 (2003)
50. R.M. Atlas, J. Philp, *Bioremediation: applied microbial solutions for real-world environmental cleanup* (ASM Press, Washington, D.C., 2005)
51. W. Chen, M. Shimazu, A. Mulchandani, ACS Sym. Ser. **863**, 25–36 (2004)
52. F. Ely, J.L. Foo, C.J. Jackson, L.R. Gahan, D. Ollis, G. Schenk, Curr. Top. Biomed. Res. **9**, 63–78 (2007)
53. CSIRO Orica Watercare, Enzyme product removes pesticides from water. http://www.csiro.au/en/Outcomes/Environment/Australian-Landscapes/Pesticide-Bioremediation.aspx. Accessed 20 Feb 2013
54. S.B. Bird, T.D. Sutherland, C. Gresham, J. Oakeshott, C. Scott, M. Eddleston, Toxicology **247**, 88–92 (2008)
55. C.J. Jackson, C. Scott, A. Carville, K. Mansfield, D.L. Ollis, S.B. Bird, Biochem. Pharmacol. **80**, 1075–1079 (2010)
56. A. Bretholz, R. Morrisey, R.S. Hoffman, Toxicology **257**, 172 (2009)
57. I. Horne, X. Qiu, J.W. Liu, P.D. Carr, J.G. Oakeshott, D.L. Ollis, H. Yang, R.J. Russell, S.Y. McLoughlin, C.M.J. Jeffries, Protein Eng. **16**, 135–145 (2003)
58. I. Horne, X. Qiu, D.L. Ollis, R.J. Russell, J.G. Oakeshott, FEMS Microbiol. Lett. **259**, 187–194 (2006)
59. Schrödinger, The PyMOL Molecular Graphics System, Version 1.3r1 (2010)
60. C. Jackson, H.-K. Kim, P.D. Carr, J.-W. Liu, D.L. Ollis, Biochim. Biophys. Acta, Proteins Proteomics **1752**, 56–64 (2005)
61. T.J. Larson, M. Ehrmann, W. Boos, J. Biol. Chem. **258**, 5428–5432 (1983)
62. C.J. Jackson, K.S. Hadler, P.D. Carr, A.J. Oakley, S. Yip, G. Schenk, D.L. Ollis, Acta Crystallogr. F **64**, 681–685 (2008)
63. C. Cojocel, *Beta-Lactam Antibiotics* (Springer, Boston, MA, 2008)
64. J.F. Fisher, S.O. Meroueh, S. Mobashery, Chem Rev. **105**, 395–424 (2005). (Washington, DC, U.S.)
65. J. Spencer, A.R. Clarke, T.R. Walsh, J. Biol. Chem. **276**, 33638–33644 (2001)
66. S.R. Palumbi, Science **293**, 1786–1790 (2001)
67. R. Breslow, Acc. Chem. Res. **28**, 146–153 (1995)
68. A.M. Barrios, S.J. Lippard, J. Am. Chem. Soc. **122**, 9172–9177 (2000)
69. T. Koike, M. Inoue, E. Kimura, M. Shiro, J. Am. Chem. Soc. **118**, 3091–3099 (1996)
70. P. Karsten, A. Neves, A.J. Bortoluzzi, M. Lanznaster, V. Drago, Inorg. Chem. **41**, 4624–4626 (2002)
71. F. Meyer, H. Pritzkow, Eur. J. Inorg. Chem. **2005**(12), 2346–2351 (2005)
72. A. Erxleben, J. Hermann, J. Chem. Soc., Dalton Trans. (4), 569–575 (2000)
73. T.W. Lane, M.A. Saito, G.N. George, I.J. Pickering, R.C. Prince, F.M.M. Morel, Nature **435**, 42–42 (2005)
74. F.B. Johansson, A.D. Bond, U.G. Nielsen, B. Moubaraki, K.S. Murray, K.J. Berry, J.A. Larrabee, C.J. McKenzie, Inorg. Chem. **47**, 5079–5092 (2008)

75. J.A. Larrabee, C.M. Alessi, E.T. Asiedu, J.O. Cook, K.R. Hoerning, L.J. Klingler, G.S. Okin, S.G. Santee, T.L. Volkert, J. Am. Chem. Soc. **119**, 4182–4196 (1997)
76. J.A. Larrabee, W.R. Johnson, A.S. Volwiler, Inorg. Chem. **48**, 8822–8829 (2009)
77. J.A. Larrabee, S.A. Chyun, A.S. Volwiler, Inorg. Chem. **47**, 10499–10508 (2008)
78. A. Neves, M. Lanznaster, A.J. Bortoluzzi, R.A. Perlata, A. Casellato, E.E. Castellano, P. Herrald, M.J. Riley, G. Schenk, J. Am. Chem. Soc. **129**, 7486–7487 (2007)
79. N.V. Kaminskaia, B. Spingler, S.J. Lippard, J. Am. Chem. Soc. **123**, 6555–6563 (2001)
80. N.V. Kaminskaia, B. Spingler, S.J. Lippard, J. Am. Chem. Soc. **122**, 6411–6422 (2000)
81. Z. Wang, W. Fast, S.J. Benkovic, J. Am. Chem. Soc. **120**, 10788–10789 (1998)

Chapter 2
Experimental

2.1 Introduction

This chapter will give an introduction of the experimental methods used in enzyme expression and purification and the synthesis of the compounds discussed in this thesis.[1] A brief introduction will be given about the concepts of ligand design and the nomenclature used throughout this thesis.

2.2 Ligand Syntheses

2.2.1 Phenolate Based Ligands

Phenolate based molecules as templates for dinucleating ligands have been around for more than 40 years [1–3]. The term di- or binucleating ligands was first used by Robson in 1970 who reported a Schiff-base copper complex with a phenol core [3]. While the numbers of applications are vast, the purpose of the phenolic

[1] Parts of this chapter appeared in: L.J. Daumann et al., *"Immobilization of the enzyme GpdQ on magnetite nanoparticles for organophosphate pesticide bioremediation"* J Inorg Biochem. **2014**, 131, 1–7; L.J. Daumann et al., "*Asymmetric Zinc(II) Complexes as Functional and Structural Models for Phosphoesterases*" Dalton Trans. **2013**; 42, 9574–9584; L.J. Daumann et al., "*Dinuclear Cobalt(II) Complexes as Metallo-β-lactamase Mimics*" Eur. J. Inorg. Chem. **2013**, 17, 3082–3089. L.J. Daumann et al. "*Synthesis, Magnetic Properties and Phosphoesterase Activity of Binuclear Cobalt(II) Complexes*" Inorg. Chem. **2013**, 52 (4), 2029–2043; L.J. Daumann et al., "*Promiscuity Comes at a Price: Catalytic Versatility vs. Efficiency in Different Metal Ion Derivatives of the Potential Bioremediator GpdQ*" BBA—Proteins and Proteomics **2013**, 1834, 1, 425–432; L.J. Daumann et al., "*Cadmium(II) Complexes; Mimics of Organophosphate Pesticide Degrading Enzymes and Metallo-β-lactamases*" Inorg. Chem. **2012**, 51, 7669-7681; L.J. Daumann et al., "*The Role of Zn-OR and Zn-OH Nucleophiles and the Influence of p-Substituents in the Reactions of Binuclear Phosphatase Mimetics*" Dalton Trans. **2012**, 41, 1695–1708. L.J. Daumann et al., "*Spectroscopic and mechanistic studies of dinuclear metallohydrolases and their biomimetic complexes*" Dalton Trans. **2014**, 43, 910–928.".

L. J. Daumann, *Spectroscopic and Mechanistic Studies of Dinuclear Metallohydrolases and Their Biomimetic Complexes*, Springer Theses, DOI: 10.1007/978-3-319-06629-5_2,

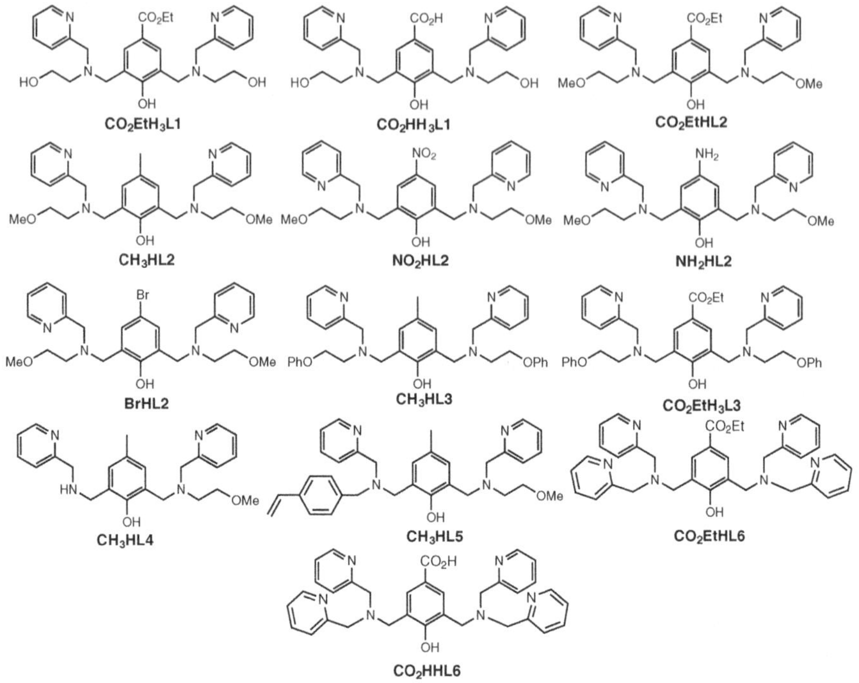

Fig. 2.1 Overview of the ligands synthesized in this work

framework that allows two metals in close proximity remains the same. The complexes of phenolate based ligands have served as biomimetic catalysts [4–12], spectroscopic models [13, 14], phosphate sensors [15], therapeutic agents [16], anionic receptors for supramolecular chemistry [17] or oxygen sensors [18] to name just a few applications. Complexes are known for almost all first row transition metals such as Co(II), Zn(II), Ni(II), Cu(II) Ti(I) V(IV)/(V) Mn(II)/(III) and Fe(II)/(III).

2.2.2 Overview of Ligands Synthesized in this Work

The new phenolate-based ligands synthesized in this work are shown in Fig. 2.1. These ligands were designed to study the influence of groups in the *para*-position to the bridging phenolic oxygen (CO_2EtH**L2**, CH_3H**L2**, NO_2H**L2** and BrH**L2**) on magnetism (Chap. 6) and phosphoesterase activity (Chaps. 4 and 6) or to compare mechanistic implications of directly bound groups like methyl ether and alcohol (CO_2EtH_3**L1** and CO_2EtH**L2**, Chaps. 5 and 6). Also a study of steric influences of bulky ligands arms is presented by comparing complexes of CH_3H**L2** and CH_3H**L3** (Chap. 4). Moreover, some ligands feature functional groups (CO_2HH_3**L1**,

NH_2H**L2**, CO_2HH**L5** and CH_3H_2**L4**) which allow attachment of the ligands to resins, inorganic supports such as silica and magnetite nanoparticles (Chap. 8). The pyridine residues are mimics for histidine, and the alcohol and ether arms are mimicking asparagine and aspartate residues, present in enzyme active sites. The bridging phenol mimics a bridging aspartate or hydroxide. The coordination sphere in the metal complexes will predominantly be completed by bridging acetato ligands.

2.2.3 Nomenclature

The nomenclature employed for the ligands is as follows: CO_2EtH_3**L1** denotes that the ligand has an ethyl ester (CO_2Et-) at the position *para* to the phenolic oxygen and that the ligand has potentially three sites for deprotonation, the phenol and two pendant alcohol donors. Two pyridines and two alcohol arms make up the **L1** donor atom set. The second ligand class **L2** reported in this thesis offers a direct comparison of methyl ether donors with the alkoxide donor in CO_2EtH_3**L1** and denotes symmetric ligands with two pendant pyridines and two methyl ethers. As only one proton can be potentially abstracted from the **L2** ligands upon deprotonation, the nomenclature was adjusted accordingly, e.g. CH_3H**L2**. **L3** ligands feature the more bulky phenyl ether instead of the methyl ether in **L2**. CH_3H**L4** is one of the two asymmetric ligands reported here. CH_3H**L5** is derived from the CH_3H**L4** ligand with an additional vinyl benzyl group. The class of **L6** ligands are lacking oxygen donor atoms but feature two pyridine moieties on either site of the dinucleating ligand instead.

2.2.3.1 Syntheses of the Precursors

The precursors for the phenolate-backbone of the ligands ethyl-4-hydroxy-3, 5-bis(hydroxymethyl)benzoate and 4-bromo-2,6-bis(hydroxymethyl)phenol were readily derived by reacting the corresponding phenols with formaldehyde and base. Both syntheses have been reported in the literature, however, the procedure for the former was altered slightly to improve yield and purity of the precursor [19, 20]. Application of lower temperatures and longer reaction times in the formaldehyde reactions seemed superior over the higher temperatures in previously reported syntheses as those caused a significant amount of polymer by-products. Also the application of freshly opened formaldehyde solution seemed to have a positive influence of the outcome of reaction. The hydroxymethyl groups were subsequently converted into bromomethyl groups by stirring the compounds with concentrated hydrogen bromide in acetic acid, as reported previously [21]. The synthesis of 2,6-bis(chloromethyl)-4-methylphenol after Paine's method [22] was straightforward and the starting material 2,6-bis(hydroxymethyl)-4-methylphenol was commercially available. 2,6-bis(bromomethyl)-4-nitrophenol was obtained in low yield after a published multistep procedure [23]. Syntheses of the ligand arms 2-methoxy-N-(pyridin-2-ylmethyl) aminoethanol, 2-phenoxy-N-(pyridin-2-ylmethyl)ethanamine and N-(2-pyridylmethyl)-2-aminoethanol followed a similar approach as previously reported [24, 25]. Typically pyridine-2-carboxaldehyde was reacted in methanol with methoxy-, phenoxy- or ethanolamine to give the corresponding Schiff base. The imine was then reduced with sodium borohydride to yield the crude precursors. Three things were crucial for a high yield: (i) Only a minimum amount of methanol had to be used (ii) freshly opened sodium borohydride was used, and (iii) the use of brine during the workup procedure was avoided for N-(2-pyridylmethyl)-2-aminoethanol as this decreased the yield of pure product significantly. The use of dry methanol or molecular sieves during the reaction had no positive effect on yield or purity of the products. The final ligands were usually synthesized by reacting two equivalents of ligand arm with the respective phenol-precursor in the presence of triethylamine as base in dichloromethane/tetrahydrofuran. After removal of the precipitated triethylamine hydrochloride/bromide by filtration and the solvent, the ligands were purified with flash column chromatography.

2.3 Materials and Methods: Model Complexes

^{1}H NMR spectra were recorded, unless otherwise stated, at room temperature with a 300, 400 or 500 MHz Bruker AV 300/400/500 spectrometer. Chemical shifts are reported in δ units relative to $CHCl_3$ ($\delta_H = 7.24$), d^4-MeOD ($\delta_H = 3.30$), CD_3CN ($\delta_H = 1.93$), d^6-acetone ($\delta_H = 3.31$) or d^6-DMSO ($\delta_H = 2.50$). The following abbreviations were used: s = singlet, d = doublet, t = triplet, dd = doublet of doublet, dt = doublet of triplet, m = multiplet. The software used for data processing was TOPSPIN 2.1 from BRUKER [26].

^{13}C NMR spectra were recorded at room temperature with a 100 MHz Bruker AV 400/500 spectrometer. Chemical shifts are given in δ units relative to $CDCl_3$ (central line of triplet: $\delta_C = 77.0$), d^4-MeOD ($\delta_C = 49.0$), CD_3CN ($\delta_C = 1.30$), D_2O with 5 % dioxane ($\delta_C = 67.2$). Two-dimensional correlation spectroscopy (COSY), heterobinuclear single quantum correlation (HSQC) and heterobinuclear multiple bond connectivity (HMBC) experiments were used to assign each signal in the spectra of the final ligands [27]. ^{13}C-NMR of $[Cd_4(CO_2EtH_2\mathbf{L1})_2(CH_3COO)_{3.75}Cl_{0.25}(H_2O)_2](PF_6)_2$ with penicillin: the solution was initially 0.075 M in penicillin G and 0.075 M of $[Cd_4(CO_2EtH_2\mathbf{L1})_2(CH_3COO)_{3.75}Cl_{0.25}(H_2O)_2](PF_6)_2$ in deuterated DMSO/acetone 1:1. The spectrum was recorded at room temperature after 24 h of mixing of both components.

^{31}P NMR spectra were recorded at room temperature with 85 % phosphoric acid as external standard ($\delta_P = 0.00$). ^{31}P NMR spectra of the Cd(II) complexes were recorded with a 600 MHz Bruker AV600 spectrometer at room temperature in the digital acquisition mode (operating frequency 242 MHz) at the NMR facilities of the Anorganisch-Chemisches Institut, University of Heidelberg.

^{113}Cd NMR was measured with a Bruker AV400 instrument and an operating frequency of 89 MHz. Cd-shifts were referenced to $Cd(OAc)_2{\cdot}2H_2O$ in D_2O (−46 ppm). The measurements for $[Cd_4(CO_2EtH_2\mathbf{L1})_2(CH_3COO)_{3.75}Cl_{0.25}(H_2O)_2](PF_6)_2$ in presence of substrates were conducted using the following procedure: to a 0.01 mM solution of the complex in CD_3CN/D_2O 1:1 were added 10 equivalents of diphenylphosphate and the solution was left to incubate for 6 h prior to spectra recording. The ^{13}C NMR spectrum of penicillin G in d^6-DMSO/d^6-acetone (1:1) with one equivalent of $[Cd_2(CO_2EtH_2\mathbf{L1})\text{-}(CH_3COO)_2]^+$ was recorded 24 h after mixing.

^{18}O-labeling studies were conducted by recording the ^{31}P NMR spectra of complex/substrate mixtures with a 400 MHz Bruker AV400 spectrometer at room temperature in the digital acquisition mode (operating frequency 161.9 MHz). Chemical shifts are reported in δ units relative to 85 % H_3PO_4 in D_2O as external reference ($\delta_P = 0.00$). For the ^{18}O-labeled sample (50 ^{18}O; 97 % purity) (Novachem, Victoria, Australia) the solution was made up from a solution of complex (0.01 mmol) in acetonitrile (0.3 mL), 100 mM HEPES buffer pH 8 (0.15 mL) and ^{18}O-water (97 %, 0.15 mL) (50:50 mixture of acetonitrile:buffer). To this, one equivalent BDNPP (5.1 mg) was added and the mixture left for one week at room temperature prior to recording the ^{31}P NMR spectra. For experiments with ^{16}O-water the solution was composed of complex (0.01 mmol) in acetonitrile (0.3 mL), 100 mM HEPES buffer pH 8 (0.15 mL) and distilled water (0.15 mL). To this, one equivalent BDNPP (5.1 mg) was added and the mixture left for one week at room temperature prior to recording the ^{31}P NMR spectra.

Magnetic Moments in Solution were determined with the Evans method [28] using

$$\chi_A^M = \frac{3M}{4\pi c}\left(\frac{\Delta v}{v}\right) \tag{2.1}$$

and

$$\mu = 2.828\sqrt{\chi_A^M T} \tag{2.2}$$

and were measured for a solution of known concentration of each complex in deuterated methanol with non-deuterated methanol in the inner capillary [28]. No diamagnetic corrections were applied. The experiments were conducted on a superconducting magnet and appropriate corrections were considered in the equation [29].

Solid State Magnetic Susceptibility Measurements were conducted with a MPMS-XL 5T superconducting quantum interference device (SQUID) from Quantum Design at the University of Heidelberg in collaboration with Professor Peter Comba and recorded as a function of the applied field (0–5 T), and at temperatures ranging from 2 to 300 K (zero field cooled method). The powdered samples were pressed into a PTFE band to avoid field-induced orientation of the powder and incorporated into two plastic straws as sample holder. The data were corrected for the diamagnetism of the PTFE band and sample holder; Pascal constant corrections for each sample were applied [30]. The program MagSaki used for the analysis of magnetic susceptibility data of the Co(II) complexes [31].

Magnetic Circular Dichroism studies were conducted at Middlebury College, Vermont, USA in collaboration with Professor Jim Larrabee. Complex samples were measured both as solid mulls (poly(dimethylsiloxane)) and as saturated ethanol solutions. Samples of air sensitive compounds were prepared in degassed ethanol. The MCD system used has a JASCO J815 spectropolarimeter and an Oxford Instruments SM4000 cryostat/magnet. Data were collected at increments of 0.5 Tesla (T) from 0 to 7.0 T and at temperatures of 1.4, 4.2, 11.3, 26 and 50 K. Each spectrum was corrected for any natural CD by subtracting the zero-field spectrum of the sample. Even when there is no sample present the instrument baseline exhibits a small deviation from zero that is both field- and wavelength-dependent. Therefore, each spectrum was also corrected by subtraction of a spectrum recorded at the same magnetic field but with no sample present. The resultant spectra were fitted to the minimum number of Gaussian peaks to achieve a satisfactory composite spectrum using the GRAMS AI software [32]. The software VTVH 2.1.1 was used to fit the data [33].

Angular Overlap Model Calculations Spectral simulations were made using the angular overlap model (AOM) using the program AOMX [34, 35]. Calculations were performed for $[Co_2(CO_2EtH_2\mathbf{L1})(CH_3COO)_2](PF_6)$ and $[Co_2(CO_2Et\mathbf{L2})(CH_3COO)_2](PF_6)$ based on the transitions obtained from the diffuse reflectance and MCD data. The coordinates were generated from the respective crystal structures, and each Co(II) metal site was treated separately. The Racah parameters C and B were fitted separately, with n in C = n B varying from 4 to 4.7.

Crystallographic Data for the complexes were collected, unless otherwise stated, at room temperature with an Oxford Diffraction Gemini Ultra dual source (Mo and Cu) CCD diffractometer with Mo ($\lambda_{K\alpha} = 0.71073$ Å) or Cu ($\lambda_{K\alpha} = 1.5418$ Å) radiation. The structures were solved by direct methods (SIR-92) and refined (SHELXL 97) by full matrix least squares methods based on F^2 [36]. These programs were accessed through the WINGX 1.70.01 crystallographic collective package [37]. All non-hydrogen atoms were refined anisotropically unless they were disordered. Hydrogen atoms were fixed geometrically and were not refined. X-ray data of the published structures were deposited with the Cambridge Crystallographic Data Center.

Positive Ion Electrospray Mass Spectrometry was carried out with a Q-Star time of flight mass spectrometer and data were processed with Bruker Compass Data Analysis 4.0 software. Mass spectral studies of the organophosphate hydrolysis reaction by $[Zn_2(CH_3\mathbf{L1})(CH_3COO)_2]^+$ were conducted by mixing one equivalent of this complex with 25 equivalents of either bis(4-nitrophenyl) phosphate (BPNPP) or 4-nitrophenyl phosphate (PNPP), the latter a hydrolysis product of the former, in 1:1 acetonitrile:water, and recording the spectrum of the resulting mixture after 1 h. BPNPP and its hydrolysis product PNPP were used as they are slower substrates (due to the lack of one nitro-group) and permitted more ready analysis than the substrate used for the kinetic studies (BDNPP). The complex $[Cd_4(CO_2EtH_2\mathbf{L1})_2\,(CH_3COO)_{3.75}Cl_{0.25}(H_2O)_2](PF_6)_2$ was dissolved in acetonitrile:water 1:1 and the mass spectra recorded with complex concentrations ranging from 0.1 mM to 10 μM. With the same solvent conditions the mass spectra were recorded in the presence of either 20 equivalents of PNPP (4-nitrophenyl phosphate), a product mimic, or 25 equivalents of DPP (diphenylphosphate), a substrate mimic lacking the activating nitro groups and is therefore not hydrolyzed by the model complex. The solutions were left to incubate at room temperature for one hour prior to spectra recording.

General FT-Infrared Spectroscopy was carried out with a Perkin Elmer FT-IR Spectrometer SPECTRUM 2000 with a Smiths DuraSamplIR II ATR diamond window.

Solution IR Measurements were recorded at the University of Heidelberg (Organisch-Chemisches Institut) with a ReactIR 45 m from Mettler Toledo between 650 and 2800 cm^{-1}. The spectrum of nitrocefin and in the presence of one equivalent $[Cd_2(CO_2EtH_2\mathbf{L1})(CH_3COO)_2]^+$ was recorded in acetonitrile:water (50:50) at 293 K every 2 min for 2 h. The initial concentrations of complex and substrate were 5 mM each. In-solution IR spectra for the Co(II) complexes were also recorded at the University of Heidelberg (Organisch-Chemisches Institut) with a ReactIR 45 m from Mettler Toledo between 650 and 2800 cm^{-1}. The spectra of penicillin G in the presence of one equivalent of $[Co_2(CO_2EtH_2\mathbf{L1})(CH_3COO)_2]^+$ or $[Co_2(CO_2Et\mathbf{L2})(CH_3COO)_2]^+$ were recorded in dimethylsulfoxide(DMSO)/water (90:10) at 293 K every 2 min for one hour. The solvent system and the substrate penicillin, instead of nitrocefin, were chosen to directly compare the results with literature values. The initial concentrations of complex and substrate were 7 mM each.

UV-Vis Studies

Phosphatase Activity Measurements were conducted with bis(2,4-dinitrophenyl) phosphate (BDNPP) as substrate with a Varian Cary50 Bio UV/Visible spectrophotometer with a Peltier temperature controller and measured in 10 mm quartz cuvettes. The initial rate method was employed and assays were measured such that the initial linear portion of the data was used for analysis. Product formation was determined at 25 °C by monitoring the formation of 2,4-dinitrophenol. Throughout the pH range studied (4.75–8.5), the extinction coefficient of this product at 400 nm varies from 7180 at pH 4.5, 10080 at 5.0; 11400 at pH 5.5 to 12000 at 6.0 and 12100 at pH 6.5–8.5 [38]. All assays were measured in 50:50 acetonitrile:buffer with the substrate and complex initially dissolved in acetonitrile. The aqueous multicomponent buffer solution was comprised of 50 mM 2-(N-morpholino)ethanesulfonic acid (MES, pH 5.50–6.70), 4-(2-hydroxyethyl)-1-piperazineethanesulfonic acid (HEPES, pH 7.00–8.50), 2-(N-cyclohexylamino)ethane sulfonic acid (CHES, pH 9.00–10.00) and *N*-cyclohexyl-3-aminopropanesulfonic acid (CAPS, pH 10.5–11) with controlled ionic strength ($LiClO_4$) 250 mM. The pH values reported refer to the aqueous component; it should, however, be noted that the pH of a solution of the buffer was the same within error as a 1:1 mixture of buffer and acetonitrile. Assays for pH dependence were 40 μM complex and 5 mM BDNPP and for substrate dependence 40 μM in complex and 1–11.5 mM in BDNPP, respectively. pH-dependence data for monoprotic (Eq. 2.3) or diprotic (Eq. 2.4) events were fit to

$$v_0 = \frac{v_{\max}}{1 + \frac{[H^+]}{K_a}} \tag{2.3}$$

or

$$v_0 = \frac{v_{\max}}{\left(1 + \frac{[H^+]}{K_{a1}} + \frac{K_{a2}}{[H^+]}\right)} \tag{2.4}$$

respectively [39]. The data derived from substrate dependence were fit to the Michaelis-Menten equation (Eq. 2.5) [39]. Here, v is the initial rate, v_{max} is the maximum rate, K_m is the Michaelis constant, and [S] is the substrate concentration.

$$v = \frac{v_{\max}[S]}{K_m + [S]} \tag{2.5}$$

[Complex] dependence of the catalytic rate was measured with a fixed substrate concentration at 5 mM and complex concentrations ranging from 20–120 μM. Background assays were conducted by measuring the autohydrolysis and hydrolysis by two equivalents of the free respective metal ($M(OAc)_2$, M = Cd, Zn, Co, 2 mM) and were subtracted from the data. Where no solid could be obtained in complex synthesis a 1 mM stock solution of the respective Co(II), Cd(II) and Zn(II) complex was generated in acetonitrile in situ by refluxing one equivalent ligand and two equivalents metal acetate for 30 min and confirming the successful formation by mass spectrometry.

For the kinetic investigation of the immobilized Zn(II) complexes on Merrifield (MR) resin, an assay was established which allowed determination of the amount of BDNPP hydrolysed by the resins. In order to obtain initial rates, the absorbance of a solution of multicomponent buffer, acetonitrile and BDNPP was determined (T = 0), and then 0.5 mg of resin added. After 15 min 1 mL of the suspension was transferred to an Eppendorf tube and centrifuged for 10 s; the absorbance of the supernatant was measured again (T = 15) and the Abs/min calculated. For every value this was done in triplicate, however, due to swelling properties of the resin and inhomogeneous catalyst loading on the Merrifield beads the errors were larger than for the free complex.

Metallo-β-lactamase-like Activity Assays were conducted in the same aqueous multi-component buffer as mentioned above with the ionic strength controlled by 250 mM $LiClO_4$. Assays were carried out at 37 °C in 50:50 acetonitrile:buffer, with nitrocefin as substrate initially dissolved in acetonitrile (10 mM) and the complex dissolved in acetonitrile:water (1 mM). Assays conducted to investigate pH dependence were 5 μM in complex and 50 μM in nitrocefin. [Substrate] dependence assays were 5 μM in complex and 10–50 μM in nitrocefin, and [complex] dependence assays were 20–120 μM in complex and 5 mM in nitrocefin. Background assays of autohydrolysis in the presence of two equivalents of cadmium(II) or cobalt(II) acetate were subtracted from the data. The assays were monitored for the hydrolysis of the substrate at 390 nm for 5 min and left to equilibrate for 2 min before each measurement. Nitrocefin and its hydrolysis product both absorb at 390 nm therefore a corrected extinction coefficient of 13415 M^{-1} cm^{-1} was used to calculate the rate of catalysis [40, 41].

Nitrocefin Hydrolysis Product Inhibition In order to elucidate how many molecules of nitrocefin one complex could hydrolyze a titration experiment was conducted where successively one to four equivalents of nitrocefin were added to the complex under the conditions used for kinetic experiments (acetonitrile/buffer pH 8, 37 °C, 300–700 nm). After each addition the solution was scanned at intervals of one minute for a total time of 2 h. Initial concentrations were in general 0.05 mM in each complex and substrate, respectively.

Synthesis of the Blue Species and UV-Vis Experiments were conducted by measuring the absorption spectrum of 0.05 mM nitrocefin in dry acetonitrile at 37 °C and then adding one equivalent of $[Cd_2(CO_2EtH_2\mathbf{L1})(CH_3COO)_2]^+$ and recording the spectrum again after 2 min. The same conditions were applied after the addition of 10 μL water to the mixture. Mass Spectra were recorded immediately after mixing equimolar amounts of nitrocefin and $[Cd_2(CO_2EtH_2L1)(CH_3COO)_2]^+$ in dry acetonitrile. Initial concentrations of complex and substrate were 0.01 mM.

Fluoride Inhibition studies were conducted by adding 2000 equivalents sodium fluoride to a stock solution of the complex $[Cd_2(CO_2EtH_2\mathbf{L1})(CH_3COO)_2]^+$ The mixture was left at room temperature to incubate for at 24 h prior to activity measurements that were conducted as described in the kinetics section for nitrocefin and BDNPP hydrolysis.

Diffuse Reflectance Spectroscopy was conducted at the Middlebury College, Vermont, using a Varian Cary 6000i and a Purged Praying Mantis Diffuse Reflection Attachment (Harrick). Magnesium oxide was used as blank and sample carrier.

Cyclic Voltammetry was carried out with a BAS 100 W potentiostat coupled with an $Ag/AgNO_3$ reference electrode in acetonitrile, platinum or glassy carbon as working and platinum wire as standard electrode. 0.1 M tetraethyl ammonium perchlorate in dry acetonitrile served as electrolyte with ferrocene as standard. Data analysis was performed with the software BAS100 W version 2.3.

Elemental Microanalyses (C, H, N, S) were performed with a Carlo Erba Elemental Analyzer model NA1500 by Mr. George Blazak at the University of Queensland.

Transmission Electron Microscopy (TEM) was conducted at the Center of Microscopy and Microanalysis, University of Queensland together with Dr. Graeme Auchterlonie. Nanoparticles were taken up in water and sonicated for 10 min after which time the suspension was transferred onto a carbon grid. A JEOL JEM2100

LaB_6 STEM analytical transmission electron microscope with an EM-21010/21020 single tilt holder was used to analyses the samples after the water was evaporated.

X-ray Photoelectron spectroscopy (XPS) was conducted with Dr. Barry Wood at the Center for Microscopy and Microanalysis, University of Queensland, with a Kratos Axis Ultra photoelectron spectrometer which uses Al K_α (1253.6 eV) X-rays. The software Casa XPS was used for data processing [42].

Hydrogenations were done with a H-cube hydrogen reactor equipped with 10 % Pd on charcoal cartridges as catalyst [43].

2.4 Materials and Methods: GpdQ

Atomic Absorption Spectroscopy was conducted for the half-apo GpdQ enzyme to confirm the presence of one equivalent of iron. Standards containing 20, 40, 60, 80, and 100 ppb Fe, Zn, Mn, and Co were prepared in 50 mM buffer (HEPES, pH 7) from analytical stock solutions. Protein samples were prepared using the same buffer.

Expression and Purification of wt-GpdQ and Mutant Forms were conducted following a modification of a previously published method [4]. In brief, DH5R Escherichia coli cells were transformed with the GpdQ/pCY76 plasmid. Cells were grown in 4 L of TB medium containing 50 μg/mL ampicillin and 0.1 mM $CoCl_2 \cdot 6H_2O$ at 30 °C for 30 h. Furthermore, cobalt is added to the growth media as this has been shown in previous experiments that it enhances the yield of expressed protein. The plasmid used expresses GpdQ in a constitutive manner, and no induction was required. The following steps were performed at 4 °C. Cells were harvested by centrifugation and then resuspended in 40 mL of 20 mM TRIS-HCl buffer, pH 8.0. Following lysis using a French press at 10,000 psi, the cell debris was pelleted by centrifugation. The supernatant was filtered using a 0.22 μm filter (Millipore) and loaded onto a HiPrep 16/10 DEAE FF column (GE Healthcare). A linear NaCl gradient (0 to 1 M) was applied to elute the protein over 10 column volumes, and the elute was collected in 8 mL fractions. A 5 μL aliquot of each fraction was assayed with 2 mM BPNPP, and the fractions with phosphodiesterase activity were combined. Other proteins were precipitated by addition of 4 M $(NH_4)_2SO_4$ to a final $(NH_4)_2SO_4$ concentration of 1.3 M. The protein solution was loaded onto a HiLoad 26/10 Phenyl Sepharose HP column (GE Healthcare). The proteins were eluted using an $(NH_4)_2SO_4$ gradient (1.5 to 0 M) over 10 column volumes and collected again in 8 mL fractions. Fractions with phosphodiesterase activity were combined, and the proteins were concentrated to approximately 4 mL. Finally, the solution was loaded onto a HiPrep 16/10 Sephacryl S-200 HR gel filtration column and eluted with 20 mM HEPES, 0.15 M NaCl, pH 8.0. Sodium dodecyl sulfate-polyacrylamide gel electrophoresis (SDS-PAGE) analysis of the purified protein generally shows a single band migrating at ~31 kDa. The final yield was ~70 mg/L. Protein concentrations were determined by measuring the absorption at the 280 nm ($\varepsilon = 39{,}880\ M^{-1}\ cm^{-1}$ for a monomer).

Expression of the 8-3 Mutant of GpdQ from competent *E. Coli* cells was conducted at the University of Queensland using same protocol as for wt-GpdQ. The plasmid was obtained from Prof. David Ollis, Research School of Chemistry, ANU, Canberra.

Purification of the GpdQ 8-3 Mutant involved two chromatographic steps using FPLC and was carried out at the Australian National University after a protocol obtained from Ms Sylvia Yip [44]: 5 g of the cells were suspended in 45 mL buffer (50 mM TRIS pH 8) and were lysed using a French Press (14,000 psi, two cycles). The cell debris was separated by centrifugation and the supernatant was kept at 4 °C at all times. Approximately 45 mL protein solution was loaded on a Q Sepharose (strong anion exchange) column using a peristaltic pump (Buffer A 50 mM TRIS pH 8; buffer B 1 M NaCl, 50 mM TRIS pH 8) and eluted with a linear NaCl gradient (0.0–0.6 M) over 10 column volumes (cv) (75 ml = cv, flow 2 mL/min). The 14 fractions containing GpdQ (8 mL, GpdQ eluted between 0.32 and 0.5 M NaCl) were collected. After SDS-page the most pure fractions were combined. From the combined fractions GpdQ was precipitated with $(NH_4)_2SO_4$ (final concentration 2.5 M). The suspension was stirred for 30 min at 4 °C and was subsequently centrifuged, the supernatant discarded and the precipitated protein resuspended in 2 mL buffer (50 mM TRIS pH 8). The suspension was subsequently loaded onto a HiPrep 16/10 Sephacryl S-200 HR gel filtration column (GE Healthcare) and the different oligomers of GpdQ eluted with 50 mM TRIS pH 8.

Site Directed Mutagenesis was conducted at the Australian National University together with Ms T. Murray. The F and R mutagenesis primers for the Y19F and S127A mutants were: GpdqY19FSDMFor—CG AGA AGC TGT TCG GCT TTA TCG ACG, GpdqY19FSDMRev—CGA TAA AGC CGA ACA GCT TCT CGC CGC; GpdqS127ASDMFor—CGC CGG CAC TGC AAA AGG CTG GCT GAC C and GpdqS127ASDMRev—GCC AGC CTT TTG CAG TGC CGG CGC GGC. The Polymerase Chain Reaction (PCR) protocol was as follows: 51 μL milliQ water; 7.5 μL 10× pfu buffer; 3 μL dNTP (10 mM); 3 μL Fi primer (10 μM); 3 μL Ri primer (10 μM); 1.5 μL pfu. From this mixture 46 μL were transferred into a PCR tube and 4 μL DNA (PCY76 wt GpdQ 141 ng/μL) were added, to the remaining blank 2 μL milliQ water were added. The PCR was conducted with a MJ Mini Personal Thermal Cycler from BioRad, PCR program: 95 °C 45 s; (95 °C 45 s, 55 °C 45 s, 72 °C 8 min) × 30; 72 °C 10 min; 4 °C hold. The purification was conducted with a kit from Promega (Wizard SV Gel + PCR cleanup kit). After digestion (50 μL purified PCR mix, 6.5 μL NEB IV, 6.5 μL 10× BSA and 1 μL Dpn I (total of 65 μL) incubate mix for 3 h at 37 °C) the PCR cleanup was repeated. The subsequent transformation of DH5α cells was as follows: 50 μL of electro competent DH5α cells were thawed on ice. 5 μL of cleaned up digestion and 5 μL water were then added. After electroporation the cells were resuspended quickly in 1 mL YenB (ampicillin) and recovered for one hour at 37 °C. After plating 50 μL of solution on Agar/Ampicillin plates they were incubated overnight at 37 °C. Three colonies were chosen and incubated overnight in LB/Ampicillin (5 mL) each. From the spun down cells DNA was extracted using the Qiagen Miniprep kit. For sequencing 1 μL BDT, 3.5 μL 5× buffer, 1 μL primer i (3.2 pmol/μL), 6 μL DNA and 8.5 μL water were combined and subjected to the following program in a MJ Mini Personal Thermal Cycler: 94 °C 5 min (96 °C 10 s, 50 °C 5 s, 60 °C 4 min) × 35, 4 °C hold.

Preparation of half-apo GpdQ An Econo-Pac 10DG (Bio-Rad) gel filtration column was used to generate the half-apo enzyme. The column was treated with 5 mL of a chelating solution, which contained 5 mM tetrasodium salt of ethylenediaminetetraacetic acid, 5 mM 1,10-phenanthroline, 5 mM 2,6-pyridine dicarboxylic acid, 5 mM 8-hydroxyquinoline-5-sulfonic acid, 5 mM 2-mercaptoethanol, and 20 mM HEPES (pH 7.0) and was subsequently equilibrated by washing with 5 column volumes of Chelex-treated buffer (50 mM HEPES, pH 7.0) prior to use. Approximately 250 μL of GpdQ (~0.35 mM) were diluted to 3 mL with the chelating solution and then loaded onto the column. The enzyme was collected in 4 mL of buffer (50 mM HEPES, pH 7.0). The concentration of the protein was calculated by measuring its absorbance at 280 nm ($\varepsilon = 39{,}880\ cm^{-1}\,M^{-1}$) [45, 46]. Atomic absorption spectroscopy was used to determine the metal ion content.

Phosphatase Activity of wt-GpdQ, and S127A and Y19F Mutants were measured spectrophotometrically with BPNPP as substrate by monitoring the formation of the product 4-nitrophenolate at 405 nm. All kinetic experiments were done in duplicate. In the assay solution the concentration of added metal ions was 20 μM. Reactions were measured to less than 5 % of total substrate hydrolysis for 2 min and the assay mix was equilibrated for 1 min before the measurement. Rates were linear in the time frame studied. The extinction coefficient of 4-nitrophenolate was determined at each relevant pH by using a standard solution of 4-nitrophenol in the same buffer used for the enzymatic assays. All reagents (buffer, substrate, etc.) were prepared in water that had been treated with Chelex. The multicomponent buffers (50 mM of each of MES, HEPES, CHES, and CAPS) that ranged from pH 5.5 to 11.0 were also treated with Chelex for 24 h. Chelex was removed by filtration with 0.45 μm Millex syringe-driven filter units. The pH-profiles of the catalytic parameters were derived from rate measurements that were conducted at various pH values between pH 5.5 and 11.0. The enzyme was stable over the entire pH range studied; autohydrolysis of the substrate was taken into account at each pH. In the substrate concentration dependence studies, the concentrations of the half-apoenzyme and M(II) remained constant at 200 nM and 20 μM, respectively, while that of the substrate increased from 0 to 10 mM. Similar studies were also conducted in the presence of mixtures of first row divalent transition metal ions (Zn(II), Mn(II) and Co(II)) at pH 9. The data were fit to Eq. 2.5 (p.17) [39].

Kinetic Assays for the 8-3 Mutant were also performed with the substrate BPNPP and hydrolysis was followed by monitoring the formation of the reaction product 4-nitrophenol at a wavelength of 405 nm over 0.5 min at 25 °C. The assays were run in 1 mL plastic cuvettes with 10 mm pathlength using 929 μL buffer (50 mM TRIS pH8), 70μL BPNPP (33.6 mM) and 1 μL of the GpdQ fraction or with addition of metal: 919 μL buffer, 1 μL enzyme fraction (final concentrations ranged from 41–67 nM), 70 μL BPNPP (33.6 mM, final concentration 2.3 mM) and 10 μL of a 2.3 mM $CoSO_4$ or $MnSO_4$ solution (final concentration 23 μM).

Magnetic Circular Dichroism Protein samples of Fe(II)Co(II) wt-GpdQ were prepared by adding one equivalent Co(II) and ten equivalents phosphate to the Fe(II) half-apo enzyme, or for the Ser127Ala mutant by adding two equivalents Co(II). The protein was dissolved in a 60 %/40 % (v/v) mixture of glycerol/buffer (50 mM HEPES, pH 7.0) and loaded in a 0.62 cm pathlength nickel-plated copper sample cell with quartz windows. The MCD system used has a JASCO J815 spectropolarimeter and an Oxford Instruments SM4000 cryostat/magnet. Data were collected at increments of 0.5 Tesla (T) from 0 to 7.0 T and at temperatures of 1.4, 4.2, 11.3, 26 and 50 K. Each spectrum was corrected for any natural CD by subtracting the zero-field spectrum of the sample. The resultant spectra were fitted to the minimum number of Gaussian peaks to achieve a satisfactory composite spectrum using the GRAMS AI software [32].

Metal Binding Affinity of Half-apo GpdQ was determined employing activity measurements by varying the amount of added M(II), where M = $CoSO_4.6H_2O$, $MnCl_2$ H_2O or $CdCl_2$ $2.5H_2O$ to the activity assay mixture. The concentrations of the half-apoenzyme and substrate remained constant at 200 nM and 10 mM, respectively, while that of M(II) increased from 0–50 μM. The addition of M(II) to wild-type and mutant half-apo forms of GpdQ at a constant pH and substrate concentration demonstrated saturation behavior similar to that described by the Michaelis-Menten Equation (Eq. 2.5). The binding function (r), which is the molar ratio of the amount of metal ion bound to the total amount of enzyme active sites, was calculated using Eq. 2.6 (p. 24), where n is the number of sites associated with K_d, the dissociation constant, and $[M]_{free}$ is the free metal ion concentration [47]. The concentration of free metal ion was calculated from Eq. 2.7 (p. 24), where $[M]_{total}$ is the total concentration of metal ions added and [E] is the concentration of the enzyme. Since the added metal ion concentration is in large excess over enzyme concentration, $[M]_{free}$ is assumed to equal $[M]_{total}$. This assumption reduces Eq. 2.6 to Eq. 2.8 [47]

$$r = \frac{n[M]_{free}}{K_d + [M]_{free}} \tag{2.6}$$

$$[M]_{free} = [M]_{total} - r[E] \tag{2.7}$$

$$r = \frac{n[M]_{total}}{K_d + [M]_{total}} \tag{2.8}$$

K_{Assoc} values, the inverse of the K_d values, were obtained from an equation similar to Eq. 2.5. The data of the pH-dependence of K_{Assoc} were fit to Eq. 2.9 in the case of Fe(II)Co(II), Fe(II)Mn(II) and Fe(II)Zn(II) wt-GpdQ and the Fe(II)Co(II) derivatives of Tyr19Phe and Ser127Ala mutants of GpdQ [47].

$$K_{Assoc} = \frac{K_{Assoc(\max)}}{\left(1 + \frac{[H^+]}{K_{A1}} + \frac{K_{A2}}{[H^+]}\right)} \tag{2.9}$$

Equation 2.10 was used to fit the data of the Fe(II)Cd(II) derivative of wt-GpdQ.

$$K_{Assoc} = \frac{K_{Assoc(\max)}}{\left(1 + \frac{[H^+]}{K_{A1}}\right)} \tag{2.10}$$

The equation to fit the data the Fe(II)Zn(II)-substituted His217Ala mutant (117) of GpdQ is based on a triprotic model and was derived elsewhere (Eq. 2.11). [48]

$$K_{Assoc} = K_{assoc1} + \left(\frac{(K_{assoc4} - K_{assoc1})K_{A1}K_{A2}K_{A3} + (K_{assoc3} - K_{assoc1})K_{A1}K_{A2}[H^+] + (K_{assoc2} - K_{assoc1})K_{A1}[H^+]^2}{[H^+]^3 + K_{A1}[H^+]^2 + K_{A1}K_{A2}[H^+] + K_{A1}K_{A2}K_{A3}}\right) \tag{2.11}$$

2.5 Syntheses of the Ligands

2.5.1 Synthesis of Ethyl-4-hydroxy-3,5-bis (hydroxymethyl)benzoate

CO_2Et / OH — 166.17 g/mol

NaOH, Formaldehyde 37 %, 55 °C, 3 d

CO_2Et / OH OH OH — 226.23 g/mol

The compound was prepared by a modification of a previously published procedure [20]. To a cool 12 % aqueous solution of sodium hydroxide (70 mL, 0.21 mol), commercially available ethyl 4-hydroxybenzoate (15 g, 0.09 mol) was added at 0 °C. Then aqueous formaldehyde solution (37 %, 60 mL, 2.16 mol) was added. The reaction was stirred at 55 °C for 3 days. The red solution was allowed to reach room temperature and subsequently ethyl acetate (100 mL) added. The organic layer was discarded and to the remaining aqueous phase ethyl acetate (100 mL) and saturated ammonium chloride solution were added. The organic phase was collected and dried over sodium sulfate. Subsequent removal of the solvent in vacuo left an orange solid which was recrystallized from chloroform/ methanol (1:1) to yield a pale yellow powder (11.7 g, 57 %).

^{1}H NMR (d^4-MeOD, 400.13 MHz); δ 1.37 (t, 3H, CH_2CH_3, J = 7.1 Hz); 4.32 (q, 2H, CH_2CH_3, J = 7.1 Hz); 4.80 (s, 4H, CH_2OH); 7.71 (s, 2H, Ar–H). **^{13}C NMR** (d^4-MeOD, 100.62 MHz) δ 14.7 (CH_2CH_3); 61.6 (CH_2CH_3); 61.8 (CH_2OH); 122.4 (arCCO_2Et); 128.4 (arCH); 129.6 (arCCH_2OH); 159.4 (arCOH); 168.5 (CO_2Et). **ESI mass spectrometry** (methanol) m/z 153.10 $[C_8H_8O_3 + H]^+$.

FT-IR spectroscopy (ν, cm^{-1}) 3428 (m, O–H str); 2985, 2942, 2871 (w, CH_2 str); 1689 (s, C=O str); 1206 (s, C–O str); 1010 (m, C–O str). **Melting Point** 137.6–139.4 °C Lit: 139.0 °C (benzene/petroleum ether) [20], 137.0–138.0 °C (chloroform) [49].

2.5.2 Synthesis of Ethyl 3,5-bis(bromomethyl)-4-hydroxybenzoate

CO2Et / OH OH OH — 30 % HBr, acetic acid, 15 min, rt → CO2Et / Br OH Br

226.23 g/mol 352.02 g/mol

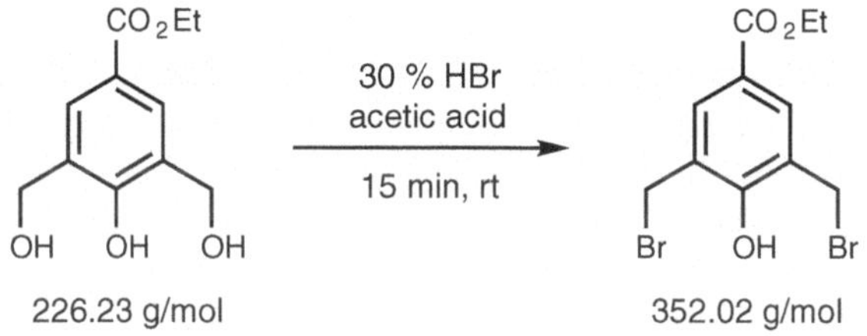

The compound was prepared by a previously published procedure [21]. 2,6-bis(hydroxymethyl)-4-hydroxybenzoate (0.58 g, 16 mmol) was added whilst stirring to a solution of hydrogen bromide in acetic acid (30 % HBr, 2 mL, 50 mmol). After 5 min a clear solution formed, from which after 10 min an orange precipitate separated. The sticky mass was cooled, filtered and washed thoroughly with cold petroleum ether until an orange solid was obtained. The resulting solid was dried in vacuo. (0.83 g, 90.45 %).

^{1}**H NMR** ($CDCl_3$, 300.13 MHz); δ 1.37 (t, 3H, CH_2CH_3, J = 7.1 Hz); 4.34 (q, 2H, CH_2CH_3, J = 7.1 Hz); 4.54 (s, 2H, CH_2Br); 6.19 (s, 1H, O*H*); 7.97 (s, 2H, ar*H*). ^{13}**C NMR** ($CDCl_3$, 100.62 MHz); δ 14.3 (CH_3); 28.4 (CH_2Br); 61.2 (CH_2CH_3); 123.4 (ar*C*CO_2Et); 124.9 (ar*C*CH_2); 132.8 (ar*C*H); 157.1 (ar*C*OH); 165.4 (*C*O_2Et). **FT-IR spectroscopy** (ν, cm^{-1}) 3291 (m, O–H str); 2985 (w, C–H str); 1881 (s, C=O str); 1235, 1191.6 (s, C–O str); 585 (s, C–Br str). **Melting point** 155.0–157.4 °C Lit: 155.0–157.0 °C (petroleum ether) [21].

2.5.3 Synthesis of 3-(chloromethyl)-2-hydroxy-5-methylbenzaldehyde

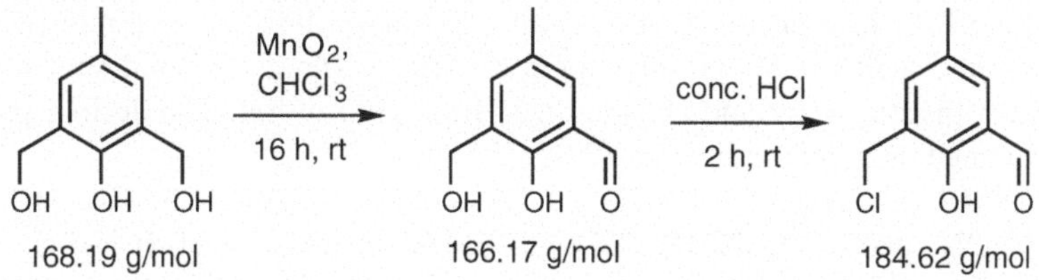

OH OH OH — MnO_2, $CHCl_3$, 16 h, rt → OH OH O — conc. HCl, 2 h, rt → Cl OH O

168.19 g/mol 166.17 g/mol 184.62 g/mol

The compound was prepared after a previously published procedure [50]. 2.6-bis(hydroxymethyl)-4-methylphenol (16 g, 0.12 mol) was stirred with activated MnO_2 (85 %, 80 g, 1.14 mol) in chloroform (250 mL) at room temperature for 16 h. The black slurry was then filtered through Celite and the yellow filtrate was concentrated to a brown oil in vacuo. The residue was taken up in hot toluene (40 mL) and filtered. A brown oil was obtained after removal of the solvent in vacuo which solidified upon standing. The solid was dissolved in ethanol (30 mL) and added dropwise to conc. hydrochloric acid (80 mL) at 40 °C. The mixture was subsequently stirred for 2 h at room temperature and the white powder obtained was filtered off and washed with water until the filtrate was pH neutral (3.94 g, 17.91 %).

^{1}H NMR ($CDCl_3$, 500.13 MHz); δ 2.33 (s, 3H, CH_3); 4.64 (s, 2H, CH_2); 7.32 (d, 1H, ar*H*, J = 1.8 Hz); 7.43 (d, 1H, ar*H*, J = 2.0 Hz), 9.84 (s, 1H, C*H*O). **^{13}C NMR** ($CDCl_3$, 100.62 MHz) δ 20.2 (CH_3); 39.8 (CH_2); 120.4 (ar*C*); 125.7 (ar*C*); 129.2 (ar*C*); 133.9 (ar*C*); 138.6 (ar*C*); 157.3 (ar*C*OH); 196.4 (*C*HO). **FT-IR spectroscopy** (ν, cm^{-1}) 3187 (s, O–H str); 2976, 2854 (w, C–H str); 1659 (s, C=O str). **Melting point** 92–94 °C Lit: 92–93 °C [50].

2.5.4 *Synthesis of 2,6-bis(bromomethyl)-4-nitrophenol*

NO_2 paraformaldehyde H_2SO_4, CH_3CO_2H 60 °C, 2 d → NO_2 … OAc; HBr rf, 5 h → NO_2 … Br OH Br

139.11 g/mol 253.21 g/mol 324.95 g/mol

The compound was prepared after a previously published procedure [23]. In a round bottom flask paraformaldehyde (95 %, 12 g, 0.4 mol), acetic acid (50 mL) and sulfuric acid (21.9 mL) were heated to 55 °C and 4-nitrophenol (13.9 g, 0.1 mol) was added. The temperature was maintained at 55 °C and the reaction was stirred for 35 h. After cooling to room temperature water (60 mL) was added and the mixture was neutralized with solid potassium carbonate (90 g). The pale yellow precipitate which resulted was filtered, washed with copious amounts of water and dried in vacuo. After recrystallization from ethanol (3×) 2 g (0.008 mol) of the 8-acetoxymethyl-6-nitro-1,3-benzodioxene were refluxed with hydrobromic acid (48 %, 60 mL) for 5 h. After cooling the solid was filtered, washed with cold water and recrystallized form chloroform. The desired product was obtained as a yellow powder (2.02 g, 40 %).

^{1}H NMR ($CDCl_3$, 300.13 MHz); δ 4.55 (s, 4H, CH_2Br); 6.58 (bs, 1H, O*H*); 8.19 (s, 2H, ar*H*). **^{13}C NMR** ($CDCl_3$, 100.62 MHz); δ 27.1 (CH_2Br); 126.6 (ar*C*CH_2Br); 127.7 (ar*C*H); 140.9 (ar*C*NO_2) 158.4 (ar*C*OH). **FT-IR spectroscopy**

(ν, cm^{-1}) 3407 (s, O–H str); 2921, 2851 (w, C–H str); 1510 (s, N=O asym. str); 1331 (s, N=O sym. str), 569 (s, C–Br str). **Melting point** 147.5–149.9 °C Lit: 149.0–150.0 °C (ethanol/hexane); [17] 146.0–147.0 °C (chloroform) [23].

2.5.5 Synthesis of 4-bromo-2,6-bis(hydroxymethyl)phenol

Br
OH
173.01 g/mol
NaOH 25 %
Formaldehyde 40 %
12 d
Br
OH OH OH
233.06 g/mol

The compound was synthesized after a previously published procedure [19]. 4-Bromophenol (17.3 g, 0.1 mol) in methanol (25 mL), aqueous sodium hydroxide solution (25 %, 50 mL) and formaldehyde were combined and left at room temperature for 12 d. Subsequently water (50 mL) and glacial acetic acid (15 mL) were added and after 2 h the solution was concentrated to 50 mL in vacuo and then cooled on ice. The solution was decanted from the orange oil/precipitate obtained on standing and the remaining oil was dissolved in aqueous sodium hydroxide solution (10 %, 50 mL). Activated charcoal was added and the suspension stirred overnight. After filtration the solution was acidified with 2 M HCl to pH 5. The resulting yellow precipitate was collected on a sintered glass funnel, recrystallized from water, dried in air and washed with cold chloroform (~20 mL) to yield the desired product as a white powder (7 g, 23 %).

^{1}H NMR (CD_3COCD_3, 300.13 MHz); δ 4.73 (s, 4H, CH_2); 7.28 (s, 2H, arCH) **^{13}C NMR** (CD_3COCD_3, 100.62 MHz); 61.4 (CH_2); 111.6 (ar$C$Br); 129.5 (ar$C$H); 130.5 (ar$C$CH2); 153.4 (ar$C$OH). **FT-IR spectroscopy** (ν, cm^{-1}) 3401, 3300 (s, O–H str); 2966, 2912, 2887 (w, CH_2 str); 1649, 1610, 1587 (w, C=C str); 1454 (s, CH_2 def); 1331 (s, O–H def); 1065 (s, C–OH def); 870 (m, Ar–H def). **Melting point** 161.7 °C Lit: 149.0–151.0 °C; [51] 164.0–168.0 °C; [52] 162.0–164.0 °C [19].

2.5.6 Synthesis of 4-bromo-2,6-bis(bromomethyl)phenol

Br
OH OH OH
233.06 g/mol
30 % HBr
in acetic acid
10 min, rt
Br
Br OH Br
358.85 g/mol

4-Bromo-2,6-bis(hydroxymethyl)phenol (3 g, 13 mmol) was added to hydrogen bromide in acetic acid (30 %, 15 mL) and stirred for 10 min until all starting material had dissolved. Upon addition of water the product precipitated form the solution and which was collected on an filter and washed thoroughly with cold petroleum spirit. The desired product was obtained as yellow powder (3.3 g, 70.5 %).

^{1}H NMR ($CDCl_3$, 300.13 MHz); δ 4.49 (s, 4H, CH_2); 5.66 (bs, 1H, O*H*); 7.39 (s, 2H, arC*H*) **^{13}C NMR** ($CDCl_3$, 100.62 MHz) 27.9 (*C*H_2); 112.7 (ar*C*Br); 127.2 (ar*C*CH_2); 133.7 (ar*C*H); 152.3 (ar*C*OH).**FT-IR spectroscopy** (ν, cm^{-1}) 3537 (s, O–H str); 3050, 2979 (w, CH_2 str); 1648, 1604 (w, C=C str); 1467 (s, CH_2 def); 869 (m, Ar–H def). **Melting point** 79.3–88.2 °C.

2.5.7 Synthesis of 2,6-bis(chloromethyl)-4-methylphenol

The compound was synthesized after a previously published procedure [22]. 2,6-bis(hydroxymethyl)-4-methylphenol (5 g, 29 mmol) was suspended in concentrated hydrochloric acid (75 mL) at room temperature and stirred overnight. The suspension was extracted with dichloromethane (3 × 50 mL) and the combined organic layers dried over sodium sulfate. After removal of the solvent in vacuo the product was obtained as white solid (5.5 g, 93.2 %).

^{1}H NMR ($CDCl_3$, 300.13 MHz); δ 2.26 (s, 3H, CH_3); 4.64 (s, 4H, CH_2Cl); 5.50 (s, 1H, O*H*); 7.07 (s, 2H, ar*H*). **^{13}C NMR** ($CDCl_3$, 100.62 MHz) 20.3 (*C*H_3); 42.5 (*C*H_2Cl); 124.6 (ar*C*CH_2); 130.5 (ar*C*CH_3); 131.6 (ar*C*H); 150.9 (ar*C*OH). **FT-IR spectroscopy** (ν, cm^{-1}) 3532 (m, O–H str); 2978, 2921 (w, CH_2 str); 1608 (w, C=C str); 1141 (m, C–O str); 869 (m, Ar–H def); 658 (m, C–Cl, skeletal vibr.). **Melting point** 81.9–84.1 °C Lit: 82.0–84.0 °C [53].

2.5.8 Synthesis of 2-methoxy-N-(pyridin-2-ylmethyl) aminoethanol

Pyridine 2-carboxaldehyde (2.25 g, 0.02 mol) in methanol (5 mL) was added dropwise to 2-methoxyethanolamine (1.58 g, 0.02 mol) in methanol (10 mL) at 0 °C. The solution was brought to room temperature on a thawing ice bath and stirred overnight. Sodium borohydride (0.83 g, 0.02 mol) was added in small portions and the solution was subsequently stirred for a further 2 h. Water (30 mL) was added and the reaction mixture was concentrated to about 30 mL in vacuo. The remaining solution was extracted with dichloromethane (3 × 20 mL) and the combined organic extracts were washed three times with brine, the solution dried over sodium sulfate and the solvent removed under vacuum. The desired product was obtained as a yellow oil (2.85 g, 90.6 %).

^{1}H NMR ($CDCl_3$, 300.13 MHz); δ 2.26 (s, 1H, N*H*); 2.72 (t, 2H, NCH_2CH_2, J = 5.4 Hz); 3.20 (s, 3H, OCH_3); 3.43 (t, 2H, CH_2OCH_3, J = 5.4 Hz); 3.79 (s, 2H, $arCH_2N$); 7.02 (dq, 1H, py*H*, J = 7.5, 4.9 Hz); 7.19 (dt, 1H, py*H*, J = 7.7, 1.0 Hz); 7.48 (td, 1H, py*H*, J = 7.6, 1.8 Hz); 8.40 (dq, 1H, py*H*, J = 4.9, 1.73, 0.9 Hz). **^{13}C NMR** ($CDCl_3$, 100.62 MHz); δ 48.6 (NCH_2CH_2); 54.9 (OCH_3); 58.8 ($arCH_2N$); 71.8 (NCH_2CH_2); 121.7 (py*C*H); 121.9 (py*C*H); 136.2 (py*C*H); 149.1 (py*C*H); 159.6 (py*C*CH_2). **ESI mass spectrometry** (methanol) m/z 167.10 $[C_9H_{14}N_2O + H]^+$. **FT-IR spectroscopy** (ν, cm^{-1}) 3319 (m, N–H str); 2889, 2830 (m, C–H str); 1592 (m, C=C str); 1435 (m, C–H def); 1357 (w, C–H sym. def); 1110 (s, C–O–C str); 757 (s, py–H).

2.5.9 *Synthesis of 2-phenoxy-N-(pyridin-2-ylmethyl) ethanamine*

MeOH, 0 °C, rt, 24 h; $NaBH_4$, 0 °C, 4 h

107.11 g/mol 137.18 g/mol 226.27 g/mol 228.29 g/mol

Pyridine 2-carboxaldehyde (0.39 g, 0.004 mol) in methanol (2 mL) was added dropwise to 2-phenoxyethanolamine (0.5 g, 0.004 mol) in methanol (3 mL) at 0 °C. The solution was brought to room temperature and stirred overnight. Subsequently sodium borohydride (0.2 g, 0.005 mol) was added in small portions and the solution stirred for a further 4 h. Water (6 mL) was added and the reaction mixture concentrated to about 6 mL in vacuo. The remaining solution was extracted with dichloromethane (3 × 10 mL) and the combined organic extracts dried over sodium sulfate and the solvent removed under vacuum to yield a yellow oil (0.75 g, 91.2 %).

^{1}H NMR ($CDCl_3$, 500.13 MHz); δ 2.14 (s, 1H, N*H*); 3.04 (t, 2H, NCH_2CH_2, J = 5.3 Hz); 3.98 (s, 2H, $arCH_2N$); 4.09 (t, 2H, NCH_2CH_2, J = 5.3 Hz); 6.89 (m, 2H, ar*H*, 6-H); 6.91 (m, 1H, *arH*); 7.13 (m, 1H, py*H*, J = 7.3, 4.9, 1.0 Hz); 7.24 (m, 2H, ar*H*, J = 7.4 Hz); 7.31 (d, 1H, py*H*, J = 7.8 Hz); 7.61 (td, 1H, py*H*, J = 7.7, 1.8 Hz); 8.54 (dq, 1H, py*H*, J = 4.9, 1.7, 0.9 Hz). **^{13}C NMR** ($CDCl_3$,

100.62 MHz); δ 48.4 (N*C*H_2CH_2); 55.1 (ar*C*H_2N); 67.3 (*C*H_2O); 114.5 (ar*C*H); 120.7 (ar*C*H); 121.9 (py*C*H); 122.1 (py*C*H); 129.3 (ar*C*H); 136.4 (py*C*H); 144.3 (py*C*H); 158.8 (ar*C*O); 159.6 (py*C*CH_2). **ESI mass spectrometry** (methanol) m/z 229.10 $[C_{14}H_{16}N_2O + H]^+$. **FT-IR spectroscopy** (ν, cm^{-1}) 3405 (m, N–H str); 2924 (m, C–H str); 1594 (m, C=C str); 1510 (s, N–H def); 1448 (m, C–H def); 1331 (w, C–H sym. def); 1212 (s, C–O–C str); 777 (w, Ar–H); 746 (s, py–H); 692 (w, Ar–H).

2.5.10 Synthesis of N-(2-Pyridylmethyl)-2-aminoethanol

MeOH, 0 °C, rt, 2 h; $NaBH_4$, 0 °C, rt, 2 h

107.11 g/mol 61.08 g/mol 150.18 g/mol 152.19 g/mol

The compound was prepared by a slight modification of previously published procedures [24, 25]. Pyridine 2-carboxaldehyde (4.42 g, 0.04 mol) in methanol (10 mL) was added dropwise to ethanolamine (2.42 mL, 0.04 mol) in methanol (20 mL) at 0 °C. The yellow solution was brought to room temperature and stirred for 2 h. Sodium borohydride (3.60 g, 0.1 mol) was added in small portions at 0 °C and the solution was stirred for a further 2 h. Water (30 mL) was added and the reaction mixture was concentrated to about 30 mL in vacuo. The remaining solution was extracted with dichloromethane (3 × 20 mL) and the combined organic phases were dried over sodium sulfate and the solvent removed under vacuum to yield a yellow oil. (2.50 g, 40.58 %).

^{1}H NMR ($CDCl_3$, 400.13 MHz); δ 2.14 (s, 1H, N*H*); 2.91 (t, 2H, NC*H_2*CH_2, J = 5.7 Hz); 3.79 (t, 2H, C*H_2*OH, J = 5.1 Hz); 5.27 (s, 2H, arC*H_2*N); 5.40 (bs, 1H, O*H*); 7.26 (m, 1H, py*H*, J = 6.2 Hz); 7.63 (d, 1H, py*H*, J = 7.7 Hz); 7.69 (tt, 1H, py*H*, J = 7.8, 1.8 Hz), 8.55 (dd, 1H, py*H*, J = 4.9 Hz). **^{13}C NMR** ($CDCl_3$, 100.62 MHz); δ 56.9 (N*C*H_2CH_2); 58.2 (ar*C*H_2N); 59.7 (*C*H_2 OH); 123.2 (py*C*H); 123.2 (py*C*H); 137.3 (py*C*H); 149.4 (py*C*H); 153.8 (py*C*CH_2). **ESI mass spectrometry** (methanol) m/z 175.07 $[C_8H_{12}N_2O + Na]^+$; 153.11 $[C_8H_{12}N_2O + H]^+$. **FT-IR spectroscopy** (ν, cm^{-1}) 3214 (m, O–H str); 2919 (m, C–H str); 2849 (m, C–H str); 1596 (m, C=C str); 1433 (m, O–H def); 832, 763 (s, py–H).

2.5.11 Synthesis of Ethyl 4-hydroxy-3,5-bis(((2-hydroxyethyl)(pyridin-2-ylmethyl)amino)methyl)benzoate (CO_2EtH_3L1)

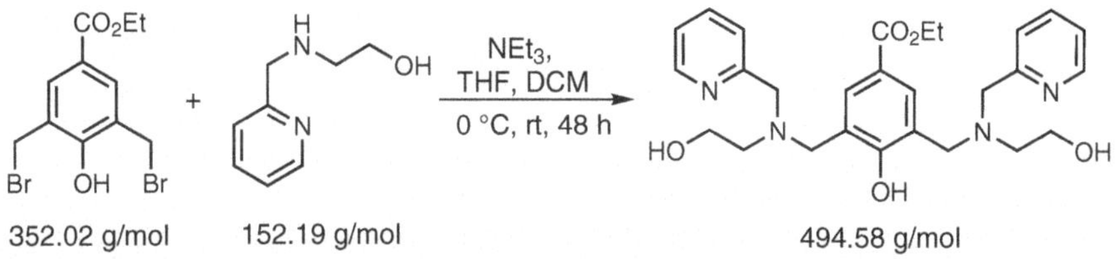

Ethyl 3,5-bis(bromomethyl)-4-hydroxybenzoate (0.50 g, 14 mmol) in dichloromethane (4 mL) was added dropwise to N-(2-pyridylmethyl)-2-aminoethanol (0.43 g, 28 mmol) and triethylamine (0.80 g) in tetrahydrofurane (4 mL) at 0 °C. The resulting yellow solution was stirred for 48 h and monitored with TLC, filtered to remove the white precipitate of triethylamine hydrobromide and the solvent removed under vacuum. The resulting brown oil was purified by flash column chromatography (ethyl acetate/methanol, 8:2, I_2 stain, $R_f = 0.44$) to yield a yellow oil (0.50 g, 72.2 %).

^{1}H NMR ($CDCl_3$, 500.13 MHz); δ 1.32 (t, 3H, CH_2CH_3, J = 7.1 Hz); 2.67 (t, 4H, NCH_2CH_2, J = 5.0 Hz); 3.67 (t, 4H, NCH_2CH_2, J = 5.0 Hz); 3.77 (s, 4H, arCH_2N); 3.83 (s, 4H, NCH_2py); 4.26 (q, 2H, CH_2CH_3, J = 7.1); 7.09 (m, 2H, py*H*, J = 5.8, 0.8 Hz); 7.20 (d, 2H, py*H*, J = 7.8 Hz); 7.54 (td, 2H, py*H*, J = 7.6, 1.7 Hz); 7.64 (s, 2H, ar*H*); 8.51 (dq, 2H, py*H*, J = 4.8, 0.8 Hz). **^{13}C NMR** ($CDCl_3$, 100.62 MHz); δ 14.4 (CH_2CH_3); 55.8 (arCCH_2), 56.5 (NCH_2CH_2); 58.5 (NCH_2CH_2); 58.8 (NCH_2py); 60.6 (CH_2CH_3); 122.5 (py*C*); 123.1 (ar*C*CO_2Et); 123.2 (ar*C*CH_2); 123.4 (py*C*); 131.8 (ar*C*H); 137.0 (py*C*); 149.1 (py*C*); 157.3 (py*C*CH_2); 160.9 (ar*C*OH); 166.3 (CO_2Et). **FT-IR spectroscopy** (*ν*, cm^{-1}) 3296 (m, O–H str); 2827 (m, C–H str); 1704 (m, C=O str); 1595 (m, C=C str); 1571 (m, C=C str); 1304 (m, C–O sym def); 1204 (s, C–O asym def); 1026 (m, C–O str); 758 (s, py–H). **ESI mass spectrometry** (methanol) m/z 495.19 $[C_{27}H_{34}N_4O_5 + H]^+$.

2.5.12 Synthesis of 4-bromo-2,6-bis(((2-methoxyethyl)(pyridin-2-ylmethyl)amino)methyl)phenol (BrHL2)

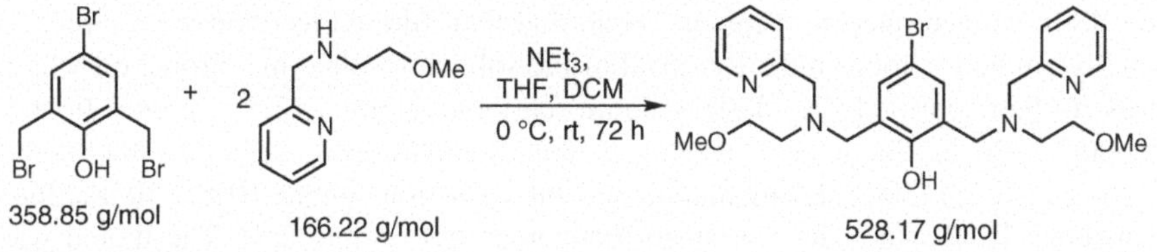

To a solution of 2-methoxy-N-(pyridin-2-ylmethyl)aminoethanol (0.925 g, 5.6 mmol) and triethylamine (1.25 g) in tetrahydrofuran (10 mL) was added dropwise a solution of 4-bromo-2,6-bis(bromomethyl)phenol (1 g, 2.7 mmol) in dichloromethane (10 mL) at 0 °C. The reaction mixture was stirred for 72 h and filtered to remove the precipitate of triethylamine hydrobromide. Removal of the solvent resulted in a yellow oil which was further purified by flash column chromatography (1 g crude ligand, length = 30 cm, diameter = 1.5 cm, ethyl acetate (until first band is eluted) then methanol/ethyl acetate 1:5, $FeCl_3$ stain, $R_f = 0.55$ in ethyl acetate). The ligand was obtained as a yellow oil (930 mg, 63 %).

^{1}H NMR ($CDCl_3$, 300.13 MHz) δ 2.79 (t, 4H, NCH_2CH$_2$); 3.26 (s, 6H, OCH_3); 3.52 (t, 4H, CH_2OCH$_3$); 3.79 (s, 4H, CH_2N); 3.96 (s, 4H, NCH_2py); 5.17 (s, 1H, OH); 7.08 (m, 2H, py*CH*); 7.19 (s, 2H, ar*CH*); 7.44 (m, 2H, py*CH*), 7.64 (m, 2H, py*CH*); 8.50 (m, 2H, py*CH*). **^{13}C NMR** ($CDCl_3$, 100.62 MHz); δ 52.8 (arCH_2N); 54.6 (NCH_2); 58.7 (OCH_3); 59.8 (NCH_2py); 69.9 (CH_2OCH$_3$); 110.7 (ar*C*Br); 122.45 (py*C*H); 123.6 (py*C*H); 123.8 (ar*C*CH$_2$); 131.9 (ar*C*H); 136.8 (py*C*H); 148.9 (py*C*H); 156.0 (ar*C*OH); 156.8 (py*C*CH$_2$). **FT-IR spectroscopy** (ν, cm^{-1}) 3231 (m, O–H str); 2927, 2879 (m, CH_2 str); 1592, 1572 (m, C=C str); 1435 (s, CH_2 def); 1112 (s, C–O str); 832, 756 (s, Ar–H). **ESI mass spectrometry** (methanol) m/z 529.10, 531.07 $[C_{26}H_{33}BrN_4O_3 + H]^+$; 551.16, 553.16 $[C_{26}H_{33}BrN_4O_3 + Na]^+$.

2.5.13 Synthesis of Ethyl 4-hydroxy-3,5-bis(((2-methoxyethyl)(pyridin-2-ylmethyl)amino)methyl)benzoate (CO_2EtHL2)

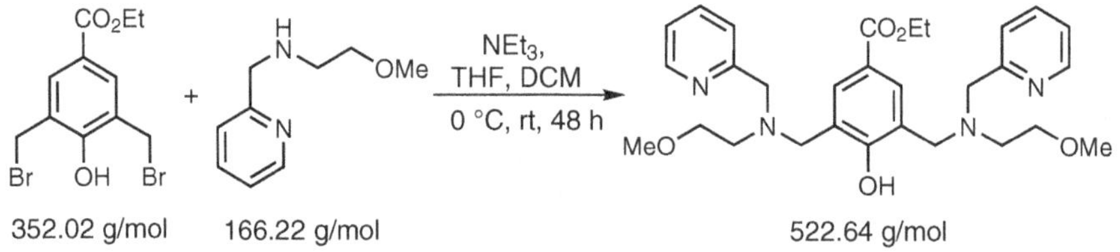

CO_2Et / Br OH Br + (pyridin-2-ylmethyl)(2-methoxyethyl)amine — NEt_3, THF, DCM, 0 °C, rt, 48 h →

352.02 g/mol 166.22 g/mol 522.64 g/mol

To a solution of 2-methoxy-N-(pyridin-2-ylmethyl)aminoethanol (2.8 g, 16.8 mmol) and triethylamine (3.96 g) in tetrahydrofuran (25 mL) was added dropwise a solution of ethyl-3,5-bis(bromomethyl)-4-hydroxybenzoate (2.5 g, 7.1 mmol) in dichloromethane (25 mL) at 0 °C. A white precipitate, identified as triethylamine hydrobromide, formed immediately. The reaction mixture was stirred for 48 h, concentrated to 4 mL in vacuo and then filtered to remove precipitated triethylamine hydrobromide. Removal of the solvent resulted in a brown oil which was further purified by flash column chromatography (2.5 g crude ligand, length = 30 cm, diameter = 1.5 cm, methanol/ethyl acetate 2:8 (1 L), methanol/ethyl acetate 2:1 (300 mL), methanol (300 mL), fraction volume 10 mL, 50 fractions collected. $FeCl_3$ stain, $R_f = 0.46$ in methanol/ethyl acetate 2:8). The ligand was obtained as a yellow oil (2.19 g, 60 %).

[1]**H NMR** ($CDCl_3$, 300.13 MHz); δ 1.34 (t, 3H, CH_2CH_3, J = 7.1 Hz); 2.73 (t, 2H, NCH_2CH_2, J = 5.7 Hz); 3.24 (s, 6H, OCH_3); 3.49 (t, 2H, NCH_2CH_2, J = 5.7 Hz); 3.81 (s, 2H, arCH_2N); 3.85 (s, 2H, NCH_2py); 4.30 (q, 2H, CH_2CH_3, J = 7.1 Hz); 7.10 (ddd, 2H, py*H*, J = 7.4, 4.9, 1.1 Hz); 7.42 (d, 2H, py*H*, J = 7.8 Hz); 7.57 (td, 4H, py*H*, J = 7.7, 1.3 Hz); 7.83 (s, 2H, ar*H*); 8.48 (dq, 2H, py*H*, J = 4.9, 1.7, 0.8 Hz). [13]**C NMR** ($CDCl_3$, 100.62 MHz); δ 14.2 (CH_2CH_3); 53.1(NCH_2CH_2), 55.0 (arCCH_2); 58.5 (OCH_3); 60.4, 60.5 (CH_2CH_3/NCH_2py); 70.7 (NCH_2CH_2); 120.4 (py*C*); 121.5 (ar*C*CO_2Et); 122.9 (ar*C*CH_2); 123.6 (py*C*); 130.4 (ar*C*H); 136.4 (py*C*); 148.6 (py*C*); 158.5 (py*C*CH_2); 160.3 (ar*C*OH); 166.5 (CO_2Et). **FT-IR spectroscopy** (ν, cm^{-1}) 2927 (m, C–H str); 1705 (m, C=O str); 1602 (m, C=C str); 1197 (m, C–O–C str): 1109 (m, C–O str); 832 (m, Ar–H); 762 (s, py–H). **ESI mass spectrometry** (methanol) m/z 523.21 $[C_{29}H_{38}N_4O_5 + H]^+$.

2.5.14 Synthesis of 2,6-bis(((2-methoxyethyl)(pyridin-2-ylmethyl)amino)methyl)-4-methylphenol (CH_3HL2)

To a solution of 2-methoxy-N-(pyridin-2-ylmethyl)aminoethanol (1.50 g, 9 mmol) and triethylamine (0.91 g, 9 mmol) in tetrahydrofuran (4.5 mL) was added dropwise a solution of 2,6-bis(chloromethyl)-*p*-cresol (0.94 g, 5 mmol) in dichloromethane (4 mL) at 0 °C. A white precipitate formed immediately. The reaction mixture was stirred for 48 h and then filtered to remove precipitated triethylamine hydrochloride. Removal of the solvents left a yellow oil which was further purified by flash column chromatography (2.17 g crude ligand, ethyl acetate/methanol 8:2 (900 mL), ethyl acetate/methanol 1:1 (400 mL), I_2 stain, R_f = 0.66 in ethyl acetate/methanol 8:2). The ligand was obtained as a yellow oil (1.35 g, 65.4 %).

[1]**H NMR** ($CDCl_3$, 500.13 MHz); δ 2.21 (s, 3H, CH_3); 2.75 (t, 2H, N–CH_2–, J = 5.8 Hz); 3.25 (s, 6H, OCH_3); 3.50 (t, 4H, CH_2OCH_3, J = 5.8 Hz); 3.77 (s, 4H, arCH_2N); 3.84 (s, 4H, NCH_2ar); 6.92 (s, 2H, arC*H*); 7.10 (ddd, 2H, pyC*H*, J = 7.4, 4.9, 1.1 Hz); 7.46 (d, 2H, pyC*H*, J = 7.8 Hz); 7.60 (td, 2H, pyC*H*, J = 7.6, 1.8 Hz); 8.47 (dq, 2H, pyC*H*, J = 4.9, 1.7 Hz). [13]**C NMR** ($CDCl_3$, 100.62 MHz); δ 20.6 (arCH_3); 53.8 (NCH_2CH_2), 55.6 (arCH_2N); 58.7 (OCH_3); 60.9 (NCH_2py); 71.0 (CH_2OCH_3); 123.0 (py*C*H); 124.0 (py*C*H); 124.7 (ar*C*CH_2); 128.2 (ar*C*CH_3); 130.3 (ar*C*H); 137.4 (py*C*H); 149.8 (pyCH); 154.6 (ar*C*OH); 159.8 (py*C*CH_2). **FT-IR spectroscopy** (ν, cm^{-1}) 2922, 2875, 2817 (m, C–H str); 1591 (m, C=C str);

1434 (m, O–H def); 1365 (m, C–O–C str): 1109 (m, C–O str): 844 (m, Ar–H); 756 (s, py–H). **ESI mass spectrometry** (methanol) m/z 487.18 $[C_{27}H_{36}N_4O_3 + Na]^+$, 465.22 $[C_{27}H_{36}N_4O_3 + H]^+$.

2.5.15 Synthesis of 4-hydroxy-3,5-bis(((2-hydroxyethyl)(pyridin-2-ylmethyl)amino)methyl)benzoic acid (CO_2HH_3L1) and 4-hydroxy-3,5-bis(((2-methoxyethyl)(pyridin-2-ylmethyl)amino)methyl)benzoic acid (CO_2HHL2)

1M HCl
80 °C, 3-4 d

R = H 494.58 g/mol
R = Me 522.64 g/mol

R = H 466.53 g/mol
R = Me 494.25 g/mol

The ethyl ester ligand (0.1 g, 0.2 mmol) was dissolved in 1 M HCl (5 mL, 5 mmol) and stirred at 80 °C for 4 days. The solvent was removed from the reaction mixture in vacuo leaving an orange solid which was redissolved twice in 20 % DCl in D_2O and repeatedly concentrated. NMR analysis showed that the compound was obtained as hydrochloride with all nitrogens protonated (60 mg, 50 %).

CO_2HHL1: ^{1}H NMR ($CDCl_3$, 300.13 MHz); δ 3.47 (t, 4H, NCH_2CH_2); 4.04 (t, 4H, NCH_2CH_2); 4.75 (s, 4H, arCH_2); 4.98 (s, 4H, NCH_2py); 7.96 (t, 2H, py*H*); 8.13 (s, 2H, ar*H*); 8.15 (d, 2H, py*H*); 8.45 (t, 2H, py*H*); 8.70 (d, 2H, py*H*). **^{13}C NMR** ($CDCl_3$, 100.62 MHz); δ 55.3 (arCCH_2); 55.7 (NCH_2CH_2), 57.1 (NCH_2CH_2, NCH_2py); 119.8 (ar*C*CH_2); 124.7 (ar*C*CO_2H); 128.3 (py*C*H); 129.9 (py*C*H); 138.1 (ar*C*H); 145.7 (py*C*H); 146.2 (py*C*H); 147.7 (py*C*); 161.6 (ar*C*OH); 167.9 (CO_2H). **ESI mass spectrometry** (methanol) m/z 489.29 $[C_{24}H_{30}N_4O_5Na]^+$; 457.20 $[C_{25}H_{31}N_4O_5]^+$; 315.15 $[C_{17}H_{19}N_2O_4]^+$; 271.17 $[C_{16}H_{18}N_2O_2]^+$. **FT-IR spectroscopy** (ν, cm^{-1}) 3362 (m, O–H str); 2954, 2851 (m, C–H str); 2499 (s, N–H str); 1720 (s, C=O str); 1435 (m, O–H def); 1236 (m, C–O–C str); 771 (s, py–H).

CO_2HHL2: The ^{1}H and ^{13}C peaks of the methyl ether group were not observed in some spectra as this group has been cleaved under the conditions. **^{1}H NMR** ($CDCl_3$, 500.13 MHz); δ 3.03 (s, 6H, OCH_3); 3.39 (t, 4H, NCH_2CH_2, J = 4.5 Hz); 3.77 (t, 4H, NCH_2CH_2, J = 4.5 Hz); 4.42 (s, 4H, arCH_2N); 4.61 (s, 4H, NCH_2py); 7.82 (t, 2H, py*H*, J = 6.1 Hz); 7.87 (s, 2H, ar*H*); 7.91 (d, 2H, py*H*, J = 7.7 Hz); 8.32 (t, 2H, py*H*, J = 7.6 Hz); 8.66 (d, 2H, py*H*, J = 5.0 Hz). **^{13}C NMR** (d^4-MeOD, 100.62 MHz) 55.2 (arCH_2N); 56.7 (NCH_2CH_2); 59.4 (NCH_2CH_2); 119.9 (ar*C*CH_2); 123.2 (ar*C*CO_2H); 127.2 (py*C*H); 127.9 (py*C*H); 136.6 (ar*C*H); 143.8

(py*C*H); 146.7 (py*C*H); 149.2 (py*C*); 161.4 (ar*C*OH); 169.1 (*C*O_2H). **FT-IR spectroscopy** (ν, cm^{-1}) 3365 (m, O–H str); 2936, 2822 (m, C–H str); 2496 (s, N–H str); 1704 (s, C=O str); 1441 (m, O–H def); 1394 (m, C–O–C str); 1104 (m, C–O str); 767 (s, py–H). **ESI mass spectrometry** (methanol) m/z 495.19 $[C_{26}H_{34}N_4O_5 + H]^+$

2.5.16 Synthesis of ethyl 4-hydroxy-3,5-bis(((2-phenoxyethyl) (pyridin-2-ylmethyl)amino)methyl)Benzoate (CO_2EtHL3)

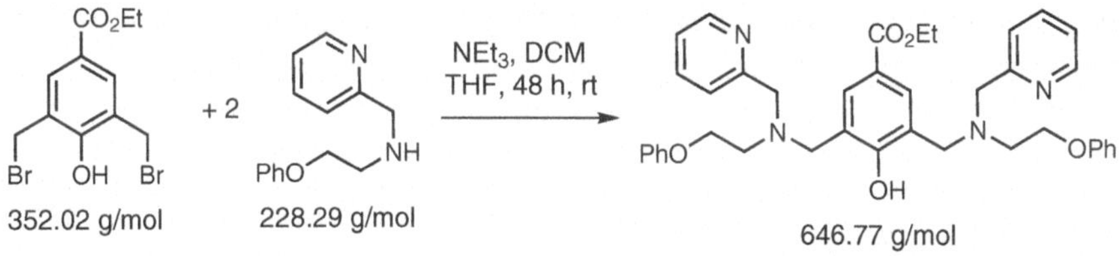

Ethyl 3,5-bis(bromomethyl)-4-hydroxybenzoate (0.19 g, 0.55 mmol) in dichloromethane (2 mL) was added dropwise to a mixture of 2-phenoxy-N-(pyridin-2-ylmethyl)ethanamine (0.25 g, 1.09 mmol) and triethylamine (0.30 g) in tetrahydrofuran (2 mL) at 0 °C. The resulting yellow solution was stirred for 48 h at room temperature and monitored with TLC. The solution was then filtered to remove the white precipitated triethylamine hydrobromide and the solvent removed under vacuum. The brown oil was purified by flash column chromatography (ethyl acetate/methanol, 1:1, $FeCl_3$ stain) to yield a yellow oil (0.18 g, 50 %).

^{1}H NMR ($CDCl_3$, 500.13 MHz) δ 1.39 (t, 3H, CH_2CH_3, J = 7.2); 3.04 (t, 4H, NCH_2CH_2, J = 5.7); 3.94 (s, 4H, arCH_2N); 3.99 (s, 4H, NCH_2py); 4.11 (t, 4H, N$CH_2$$CH_2$, J = 5.7); 4.35 (q, 2H, CH_2CH_3, J = 7.2); 6.86 (m, 4H, ar*H*, J = 8.7, 1.0); 6.92 (tt, 2H, ar*H*, J = 7.4 1.0); 7.16 (ddd, 2H, py*H*, J = 7.5, 4.9, 1.0); 7.25 (m, 4H, ar*H*, J = 7.3); 7.58 (d, 2H, py*H*, J = 7.8); 7.63 (td, 2H, py*H*, J = 7.6, 1.7); 7.90 (s, 2H, ar*H*); 8.49 (dq, 2H, py*H*, J = 4.9, 1.7. 0.8). **^{13}C NMR** ($CDCl_3$, 100.62 MHz) δ 14.4 (CH_2CH_3); 52.4 (NCH_2CH_2), 54.9 (arCCH_2N); 60.3 (NCH_2py/CH_2CH_3); 65.5 (NCH_2CH_2); 114.4 (ar*C*H); 120.7 (ar*C*H); 122.1 (py*C*H); 122.2 (arCCH_2); 122.7 (py*C*H); 123.1 (ar*C*H); 123.7 (arCCO_2Et); 129.3 (ar*C*H); 130.8 (ar*C*H); 136.6 (py*C*H); 148.8 (py*C*H); 158.5 (py*C*); 160.6 (ar*C*OH); 166.7 (*C*O_2Et). **ESI mass spectrometry** (methanol) m/z 647.26 $[C_{39}H_{42}N_4O_5 + H]^+$. **FT-IR spectroscopy** (ν, cm^{-1}) 2924, 2850 (m, C–H str); 1704 (m, C=O str); 1597 (m, C=C str); 1241 (m, C–O–C str): 1199 (m, C–O str): 752 (s, py–H).

2.5.17 Synthesis of 4-methyl-2,6-bis(((2-phenoxyethyl)(pyridin-2-ylmethyl)amino)methyl)phenol (CH_3HL3)

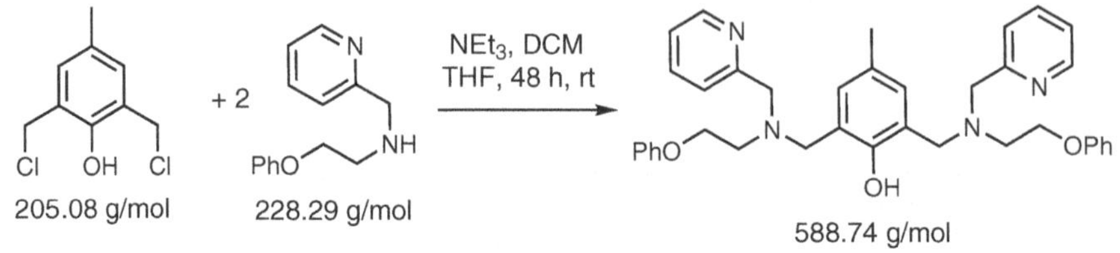

2,6-bis(chloromethyl)-*p*-cresol (0.16 g, 0.55 mmol) in dichloromethane (2 mL) was added dropwise to a mixture of 2-phenoxy-N-(pyridin-2-ylmethyl)ethanamine (0.25 g, 1.09 mmol) and triethylamine (0.30 g) in tetrahydrofuran (2 mL) at 0 °C. The resulting yellow solution was stirred for 48 h at room temperature. It was then filtered to remove the white precipitate of triethylamine hydrobromide and the solvent removed under vacuum. The yellow oil was purified by flash column chromatography (ethyl acetate/methanol, 9:1, $FeCl_3$ stain) to yield the ligand as yellow oil (0.17 g, 52 %).

^{1}H NMR ($CDCl_3$, 500.13 MHz) δ 2.26 (s, 3H, CH_3); 3.03 (t, 4H, CH_2O, J = 5.8 Hz); 3.88 (s, 4H, arCH_2N); 3.98 (s, 4H, NCH_2py); 4.10 (t, 4H, N$CH_2$$CH_2$, J = 5.8 Hz); 6.85 (m, 4H, ar*CH*, J = 8.7, 1.0 Hz); 6.92 (tt, 2H, ar*CH*, J = 7.4, 1.0 Hz); 6.99 (s, 2H, ar*CH*); 7.13 (ddd, 2H, py*CH*, J = 7.4, 4.9, 1.1 Hz); 7.24 (m, 4H, *CH*, J = 7.4 Hz); 7.49 (d, 2H, py*CH*, J = 7.8 Hz); 7.60 (td, 2H, py*CH*, J = 7.6, 1.8 Hz); 8.53 (dq, 2H, py*CH*, 4.8, 1.7, 0.8 Hz). **^{13}C NMR** ($CDCl_3$, 100.62 MHz); δ 20.5 (CH_3); 52.3 (CH_2OPh), 55.2 (arCH_2N); 60.4 (NCH_2py); 65.5 (NCH_2); 114.4 (ar*C*H); 120.6 (ar*C*H); 122.0 (py*C*H); 123.2 (py*C*H); 123.3 (ar*C*); 127.6 (ar*C*CH_3); 129.3 (ar*C*H); 136.5 (py*C*H); 148.8 (py*C*H); 153.6 (ar*C*OH); 158.6 (py*C*). **FT-IR spectroscopy** (ν, cm^{-1}) 2922, 2831 (m, C–H str); 1597 (m, C=C str); 1475 (m, C–H def); 1242 (m, C–O–Ph str); 752 (s, py–H); 729, 691 (s, Ar–H). **ESI mass spectrometry** (methanol) m/z 611.23 $[C_{37}H_{40}N_4O_3 + Na]^+$; 589.25 $[C_{37}H_{40}N_4O_3 + H]^+$.

2.5.18 Synthesis of 2,6-bis(((2-methoxyethyl)(pyridin-2-ylmethyl)amino)methyl)-4-nitrophenol (NO_2HL2)

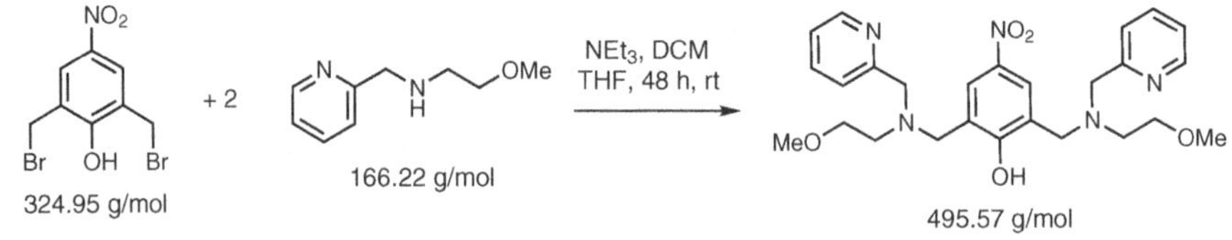

2-methoxy-N-(pyridin-2-ylmethyl)aminoethanol (0.51 g, 31 mmol) was dissolved together with triethylamine (0.7 g) in tetrahydrofuran (5 mL) and cooled to 0 °C. Upon dropwise addition of 2,6-bis(bromomethyl)-4-nitrophenol (0.5 g, 15 mmol) in dichloromethane (7 mL) the solution turned bright yellow and a precipitate emerged immediately. The mixture was stirred for 3 days at room temperature, filtered and concentrated in vacuo. After further purification with column chromatography (ethyl acetate/methanol, 8:1, $FeCl_3$ stain, $R_f = 0.76$) a yellow oil was obtained (0.5 g, 68 %).

^{1}H NMR ($CDCl_3$, 500.13 MHz) δ 2.80 (t, 3H, NCH_2CH_2, J = 5.6 Hz); 3.29 (s, 6H, OCH_3); 3.52 (t, 4H, NCH_2CH_2, J = 5.6 Hz); 3.85 (s, 4H, arCH_2N); 3.91 (s, 4H, NCH_2py); 7.16 (dd, 2H, py*CH*, J = 6.9 Hz); 7.42 (d, 2H, py*CH*, J = 7.8 Hz); 7.65 (td, 2H, py*CH*, J = 7.8, 1.7 Hz); 8.14 (s, 2H, ar*CH*); 8.52 (d, 2H, py*CH*, J = 4.4 Hz). **^{13}C NMR** ($CDCl_3$, 100.62 MHz) δ 53.2 (NCH_2); 54.6 (arCH_2N); 58.8 (OCH_3); 60.3 (NCH_2py); 70.6 (CH_2OCH_3); 122.2 (py*C*H), 123.0 (py*C*H); 124.5 (py*C*H); 125.0 (ar*C*H); 136.7 (py*C*H); 139.7 (*C*NO_2); 148.9 (py*C*H); 158.5 (py*C*); 162.5 (*C*OH). **FT-IR spectroscopy** (ν, cm^{-1}) 3372 (m, O–H str); 2927, 2878, 2821 (m, C–H str); 1592 (m, C=C str); 1514 (m, N=O asym. str); 1471 (m, C–H def); 1329 (s, N=O sym. str); 1095 (m, C–O str); 845 (w, ar–H); 752 (s, py–H). **ESI mass spectrometry** (methanol) m/z 496.25 $[C_{26}H_{33}N_5O_5 + H]^+$.

2.5.19 Synthesis of 4-amino-2,6-bis(((2-methoxyethyl)(pyridin-2-ylmethyl)amino)methyl)phenol (NH_2HL2)

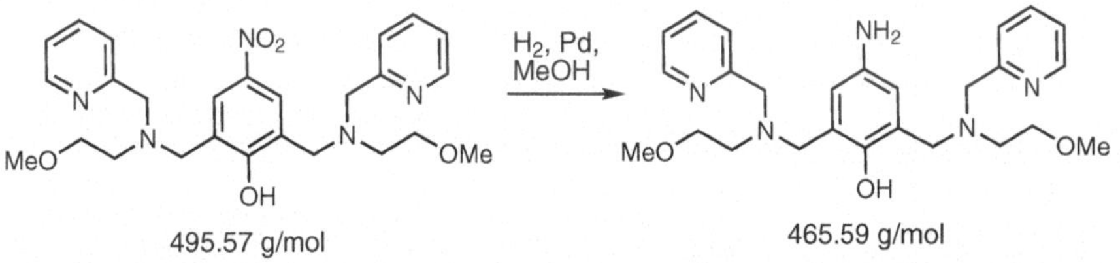

NO2 … OH, MeO, OMe — 495.57 g/mol; H_2, Pd, MeOH →; NH2 … OH — 465.59 g/mol

A 0.05 M solution of NO_2**HL2** (130 mg) in methanol was cycled twice in an H-cube Hydrogen reactor equipped with a palladium cartridge (full H_2 mode, 1 mL/min flow) at room temperature and atmospheric pressure. After one cycle the orange color of the solution had faded to yellow. After removal of the solvent the desired product was obtained as yellow oil (105 mg, 85 %).

^{1}H NMR ($CDCl_3$, 500.13 MHz) δ 2.82 (t, 3H, NCH_2CH_2, J = 5.5 Hz); 3.24 (s, 6H, OCH_3); 3.50 (t, 4H, NCH_2CH_2, J = 5.5 Hz); 3.79 (s, 4H, arCH_2N); 3.91 (s, 4H, NCH_2py); 6.56 (s, 2H, ar*CH*); 7.13 (t, 2H, py*CH*, J = 5.7 Hz); 7.42 (d, 2H, py*CH*, J = 7.8 Hz); 7.60 (t, 2H, py*CH*, J = 3.8 Hz); 8.47 (d, 2H, py*CH*, J = 3.9 Hz). **^{13}C NMR** ($CDCl_3$, 100.62 MHz) δ 52.9 (NCH_2); 55.2 (arCH_2N); 58.6 (OCH_3); 59.9 (NCH_2py); 70.0 (CH_2OCH_3); 116.63 (ar*C*H); 122.2 (py*C*H);

123.4 (py*C*H); 123.6 (py*C*H); 136.6 (py*C*H); 138.1 (*C*NH_2); 148.6 (*C*OH); 148.8 (py*C*H); 158.0 (py*C*). **FT-IR spectroscopy** (ν, cm^{-1}) 3349 (m, N–H/O–H str); 2928, 2881, 2824 (m, C–H str); 1622 (m, N–H def); 1572 (m, C=C str); 1476 (m, C–H def); 1100 (m, C–O str); 760 (s, py–H). **ESI mass spectrometry** (methanol) m/z 466.15 $[C_{26}H_{35}N_5O_3 + H]^+$.

2.5.20 Synthesis of 2-(((2-methoxyethyl)(pyridin-2-ylmethyl) amino)methyl)-4-methyl-6-(((pyridin-2-ylmethyl) amino)methyl)phenol (CH_3HL4)

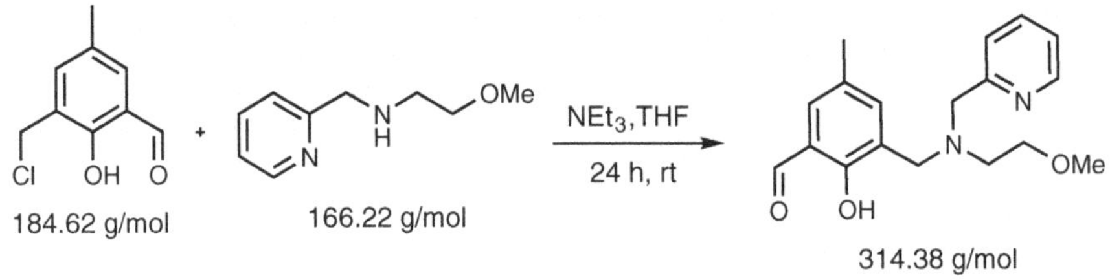

Triethylamine (2.6 mL) was added dropwise to a mixture of 3-(chloromethyl)-2-hydroxy-5-methylbenzaldehyde (1.0 g, 6.0 mmol) and 2-methoxy-N-(pyridin-2-ylmethyl)ethanamine (1.0 g, 6.0 mmol) in tetrahydrofurane (45 mL) at room temperature and the mixture stirred for 24 h. After filtration and concentration in vacuo the residue was taken up in water (30 mL) and extracted with dichloromethane (3· × 30 mL). The combined organic layers were dried over sodium sulfate and concentrated in vacuo to yield 1.69 g (89 %) of crude 2-hydroxy-3-(((2-methoxyethyl)(pyridin-2-ylmethyl)amino)methyl)-5-methylbenzaldehyde as an orange oil which was used in the next step without further purification.

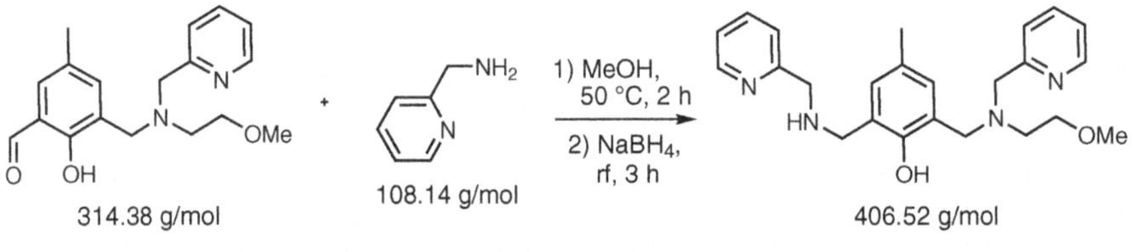

2-hydroxy-3-(((2-methoxyethyl) (pyridin-2-ylmethyl)amino)methyl)-5-methylbenzaldehyde (1.69 g, 5.37 mmol) was dissolved in methanol (50 mL) and 2-aminomethylpyridine (0.58 g, 5.37 mmol) in methanol (25 mL) was added dropwise at room temperature. The resulting mixture was stirred at 50 °C for 2 h. The mixture was subsequently cooled to 0 °C and sodium borohydride (0.72 g, 19.0 mmol) was added in small portions. After heating to reflux for 3 h the crude ligand solution was concentrated in vacuo, taken up in acidified water (100 mL, pH 2) and extracted with dichloromethane (3 × 35 mL). The combined organic layers were washed with saturated sodium bicarbonate solution (3 × 50 mL) and dried over sodium sulfate. After removal of the solvent in vacuo the crude ligand (1.53 g, 70 %) was purified with flash column chromatography (ethyl acetate until first band is eluted, methanol/ethyl acteate 9:1, $FeCl_3$ stain, $R_f = 0.48$ in ethyl acetate) to yield the ligand as a yellow oil (900 mg, 41 %).

^{1}H NMR ($CDCl_3$, 500.13 MHz) δ 2.20 (s, 3H, arCCH_3); 2.76 (t, 2H, NCH_2CH_2, J = 5.5 Hz); 3.26 (s, 3H, OCH_3); 3.50 (t, 2H, NCH_2CH_2, J = 5.5 Hz); 3.77 (s, 2H, arCCH_2N); 3.85 (s, 2H, arCCH_2NH); 3.92 (s, 2H, NCH_2py); 3.96 (s, 2H, NHCH_2py); 6.78 (d, 1H, arC*H*, J = 1.8 Hz); 6.92 (d, 1H, arC*H*, J = 1.8 Hz); 7.13 (qd, 2H, py*H*, J = 7.4, 1.2 Hz); 7.34 (dd, 2H, py*H*, J = 5.1, 1.5 Hz); 7.62 (tt, 2H, py*H*, J = 7.6, 1.9 Hz); 8.51 (ddt, 2H, py*H*, J = 6.6, 1.7, 0.9 Hz). **^{13}C NMR** ($CDCl_3$, 100.62 MHz) δ 20.4 (arCCH_3); 49.2 (NCH_2py); 52.6 (N$CH_2$$CH_2$); 53.7 (NH$CH_2$py); 56.9 (arC$CH_2$N); 58.6 (O$CH_3$); 60.0 (arC$CH_2$NH); 70.1 (N$CH_2$$CH_2$); 122.0 (py*C*H); 122.2 (py*C*H); 122.3 (py*C*H); 122.4 (ar*C*); 123.4 (py*C*H); 124.4 (ar*C*); 127.8 (ar*C*CH_3); 129.2 (ar*C*H); 129.7 (ar*C*H); 136.5 (py*C*H); 136.6 (py*C*H); 148.9 (py*C*H); 149.0 (py*C*H); 153.5 (ar*C*OH); 158.0 (py*C*CH_2); 158.6 (py*C*CH_2). **FT-IR spectroscopy** (*ν*, cm^{-1}) 3402 (m, O–H str); 2917, 2819 (m, C–H str); 1108 (m, C–O str); 863 (w, ar–H); 755 (s, py–H). **ESI mass spectrometry** (methanol) m/z 407.23 $[C_{24}H_{30}N_4O_2 + H]^+$.

2.5.21 Synthesis of 2-(((2-methoxyethyl)(pyridine-2-ylmethyl) amino)methyl)-4-methyl-6-(((pyridin-2-ylmethyl) (4-vinylbenzyl)amino)methyl)phenol (CH_3HL5)

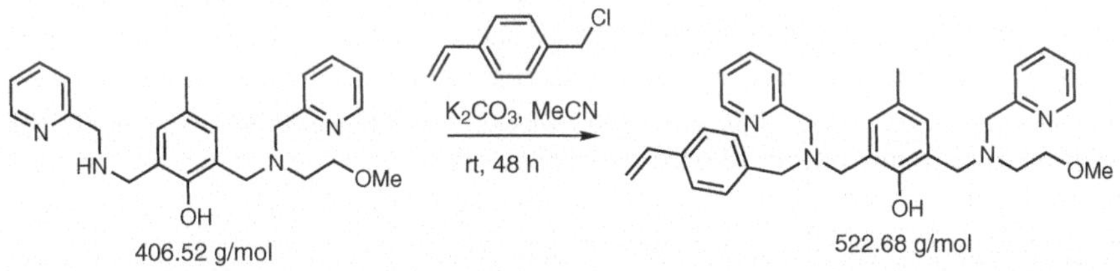

CH_3**HL4** (500 mg) and vinylbenzylchloride (0.174 mL) were stirred with dry powdered potassium carbonate (0.256 g) in acetonitrile (5 mL) at room temperature for two days. After this time the mixture was filtered and concentrated in vacuo. The crude ligand was purified by flash column chromatography (ethyl acetate/methanol 5:2) and obtained as orange oil (Yield: 38 %, 0.24 g).

^{1}H NMR ($CDCl_3$, 500.13 MHz) δ 2.24 (s, 3H, arCCH_3); 2.77 (t, 2H, NCH_2CH_2, J = 5.7 Hz); 3.27 (s, 3H, OCH_3); 3.52 (t, 2H, NCH_2CH_2, J = 5.7 Hz); 3.65 (s, 2H, CH_2); 3.71 (s, 2H, CH_2); 3.77 (s, 2H, CH_2); 3.79 (s, 2H, CH_2); 3.86 (s, 2H, CH_2); 5.19 (dd, 1H, CH=CH_2 cis, J = 10.9, 0.6 Hz); 5.69 (dd, 1H, CH=CH_2 trans, J = 17.6, 0.8 Hz); 6.67 (dd, 1H, C*H*=CH_2, J = 17.6, 10.9 Hz); 6.89 (d, 1H, arC*H*, J = 1.7 Hz); 7.03 (d, 1H, arC*H*, J = 1.8 Hz); 7.13 (m, 2H, pyC*H*, J = 6.4, 0.9 Hz);

7.34 (s, 4H, vinylbenzyl*H*); 7.43 (d, 1H, pyC*H*, J = 7.8 Hz); 7.51 (d, 1H, pyC*H*, J = 7.8 Hz); 7.62 (m, 2H, pyC*H*, J = 7.7, 1.8 Hz); 8.51 (m, 2H, pyCH, J = 0.9 Hz). **^{13}C NMR** ($CDCl_3$, 100.62 MHz) δ 20.6 (arCCH_3); 52.9 (N$CH_2$$CH_2$); 53.8 ($CH_2$); 55.7 ($CH_2$); 57.9 ($CH_2$); 58.6 (O$CH_3$); 59.6 ($CH_2$); 60.4 ($CH_2$); 70.4 (N$CH_2$*C*$H_2$); 113.4 (CH=*C*$H_2$); 121.8 (py*C*H); 122.0 (py*C*H); 122.9 (py*C*H); 123.2 (py*C*H); 123.9 (ar*C*–CH=CH_2); 126.1 (vinylbenzyl ar*CH*); 127.5 (ar*C*CH_3); 129.0 (ar*C*H); 129.1 (vinylbenzyl ar*CH*); 129.3 (ar*CH*); 136.4 (pyCH); 136.5 (py*C*H); 136.6 (*C*H=CH_2); 138.5 (arC); 148.8 (py*C*H); 148.9 (py*C*H); 153.5 (ar*C*OH); 158.0 (py*C*CH_2); 158.6 (py*C*CH_2). **FT-IR spectroscopy** (ν, cm^{-1}) 2918, 2816 (m, C–H str); 1109 (m, C–O str); 861 (w, ar–H); 755 (s, py–H). **ESI mass spectrometry** (methanol) m/z 523.20 [$C_{33}H_{38}N_4O_2$ + H]$^+$.

2.5.22 Synthesis of Ethyl 4-hydroxy-3,5-bis((bis(pyridine-2-ylmethyl)amino)methyl)benzoate (CO_2EtHL6)

CO_2Et / Br OH Br — 352.02 g/mol + 2 (bis(pyridin-2-ylmethyl)amine) — 199.25 g/mol → NEt₃, DCM, THF, 48 h, rt → 588.70 g/mol

Bis(pyridin-2-ylmethyl)amine (0.57 g, 28 mmol) was dissolved in tetrahydrofuran (4 mL) together with triethylamine (0.8 g) and cooled to 0 °C. Upon addition of ethyl 3,5-bis(bromomethyl)-4-hydroxybenzoate (0.50 g, 14 mmol) in dichloromethane (4 mL) a precipitate emerged immediately. The mixture was stirred for 48 h at room temperature, filtered and concentrated in vacuo. After further purification with column chromatography (ethyl acetate/methanol, 8:1, $FeCl_3$ stain, R_f = 0.55) a brown oil was obtained (0.5 g, 61 %).

^{1}H NMR ($CDCl_3$, 500.13 MHz) δ 1.31 (t, 3H, CH_2CH_3, J = 7.1 Hz); 3.77 (s, 4H, arCH_2N); 3.82 (s, 8H, NCH_2py); 4.25 (q, 2H, C$H_2$$CH_3$, J = 7.1 Hz); 7.06 (ddd, 4H, py*H*, J = 7.4, 4.9, 1.0 Hz); 7.42 (d, 4H, py*H*, J = 7.8 Hz); 7.54 (td, 4H, py*H*, J = 7.7, 1.8 Hz); 7.88 (s, 2H, ar*H*); 8.45 (d, 4H, py*H*, J = 4.8 Hz). **^{13}C NMR** ($CDCl_3$, 100.62 MHz) δ 14.3 (CH_3); 50.1 (arC*C*H_2N); 60.2 (py*C*H); 60.3 (*C*$H_2$$CH_3$); 120.3 (ar*C*$CO_2$Et); 121.9 (py*C*H); 122.8 (py*C*H); 124.1 (ar*C*CH_2), 130.9 (ar*C*H); 136.5 (py*C*H); 148.6 (py*C*H); 158.8 (py*C*); 160.4 (ar*C*OH); 166.6 (*C*O_2Et). **ESI mass spectrometry** (methanol) m/z 589.25 [$C_{35}H_{36}N_6O_3$ + H]$^+$. **FT-IR spectroscopy** (ν, cm^{-1}) 2982, 2928, 2822 (m, C–H str); 1704 (m, C=O str); 1592 (m, C=C str); 1199 (m, C–O str); 758 (m, py–H); 728 (s, py–H).

2.5.23 *Synthesis of 2-(1,3-dioxoisoindolin-2-yl)ethyl Benzenesulfonate*

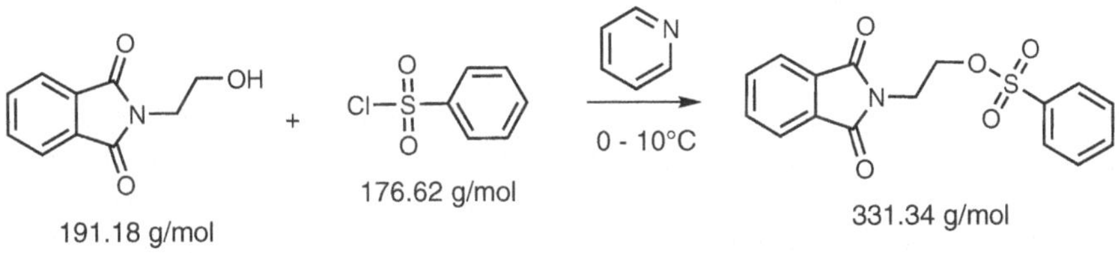

191.18 g/mol + 176.62 g/mol → (pyridine, 0 - 10°C) 331.34 g/mol

N-(2-hydroxyethyl)-phthalimide (5 g, 26 mmol) was dissolved in pyridine (30 mL) in a dry 3-necked flask equipped with a condenser, pressure equalizing dropping funnel, thermometer and nitrogen inlet and cooled to 0 °C. Benzenesulfonylchloride (10.5 mL, 79 mmol) was added dropwise such the temperature did not rise above 10 °C. Upon completion of the addition, the slurry was stirred at 50 °C for 30 min, cooled to room temperature and poured into a mixture of hydrochloric acid (40 mL), water (50 mL), methanol (100 mL) and ice (25 g). The white precipitate was dried under vacuum, washed twice with cold methanol (50 mL) and recrystallized from ethanol ($\sim$140 mL) to yield the desired product as white needles (7 g, 80 %).

^{1}H NMR ($CDCl_3$, 500.13 MHz) δ 3.91 (t, 2H, J = 5.4 Hz, CH_2OSO_2); 4.32 (t, 2H, J = 5.3 Hz, NCH_2CH_2); 7.36 (tt, 2H, J = 8.0, 1.7 Hz, ar*H*); 7.46 (tt, 1H, J = 7.6, 1.2 Hz, ar*H*); 7.68–7.72 (m, 2H, ar*H*); 7.76–7.81 (m, 4H, ar*H*). **^{13}C NMR** ($CDCl_3$, 100.62 MHz) δ 36.8 (NCH_2); 66.5 (CH_2OSO_2); 123.4 (arCH); 127.7 (ar*H*); 129.0 (ar*H*); 131.7 (arCCO); 133.6 (ar*H*); 134.1 (ar*H*); 135.6 ($arCSO_3$); 167.6 (CON). **ESI mass spectrometry** (methanol) m/z 353.96 $[C_{16}H_{13}NO_5S + Na]^+$. **FT-IR spectroscopy** (ν, cm^{-1}) 2949 (w, C–H str); 1773 (m, $C{=}O_{Imid}$); 1714 (s, $C{=}O_{Imid}$); 1390, 1356 (s, CH_2OSO_2–Ph); 1185 (s, SO_2–Ph); 1045 (m, S=O); 748, 722, 679 (m, ar–H). **Melting point** 138.2 °C.

2.5.24 *Synthesis of 4-((4-vinylbenzyl)oxy)phenol*

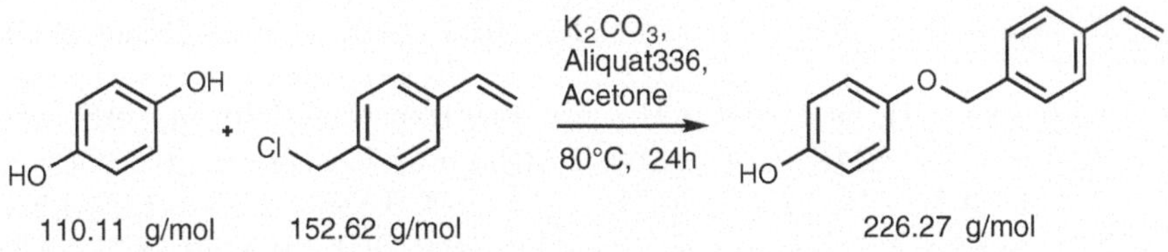

110.11 g/mol + 152.62 g/mol → (K_2CO_3, Aliquat336, Acetone, 80°C, 24h) 226.27 g/mol

The compound was prepared after a slight modification of a previously published protocol [54]. 1,4-hydroquinone (8.8 g, 80 mmol), 4-vinylbenzyl chloride (3.0 g, 20 mmol), dried, powdered potassium carbonate (2.8 g, 20 mmol) and Aliquat336 (0.8 g, 2 mmol) were dissolved in acetone (75 mL) and the mixture heated to

80 °C for 24 h. After cooling to room temperature the mixture was concentrated in vacuo, chloroform (30 ml) added and the mixture filtered through Celite. The crude product was purified with flash column chromatography (ethyl acetate/hexane 9:1, $R_f = 0.76$, I_2 stain) to yield an orange solid (1.66 g, 36 %).

^{1}H NMR ($CDCl_3$, 500.13 MHz) δ 5.01 (s, 2H, C*H*$_2$); 5.26 (dd, 1H, CH=C*H*$_2$, J = 10.8 Hz); 5.79 (dd, 1H, CH=C*H*$_2$, J = 17.6, 1.5 Hz); 6.73–6.88 (m, 5H, ar*H*, C*H*=CH$_2$); 7.37–7.43 (q, 4H, ar*H*, J = 10.8 Hz). **^{13}C NMR** ($CDCl_3$, 100.62 MHz) δ 70.5 (O*C*H$_2$); 114.0 (CH=*C*H$_2$); 116.0 (ar*C*H); 116.1 (ar*C*H); 126.4 (ar*C*H); 127.7 (ar*C*H); 136.5 (*C*H=CH$_2$); 136.8 (ar*C*); 137.3 (ar*C*); 149.7 (ar*C*OH); 153.0 (ar*C*OCH$_2$). **ESI mass spectrometry** (methanol) m/z 249.12 $[C_{15}H_{14}O_2 + Na]^+$. **FT-IR spectroscopy** (ν, cm^{-1}) 3422 (m, O–H str); 3032 (w, C=C str); 2942, 2864 (w, CH_2 str); 1236 (s, C–O str); 1013 (s, C–O–C str); 991 (m, CH=CH$_2$ def); 818 (s, ar–H). **Melting point** 127.6–130.5 °C Lit:143.0–144.0 °C [54].

2.5.25 Synthesis of 2,6-bis(hydroxymethyl)-4-[(4-vinylbenzyl)oxy]phenol

This compound was prepared after a previously published procedure [55]. The 4-((4-vinylbenzyl)oxy)phenol (1 g, 4.4 mmol) was dissolved in tetrahydrofuran (2 mL) and formaldehyde (37 %, 1 mL) and aqueous sodium hydroxide solution (2.6 M, 2.5 mL) were added. The brown suspension was stirred for two days at room temperature, subsequently hydrochloric acid (0.1 M, 5 mL) were added and the aqueous phase extracted with ethyl acetate (2 × 15 mL). The combined organic layers were dried over sodium sulfate and the solvent removed in vacuo. After purification with column chromatography (hexane/ethyl acetate 1:1, $FeCl_3$ stain, $R_f = 0.3$) a pale yellow powder was obtained (372 mg, 59 %).

^{1}H NMR ($CDCl_3$, 500.13 MHz) δ 4.75 (s, 4H, C*H*$_2$OH); 4.96 (s, 2H, arC*H*$_2$O); 5.22 (dd, 1H, CH=C*H*$_2$, J = 10.8 Hz); 5.75 (dd, 1H, CH=C*H*$_2$, J = 17.6, 0.6 Hz); 6.67–6.76 (m, 3H, ar*H*, C*H*=CH$_2$, J = 10.9 Hz); 7.32–7.40 (q, 4H, ar*H*, J = 12.4 Hz). **^{13}C NMR** ($CDCl_3$, 100.62 MHz) δ 63.7 (*C*H$_2$OH); 70.6 (O*C*H$_2$); 114.2, 114.4 (CH=*C*H$_2$); 126.5 (ar*C*H); 127.1 (ar*C*CH$_2$OH); 127.7 (ar*C*H); 136.5 (arCH); 136.7 (ar*C*); 137.4 (ar*C*CH=CH$_2$); 148.7 (ar*C*OH); 151.8 (ar*C*OCH$_2$). **ESI mass spectrometry** (methanol) m/z 309.20 $[C_{17}H_{18}O_4 + Na]^+$. **FT-IR spectroscopy** (ν, cm^{-1}) 3423 (m, O–H str); 3032 (w, C=C str); 2942, 2864 (w, CH_2 str); 1236 (s, C–O str); 1013 (s, C–O–C str); 991 (m, CH=CH$_2$ def); 818 (s, ar–H). **Melting point** 100.2–105.5 °C.

2.6 Substrate Synthesis

2.6.1 Synthesis of bis(2,4-dinitrophenyl)Phosphate (BDNPP)

pyridine, CH_3CN, 0 °C, 15 min

$POCl_3$

184.11 g/mol 151.33 g/mol 509.28 g/mol

This compound was prepared with slight modifications of a previously published procedure [56]. 2,4-dinitrophenol was recrystallized from ethanol 3 times and stored under ethanol prior to use. Acetonitrile was distilled and dried with activated 3 Å molecular sieves. Pyridine was distilled and dried over 4 Å molecular sieves. 2,4-Dinitrophenol (2.76 g, 15 mmol) was dried and transferred into a Schlenk flask which was carefully evacuated and flushed with nitrogen twice. Acetonitrile (30 mL) and pyridine (2.5 mL) were added via a syringe/septum. Freshly distilled $POCl_3$ (0.5 mL, 5 mmol) was added after the mixture was cooled to 0 °C on an ice bath. The mixture was stirred for 15 min and subsequently poured on ice water (200 mL). A yellow oil released from the reaction mixture which crystallized overnight in the fridge. The precipitate was dried in vacuo and recrystallized twice from acetone/diethyl ether to yield a white powder (1.6 g, 60 %).

^{1}H NMR (d^4-MeOD, 500.13 MHz); δ 7.85 (d, 2H, ar*H*, J = 9.1 Hz); 8.04 (t, 2H, py*H*, J = 6.8 Hz); 8.40 (dd, 2H, ar*H*, J = 9.2, 2.8 Hz); 8.59 (tt, 1H, py*H*, J = 7.8, 2.9 Hz); 8.66 (dd, 2H, ar*H*, J = 1.3, 0.5 Hz); 8.79 (d, 2H, py*H*, J = 5.2 Hz). **^{13}C NMR** (d^4-MeOD, 100.62 MHz); δ 122.1 (ar*C*H); 124.4 (ar*C*H); 128.7 (ar*C*H); 129.5 (ar*C*NO_2); 143.1 (ar*C*NO_2); 144.2 (ar*C*NO_2); 148.2 (ar*C*NO_2); 151.2 (ar*C*O); 151.3 (ar*C*O). **^{31}P NMR** (d^4-MeOD, 161.9 MHz) δ -14.19 **FT-IR spectroscopy** (ν, cm^{-1}): 2569 (b, P–O str); 2160 (w, py, N^+–H str); 1602 (m, C=C str), 1526 (s, C–NO_2 asym str); 1341 (s, C–NO_2 sym str); 1258 (s, P = O str); 1239 (m, P–O–Ar str); 874, 758, 739, 683 (m, Ar–H str). **Microanalysis** $C_{17}H_{12}N_5O_{12}P$: calc. C: 40.09, H: 2.37, N: 13.75; found: C 40.08, H 2.24, N 13.62 %. **Melting Point** 166.5–170.9 °C Lit: 157.0–160.0 °C [56].

2.7 Nanoparticle Synthesis and Immobilization Procedures

2.7.1 Synthesis of Functionalized Magnetite Nanoparticles

$$\underset{270.30\ \text{g/mol}}{FeCl_3 \times 6\ H_2O} + \underset{278.01\ \text{g/mol}}{FeSO_4 \times 7\ H_2O} \xrightarrow{1.5\ \text{M aq.}\ NH_3} \underset{231.53\ \text{g/mol}}{Fe_3O_4} = \text{MNP}$$

The magnetite nanoparticles were synthesized after a previously published procedure [57, 58]. Solutions of ferric trichloride (1.15 g in 50 mL degassed water) and ferrous sulfate (0.69 g, in 50 mL degassed water) were combined under nitrogen and 1.5 M ammonia solution was added dropwise with vigorous stirring. As the pH reached 9, the black precipitated magnetite was collected with a magnet and washed with water (5 × 50 mL) and ethanol (2 × 50 mL). The black powder was dispersed in ethanol to give a 5 wt % suspension.

FT-IR spectroscopy (ν, cm^{-1}) 560 (m, Fe–O str). **Microanalysis** Fe_3O_4: found C 0.24, H 0.35, N 0.05, S 0.75 % (slight sulfur contamination due to sulfate residues)

APTS; methyl acrylate; ethylene-diamine; G3-PAMAM modified MNP

Magnetite stock solution (100 mL) was sonicated for 30 min and diluted with ethanol (300 mL) in a two-necked flask equipped with a reflux condenser and a pressure equalized dropping funnel. 3-aminopropyltrimethoxysilane (APTS, 20 mL) was added over 7 h at 60 °C and the nanoparticles were subsequently washed with methanol (3 × 50 mL) by magnetic separation and dried in vacuo.

FT-IR spectroscopy (ν, cm^{-1}) 2949, 2866 (w, C–H str); 1015 (m, Si–O str); 569 (m, Fe–O str). **Microanalysis** found C 1.69, H 0.64, N 0.47, S 0.69 % (slight sulfur contamination due to sulfate residues).

1.16 g of APTS modified magnetite nanoparticles were dispersed in methanol (100 mL) and methyl acrylate (40 mL) was added. After sonication for 5 min the mixture was stirred overnight at room temperature and washed with methanol (5 × 20 mL). Ethylenediamine (8 mL) and methanol (10 mL) were added and stirred at 50 °C for 5 h. This step was repeated twice with increased amounts of methyl acrylate (60 and 80 mL) and ethylenediamine (12 and 20 mL) and the final product was washed with methanol (3 × 50 mL) and water (3 × 50 mL).

FT-IR spectroscopy (ν, cm^{-1}) 3314 (b, N–H); 2668 (w, C–H str); 1625, 1551 (m, $C{=}O_{Amide}$); 1017 (m, Si–O str); 715 (w, CH_2); 560 (m, Fe–O str). **Microanalysis** found C 3.53, H 0.75, N 1.44, S 0.00 %.

2.7.2 Immobilization of CH_3HL4 on Merrifield Resin

Cl
N
H
N
N
N
O
OH
406.52 g/mol
K_2CO_3,
CH_3CN
rt, 5 d
N
N
N
O
OH
N
M-CH_3HL4

To Merrifield resin (1 % cross-linked, 3.5 mmol/g Cl, 500 mg) in acetonitrile (10 mL) was added CH_3**HL4** (320 mg, 0.78 mmol) and dried potassium carbonate (250 mg) and the mixture was stirred for 10 days. After this time the orange beads of the resin were suction dried on a sintered glass funnel (the filtrate was collected to recover any unbound ligand) and washed with water (20 mL) and methanol (20 mL). The resin was further dried under high vacuum to yield 646 mg of M-CH_3**HL4**. 26 mg of ligand were recovered from the filtrate.

FT-IR spectroscopy (ν, cm^{-1}) 3352 (b, O–H str); 3025, 2922 (w, C–H str); 1110 (m, C–O–C str); 814, 756, 699 (m, Ar–H/Py–H str). **Microanalysis** found C 77.39, H 7.22, N 4.86, S 0.00 %.

2.7.2.1 Zinc Complex of MR-CH_3HL4

To MR-CH_3**HL4** (400 mg) in methanol (10 mL) was added zinc(II) acetate dihydrate (400 mg) and the mixture was refluxed for 30 min and then stirred for 24 h. Subsequently the resin was suction dried and washed with water (10 mL) and methanol (20 mL) to yield 505 mg of the immobilized complex MR-$[Zn_2(CH_3$**L4**$)(CH_3COO)_2]$.

FT-IR spectroscopy (ν, cm^{-1}) 3410 (b, O–H str); 3025, 2924 (w, C–H str); 1596 (s, bridging acetate asym. str); 1421 (s, bridging acetate sym. str); 809, 761, 700, 664 (m, Ar–H/Py–H str). **Microanalysis** found C 64.42, H 6.18, N 3.58, S 0.00 %.

2.7.3 Immobilization of NH_2HL2 on Merrifield Resin

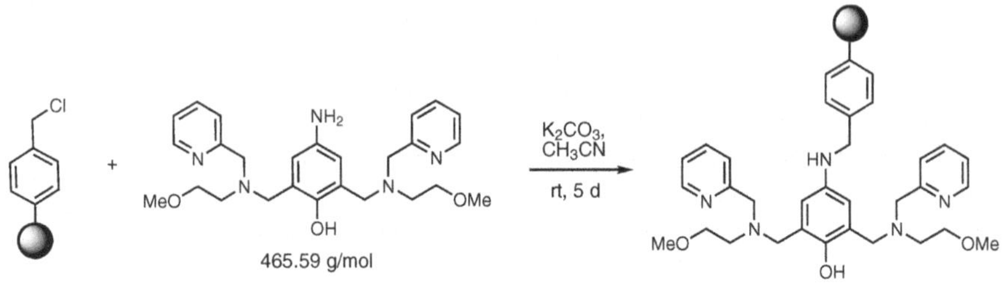

To Merrifield resin (1 % cross-linked, 3.5 mmol/g Cl, 150 mg) in acetonitrile (5 mL) was added NH_2**HL2** (100 mg, 0.21 mmol) and dried potassium carbonate (75 mg) and the mixture was stirred for 10 days. After this time the brown beads were suction dried on a sintered glass funnel (the filtrate was collected to recover any unbound ligand) and washed with water (10 mL) and methanol (10 mL). The resin was further dried under high vacuum to yield 80 mg of MR-NHH**L2**. 36 mg of ligand were recovered from the filtrate.

FT-IR spectroscopy (ν, cm^{-1}) 3359 (b, O–H str); 3025, 2921 (w, C–H str); 1111 (m, C–O–C str); 818, 757, 699 (s, Ar–H/Py–H str). **Microanalysis** found C 78.41, H 7.35, N 5.23, S 0.00 %.

2.7.3.1 Zinc Complex of MR-NHHL2

To MR-NHH**L2** (58 mg) in methanol (3 mL) was added zinc(II) acetate dihydrate (58.8 mg) and the mixture was refluxed for 30 min. Subsequently the resin was washed with methanol (20 mL) and suction dried. After removal of residual solvent in high vacuum 69 mg of the immobilized complex MR-$[Zn_2$(NH**L2**)$(CH_3COO)_2]$ were obtained.

FT-IR spectroscopy (ν, cm^{-1}) 3417 (b, O–H str); 3027, 2921 (w, C–H str); 1599 (s, bridging acetate antisym. str); 1414 (s, bridging acetate sym. str); 704 (m, Ar–H/Py–H str). **Microanalysis** found C 64.56, H 6.43, N 3.84, S 0.00 %.

2.7.4 Functionalizing G3-MNP with Glutaraldehyde

G3-MNP (150 mg) were suspended in glutaraldehyde (25 %, 20 mL) and sonicated for one hour. After 12 h standing the nanoparticles were separated from glutaraldehyde and washed with TRIS buffer pH 8 (5 × 15 mL). G3-MNP-glutaraldehyde were used in the next step without further purification for the GpdQ immobilization and for the ligand immobilization they were washed with methanol (1 mL).

2.7.4.1 Immobilization of $[Zn_2(NH_2HL2)(CH_3COO)_2]$ on G3-MNP-glutaraldehyde

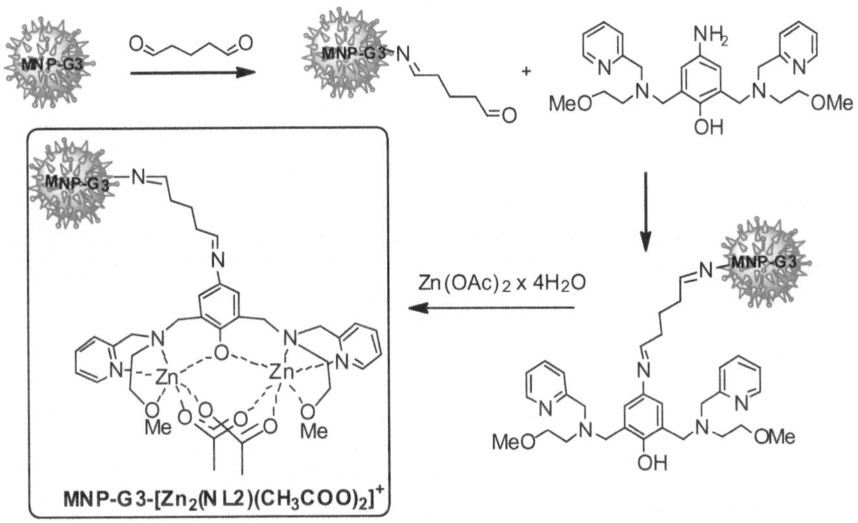

To G3-MNP-glutaraldehyde (75 mg) in methanol (1 mL) was added NH_2**HL2** (50 mg) and dried potassium carbonate (75 mg) and the mixture was left standing at room temperature for five days. After this time the nanoparticles were washed by magnetic separation until the solution was clear. The nanoparticles were then resuspended in methanol (1.5 mL) and zinc acetate dihydrate (50.2 mg) was added. After stirring for 30 min the MNP were washed with methanol (5 × 1.5 mL) and dried in vacuo.

FT-IR spectroscopy (v, cm^{-1}) 3359 (b, O–H str); 3025, 2921 (w, C–H str); 1111 (m, C–O–C str); 818, 757, 699 (s, Ar–H/Py–H str). **Microanalysis** found C 78.41, H 7.35, N 5.23, S 0.00 %.

2.7.4.2 Immobilization of Ser127Ala Mutant GpdQ on PAMAM MNP

The G3-MNP (50 mg) that had been functionalized with glutaraldehyde were added to a GpdQ stock solution (0.17 mM, 2 mL) and the mixture incubated 24 h at 4 °C. After this time a sufficiently high amount of enzyme has bound to the nanoparticles via amide formation between the lysine residues of the enzyme and the pendant aldehydes on the particles (Initially the enzyme solution was incubated for five days with the nanoparticles but analysis showed that the amount of enzyme immobilized did not increase significantly after 24 h). This was confirmed by the decrease of absorbance at 280 nm of the supernatant. After a thorough washing process (5 × 1.5 mL 50 mM TRIS pH 8) the GpdQ-nanoparticles were suspended in 0.5 mL buffer (50 mM TRIS pH 8) and stored at 4 °C until further use. The amount of GpdQ that was immobilized with this procedure was 1.488 μmol per g MNP.

2.8 Syntheses of the Metal Complexes

2.8.1 Synthesis of $[Cd_4(CO_2EtH_2L1)_2(CH_3COO)_{3.75}Cl_{0.25}(H_2O)_2](PF_6)_2$

Methanol solutions of CO_2EtH**L1** (0.1 M, 1 mL), cadmium acetate (0.1 M, 2 mL), sodium acetate (0.1 M, 1 mL) and sodium hexafluorophosphate (0.1 M, 3 mL) were combined with methanol (2 mL) and the yellow solution was left to evaporate slowly at room temperature. White crystals were obtained after 5 days. After recrystallization from a minimum of hot methanol the complex was obtained in a 66 % yield (40 mg).

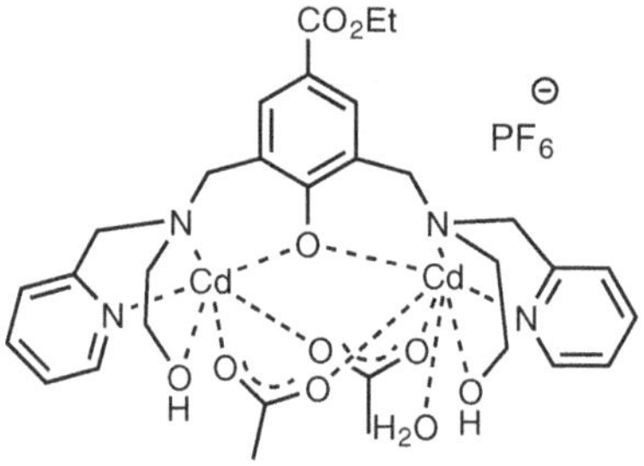

monomeric unit

ESI mass spectrometry (methanol) m/z: 837.0 $[C_{31}H_{29}Cd_2N_4O_9]^+$, 777.0 $[C_{29}H_{35}Cd_2N_4O_7]^+$, 606.1 $[C_{54}H_{66}Cd_2N_4O_{10}]^{2+}$. **Microanalysis** Anal calc. for $C_{61.5}H_{81.25}Cd_4N_8Cl_{0.25}O_{19.5}P_2F_{12}$ C 37.06, H 4.11, N 5.62 % Found: C 37.43, H 4.28, N 5.63 %. **FT-IR spectroscopy** (ν, cm^{-1}): 3173 (b, O–H str in water) 2906 (w, C–H str), 1670 (m, C=O str), 1604 (m, asym str OAc); 1547 (m, asym str OAc); 1413 (m, sym str OAc); 1373 (m, sym str OAc). 1015 (w, C–OH str), 831 (s P–F str), 766 (m, py C–H def), 556 (s, P–F). **^{1}H NMR** (CD_3CN/D_2O 1:1, 500.13 MHz, referenced to D_2O); δ 1.70 (t, 3H, CH_2CH_3, J = 7.12 Hz); 2.39 (s, 6H, acetateCH_3); 3.28 (t, 4H, NCH_2CH_2, J = 8.40 Hz); 3.43 (m, 2H, CH_2, J = 11.60 Hz); 4.13 (m, 4H, CH_2CH_2OH); 4.15–4.23 (m, 2H, CH_2); 4.45 (m, 2H, CH_2, J = 5.85 Hz); 4.59–4.66 (m, 4H, CH_2/CH_2CH_3), 7.39 (d, 2H, py*CH*, J = 7.85 Hz); 7.58 (t, 2H, py*CH*, J = 6.10 Hz); 7.77 (s, 2H, ar*CH*); 8.00 (t, 2H, py*CH*, J = 7.72 Hz); 8.70 (d, 2H, py*CH*, J = 4.70 Hz). **^{13}C NMR** (CD_3CN, /D_2O 1:1, 100.62 MHz, referenced to CD_3CN); δ The arCCO_2Et signal was not assigned due to overlapping with the solvent signal. 14.5 (CH_2CH_3); 22.7 (acetateCH_3); 56.9 (CH_2); 57.7 (CH_2); 58.3 (CH_2); 58.7 (CH_2); 61.6 (CH_2); 124.4 (py*C*H); 124.5 (py*C*H); 125.5 (ar*C*); 134.4 (ar*C*H); 140.1 (py*C*H); 148.8 (py*C*H); 156.1 (py*C*); 161.1 (ar*C*OH); 181.8 (CO_2^-). **^{113}Cd NMR** (CD_3CN/D_2O 1:1, 89 MHz, referenced to $Cd(OAc)_2 \times 2H_2O$ in D_2O) δ 36.4.

2.8.2 Synthesis of $[Cd_2(CO_2EtL2)(CH_3COO)_2H_2O](PF_6)$

CO_2Et**HL2** (180 mg, 0.34 mmol) was dissolved in methanol and cadmium(II) acetate dihydrate (183 mg, 0.68 mmol) and sodium acetate (56 mg, 0.68 mmol) added. The yellow solution was subsequently refluxed for 30 min and then allowed to cool to room temperature and sodium hexafluorophosphate (115 mg, 0.68 mmol) added. After filtration the yellow solution was left to evaporate at room temperature to leave a yellow oil. Attempts to crystallize the complex from a range of solvents were unsuccessful. The complex was, however, readily formed as confirmed by mass spectroscopic measurements. After repeated (5×) evaporation of the methanolic complex solution a white powder was obtained.

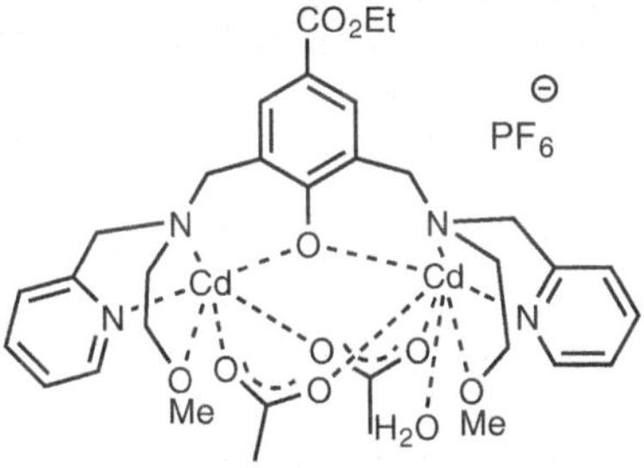

FT-IR spectroscopy (ν, cm^{-1}): 3385 (b, O–H str in water) 2927 (w, C–H str); 1702 (m, C=O str); 1603 (m, asym str OAc); 1574 (m, asym str OAc); 1441 (m, sym str OAc); 1368 (m, sym str OAc). 834 (s, P–F str). **ESI mass spectrometry** (methanol) found m/z 864.0 (100 %), 863.0 (90 %) $[C_{33}H_{43}Cd_2N_4O_9]^+$ calc. m/z 865.11 (100.0 %), 863.11 (88.3 %), 864.11 (86.7 %).**^{1}H NMR** (CD_3CN/D_2O 1:1, 500.13 MHz, referenced to D_2O); δ 1.84 (t, 3H, CH_2CH_3, J = 7.12 Hz); 2.46 (s, 6H, acetateCH_3); 3.33 (m, 2H, CH_2); 3.43 (m, 1H, CH_2); 3.63–3.77 (m, 7H, CH_2); 3.79 (s, 6H, OCH_3); 3.93–4.49 (m, 6H, CH_2); 4.75 (q, 2H, CH_2CH_3, J = 7.16 Hz); 7.57 (m, 1H, py*CH*); 7.76 (m, 2H, py*CH*); 7.87 (t, 1H, py*CH*, J = 6.79 Hz); 7.97, 8.12 (s, 2H, ar*CH*); 8.19 (m, 1H, py*CH*); 8.32 (m, 1H, py*CH*); 8.83 (m, 1H, py*CH*); 8.94 (m, 1H, py*CH*). **^{13}C NMR** (CD_3CN/D_2O 1:1, 100.62 MHz, referenced to CD_3CN); δ 14.6 (CH_2CH_3); 22.8 (acetateCH_3); 56.3 (CH_2); 59.9 (OCH_3); 60.5 (CH_2); 60.7 (CH_2); 61.8 (CH_2); 68.2 (CH_2); 68.5 (CH_2); 124.6 (pyCH); 124.9 (py*C*H); 125.5 (ar*C*); 125.7 (arC); 134.4 (ar*C*H); 134.8 (arCH); 140.3 (py*C*H); 140.6 (pyCH); 148.8 (py*C*H); 149.3 (pyCH); 155.9 (py*C*); 168.5 (arCOH); 171.0 (CO_2Et); 181.1 (CO_2^-); **^{113}Cd NMR** (CD_3CN/D_2O 1:1, 89 MHz) δ 23.4.

2.8.3 Synthesis of $[Co_2(CO_2EtL1)(CH_3COO)_2](PF_6)$

Method 1 The methanol ligand solution of CO_2Et**HL1** (0.1 M, 1 mL) was combined with a solution of cobalt(II) acetate tetrahydrate (0.1 M, 2 mL) in methanol. Subsequently sodium hexafluorophosphate (0.1 M, 3 mL) was added. The solution was left at room temperature to evaporate. Pink clusters of crystals were obtained after two days in 88 % yield (310 mg).

Method 2 (X-ray suitable crystals were obtained with this procedure) All solutions were made up in degassed methanol and combined under nitrogen in a Schlenk flask. A flask containing diethyl ether (10 mL) was attached to the Schlenk flask. When the diffusion was complete, the pink solution was poured in a beaker and left to evaporate for 3 days to yield pink needles which were subsequently analyzed by X-ray crystallography. The mass spectra of the crystals obtained by methods 1 and 2 were identical.

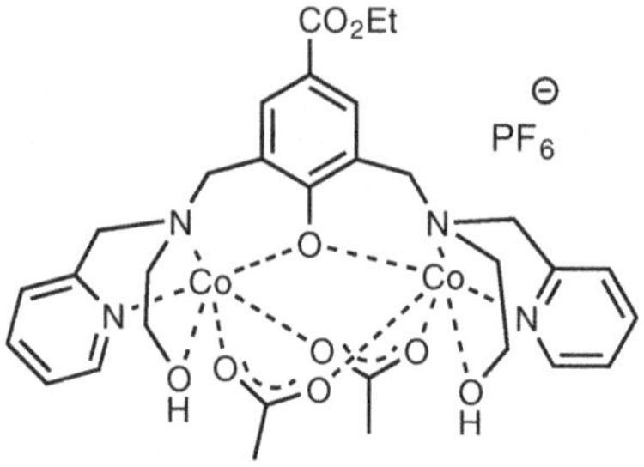

ESI mass spectrometry (methanol) m/z: 729.2 $[C_{31}H_{29}Co_2N_4O_9]^+$, 669.1 $[C_{29}H_{35}Co_2N_4O_7]^+$, 609.1 $[C_{27}H_{31}Co_2N_4O_5]^+$, 305.1 $[C_{27}H_{31}Co_2N_4O_5]^{2+}$. **Microanalysis** Anal calc. for $C_{31}H_{39}Co_2N_4O_9PF_6$ C 42.58, H 4.50, N 6.41; found: C 42.16, H 4.40, N 6.39 %. **FT-IR spectroscopy** (ν, cm^{-1}) 1605 (m, asym str acetate); 1442 (s, sym str acetate) 1022 (m, O–H); 834 (s, P–F str); 767 (m, Py–H def); 555 (m, P–F). **Magnetic moment** in solution by Evans method [28]. 6.14 BM. **UV-Vis spectroscopy** Acetonitrile, 0.02 M, $\lambda_1 = 493$ nm ($\varepsilon = 63.2$ L mol^{-1} cm^{-1}).

2.8.4 Synthesis of $[Co_2(CO_2EtL2)(CH_3COO)_2](PF_6)$

CO_2Et**HL2** (100 mg, 0.19 mmol) was dissolved in methanol (4 mL) and was combined with a solution of cobalt(II) acetate tetrahydrate (95 mg, 0.38 mmol) in methanol (6 mL). Subsequently solid sodium hexafluorophosphate (95 mg, 0.57 mmol) was added. The solution was left in a beaker at room temperature to evaporate. Pink crystals which were suitable for X-ray crystallography were obtained in 70 % yield (120 mg).

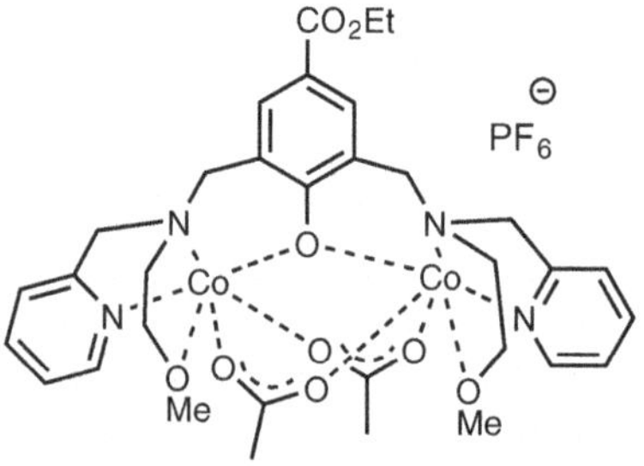

ESI mass spectrometry (methanol) m/z: 757.36 $[C_{33}H_{43}Co_2N_4O_9]^+$, 729.37 $[C_{32}H_{43}Co_2N_4O_8]^+$. **Microanalysis** Anal calc. for $C_{33}H_{43}Co_2N_4O_9PF_6$ C 43.91, H 4.80, N 6.21; found: C 43.21, H 4.61, N 6.10 %. **FT-IR spectroscopy** (ν, cm^{-1}) 1601 (s, asym str acetate); 1421 (s, sym str acetate); 1022 (m, O–H); 830 (s, P–F str); 762 (m, Py–H def); 556 (m, P–F). **Magnetic moment** (d^4-MeOD) in solution by Evans method [28]. 6.18 BM. **UV-Vis spectroscopy** Acetonitrile, 0.02 M, $\lambda_1 = 508$ nm ($\varepsilon = 73.5$ L mol^{-1} cm^{-1}).

2.8.5 Synthesis of $[Co_2(CH_3L2)(CH_3COO)_2](PF_6)$

CH_3**HL2** (125 mg, 0.25 mmol) was dissolved in methanol (2 mL) and a methanol (3 mL) cobalt(II) acetate tetrahydrate (125 mg, 0.5 mmol) solution was added dropwise at room temperature. Subsequently sodium hexafluorophosphate (140 mg, 0.8 mmol) dissolved in methanol (3 mL) was added. The pink crystals which emerged upon standing after a couple of were filtered and dried in air (150 mg, 71 %). Crystals were obtained suitable for X-ray crystallography after 2 days by using only one equivalent Co(II) and adding the hexafluorophosphate as solid.

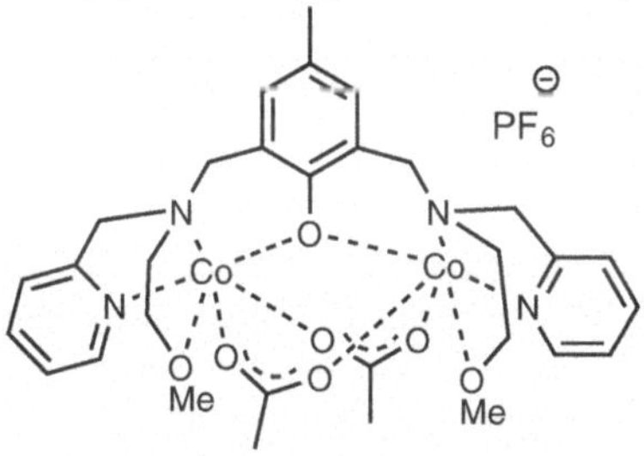

ESI mass spectrometry (methanol) m/z: 699.2 $[C_{31}H_{41}Co_2N_4O_7]^+$. **FT-IR spectroscopy** (ν, cm^{-1}): 2925 (w, C–H str), 1596 (m, C=O asym str, acetate); 1475 (m, C=O sym str, acetate); 1423 (m, C–H def); 1085 (w, C–O str), 829 (s P–F str), 555 (s, P–F str). **Microanalysis** Anal calc. for $C_{31}H_{41}Co_2N_4O_7PF_6$ C 44.09, H 4.89, N 6.63; found: C 44.09, H 4.96, N 6.58 %. **Magnetic moment** (d^4-MeOD) in solution by Evans method [28]. 6.09 BM. **UV-Vis spectroscopy** Acetonitrile, 0.02 M, $\lambda_1 = 471$ nm ($\varepsilon = 46.2$ L mol^{-1} cm^{-1}), $\lambda_2 = 519$ nm ($\varepsilon = 50.5$ L mol^{-1} cm^{-1}).

2.8.6 Synthesis of $[Co_2(NO_2L2)(CH_3COO)_2](PF_6)$

NO_2**HL2** (50 mg, 0.1 mmol) was dissolved in methanol (2 mL) and cobalt(II) acetate tetrahydrate (49 mg, 0.1 mmol) in methanol (3 mL) was added dropwise at room temperature. Subsequently sodium hexafluorophosphate (50 mg, 0.3 mmol) was added. The orange crystals which emerged upon standing were filtered and dried in air (60 mg, 66 %).

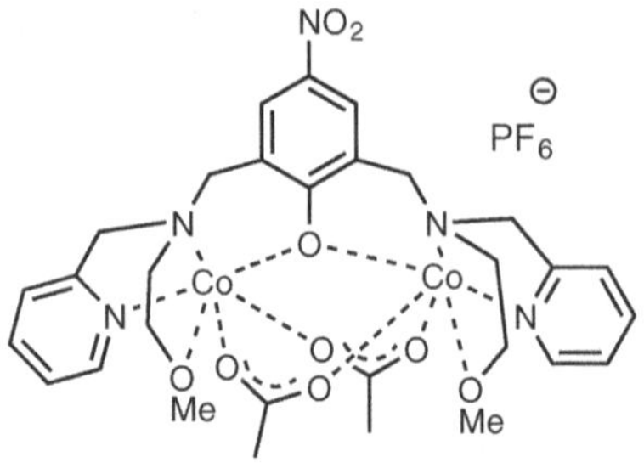

ESI mass spectrometry (methanol) m/z 784.1 $[C_{30}H_{44}Co_2N_5O_{12}]^+$. **FT-IR spectroscopy** (ν, cm^{-1}) 2933 (w, C–H str); 1595 (s, acetate asym str); 1506 (w, NO_2 asym str); 1429 (s, acetate sym str); 1317 (s, NO_2 sym str); 1086 (m, C–O str); 829 (s, P–F str); 752 (m, py C–H def); 658 (w, Ar–H def); 556 (s, P–F). **Microanalysis** Anal calc. for $C_{30}H_{40}Co_2N_5O_{10}PF_6$ C 40.33, H 4.51, N 7.84; found: C 40.41, H 4.42, N 7.92 %. **Magnetic moment** (d^4-MeOD) in solution by Evans method [28]. 7.31 BM. **UV-Vis spectroscopy** Acetonitrile, 0.02 M, $\lambda_1 = 466$ nm ($\varepsilon = 77.7$ L mol^{-1} cm^{-1}).

2.8.7 Synthesis of $[Co_2(BrL2)(PO_3F)](PF_6)$

BrH**L2** (74 mg, 0.14 mmol) was dissolved in methanol (3 mL) and combined with a solution of cobalt(II) acetate tetrahydrate (65 mg, 0.28 mmol) in methanol (3 mL). Sodium hexafluorophosphate (70 mg, 0.4 mmol) was added and, after filtration, the solution was left in a beaker to evaporate at room temperature. Pink crystals were obtained in low yield (25 mg) after 2 days. As the sodium hexafluorophosphate employed was wet, the complex crystallized with PO_3F^{2-} instead of acetates as suggested by mass spectrometry, infrared spectroscopy and microanalysis.

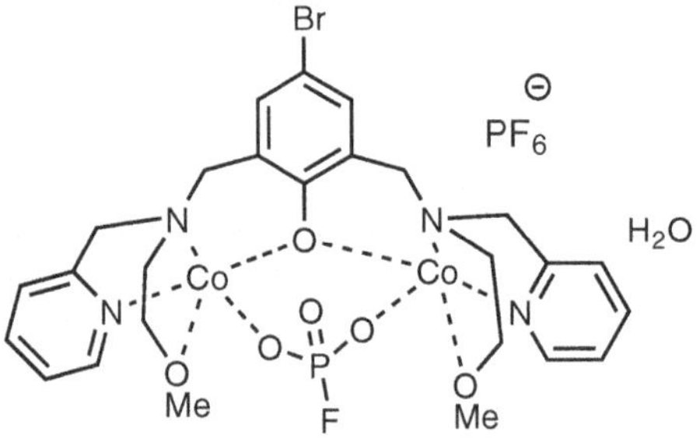

ESI mass spectrometry (methanol) m/z: 679.08 $[C_{26}H_{34}BrN_4O_5Co_2]^+$, 779.10 $[C_{26}H_{36}BrN_4O_5Co_2FO_3P]^+$. **Microanalysis** Anal calc. for $C_{26}H_{36}BrN_4O_8Co_2P_2F_7$ C 33.75, H 3.92, N 6.06; found: C 33.70, H 3.68, N 6.05 %. **FT-IR spectroscopy** (ν, cm^{-1}) 2932, 2855 (w, CH_2 str); 1435 (m, CH_2 def); 1186 (m, P = O, str); 1117 (s, C–O–C str); 830 (s, P–F str); 774 (m, Py–H def); 555 (m, P–F).

2.8.8 Synthesis of $[Co_2(BrL2)(CH_3COO)_2](PF_6)$

BrHL2 (110 mg, 0.2 mmol) was dissolved in methanol (5 mL) and was combined with a solution of cobalt(II) acetate tetrahydrate (103 mg, 0.4 mmol) in methanol (5 mL). Solid sodium hexafluorophosphate (104 mg, 0.6 mmol) was added and after filtration the solution was left in a beaker to evaporate at room temperature. Pink crystals were obtained after two days that were suitable for X-ray crystallography in 32 % yield (62 mg).

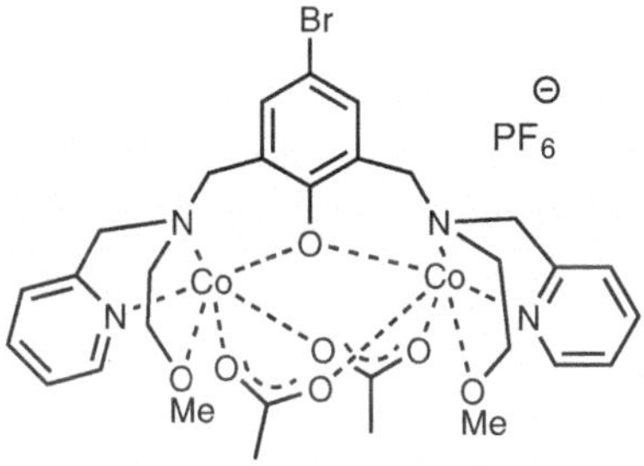

ESI mass spectrometry (methanol) m/z 734.99 $[C_{29}H_{38}BrCo_2N_4O_6]^+$. **FT-IR spectroscopy** (ν, cm^{-1}) 2931 (m, C–H); 1599 (m, C=O asym str, acetate); 1421 (s, C–O sym str, acetate); 831 (s, P–F str); 759 (m, Py–H def); 556 (m, P–F). **Microanalysis** Anal calc. for $C_{30}H_{38}BrCo_2N_4O_7PF_6$ C 39.62, H 4.21, N 6.16; found: C 39.35, H 4.20, N 6.18 %. **Magnetic moment** (d^4-MeOD) in solution by Evans method [28]. 6.09 BM. **UV-Vis spectroscopy** Acetonitrile, 0.01 M, $\lambda_1 = 460$ nm ($\varepsilon = 4.0$ L mol^{-1} cm^{-1}), $\lambda_2 = 509$ nm ($\varepsilon = 4.1$ L mol^{-1} cm^{-1}).

2.8.9 Attempted Synthesis of $[Co(II)Co(III)(CH_3L4)(CH_3CO_2)_2](PF_6)_2{\cdot}MeOH$

CH_3**HL4** (32 mg, 0.079 mmol) was dissolved in degassed methanol and solid cobalt(II) acetate tetrahydrate (39 mg, 0.15 mmol) was added under a nitrogen stream. N_2 was continuously bubbled through the solution until all starting material had dissolved and subsequently solid sodium hexafluorophosphate (39 mg, 0.24 mmol) was added and after filtration the solution was left in a beaker to evaporate at room temperature in air. After two days the solution had undergone a color change form pink to brown and after a further two days a brown solid was obtained which was dried in air (42 mg, 58 %).

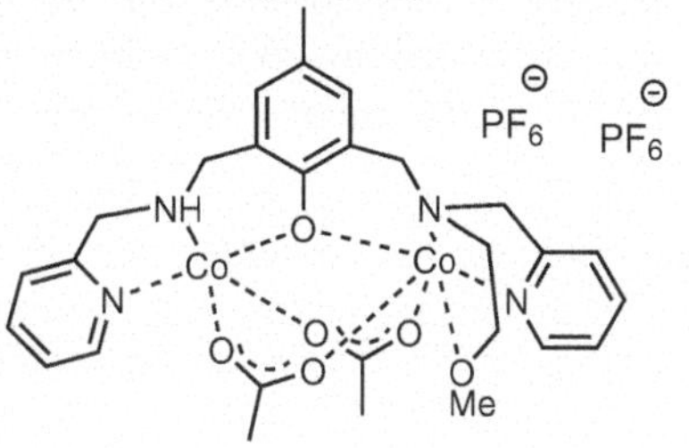

FT-IR spectroscopy (ν, cm^{-1}) 2929 (m, C–H); 1601 (m, C=O asym str, acetate); 1415 (s, C=O sym str, acetate); 830 (s, P–F str); 765, 739 (m, Py–H def); 555 (m, P–F). **Microanalysis** Anal calc. for $C_{29}H_{39}Co_2N_4O_7P_2F_{12}$ C 36.15, H 4.08, N 5.82; found: C 36.70, H 4.30, N 5.92 %.

2.8.10 *Attempted Synthesis of [Co(II)Co(III)(CH$_3$L5)(CH$_3$CO$_2$)$_2$](PF$_6$)$_2$*

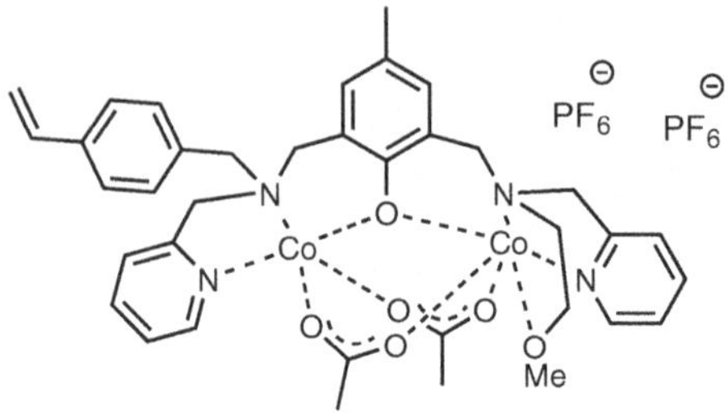

CH_3H**L5** (22 mg, 0.041 mmol) was dissolved in degassed methanol and solid cobalt(II) acetate tetrahydrate (21 mg, 0.082 mmol) was added under a nitrogen stream. N_2 was continuously bubbled through the solution until all starting material had dissolved and subsequently solid sodium hexafluorophosphate (20 mg, 0.12 mmol) was added and after filtration the solution was left in a beaker to evaporate at room temperature in air. After two days the solution had undergone a color change form light pink to dark-pink/red and after a further two days a pink solid was obtained which was dried in air (15 mg, 37 %).

FT-IR spectroscopy (ν, cm^{-1}) 2925 (m, C–H); 1602 (m, C=O asym str, acetate); 1419 (s, C=O sym str, acetate); 833 (s, P–F str); 762 (m, Py–H def); 556 (m, P–F). **Microanalysis** Anal calc. for $C_{37}H_{43}Co_2F_{12}N_4O_6P_2$ C 42.42, N 5.35, H 4.14 %, found C 42.40, N 5.34, H 4.13 %.

2.8.11 *Synthesis of [Zn$_2$(CH$_3$L2)(CH$_3$COO)$_2$](PF$_6$)*

CH_3H**L2** (0.250 g, 0.5 mmol) was dissolved in methanol (20 mL), and a methanol solution (10 mL) of zinc(II) acetate dihydrate (0.240 g, 0.9 mmol) was added dropwise. The resulting pale yellow solution was then heated under reflux for 0.5 h. The solution was permitted to cool to room temperature and sodium hexafluorophosphate (0.140 g, 0.8 mmol) was added. Upon standing colorless crystals were deposited and these were collected by filtration (0.31 g, 71 %).

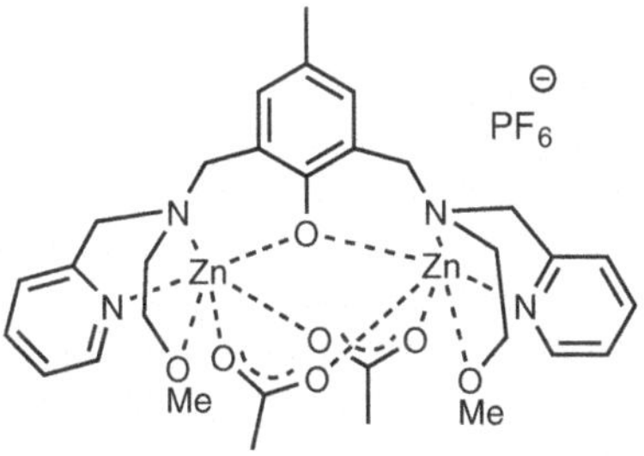

ESI mass spectrometry (methanol) m/z: 711.2 $[C_{31}H_{41}N_4O_7Zn_2]^+$. **Microanalysis** Anal. Calc. for $C_{31}H_{41}Zn_2N_4O_7$ PF_6 C 43.42, H 4.82, N 6.53 % Found: C 43.25, H 4.79, N 6.53 %. **FT-IR spectroscopy** (ν, cm^{-1}): 2924, 2851 (w, C–H str), 1594 (m, C=O asym str, acetate); 1476 (m, C=O sym str, acetate); 1428 (m, C–H def); 1093 (w, C–O str), 831 (s P–F str), 555 (s, P–F str). **^{1}H NMR** (CD_3CN, 500.13 MHz); δ 1.90 (s, 6H, acetateCH_3); 2.22 (s, 3H, CH_3); 2.52 (m, 4H, NCH_2, J = 5.7 Hz); 2.71 (bm, 4H, CH_2CH_2O); 2.98 (s, 6H, OCH_3); 3.63 (d, 2H, arCH_2N, J = 11.9 Hz); 3.85 (d, 2H, NCH_2py, J = 14.7 Hz); 4.39 (d, 2H, NCH_2py, J = 13.8 Hz); 4.48 (d, 2H, arCH_2N, J = 9.3 Hz); 7.03 (s, 2H. arC*H*); 7.49–7.51 (m, 4H, pyC*H*); 7.99 (t, 2H, pyC*H*, J = 7.4 Hz); 8.74 (d, 2H, pyC*H*, J = 4.6 Hz). **^{13}C NMR** (CD_3CN, 100.62 MHz); 20.4 (CH_3); 24.6 (acetateCH_3); 55.1 (NCH_2); 59.0 (OCH_3); 61.3 (arCH_2N), 61.6 (NCH_2py); 69.7 (CH_2O); 125.2 (pyCH); 125.8 (arC); 126.8 (CCH_3); 132.8 (arCH); 140.8 (pyCH); 148.6 (pyCH); 155.8 (pyC); 161.1 (COH); 177.7 (CO_2^-). Second set of low intensity (25 %) signals: **^{1}H NMR** (CD_3CN, 500.13 MHz) H,H-COSY; δ 2.07 (s, 6H); 3.12 (s, 6H); 3.27 (d, 2H, J = 5.4 Hz); 3.74 (m, 4H); 3.74 (s, 2H); 3.86 (d, 2H, J = 15.8 Hz); 4.01 (s, 2H); 6.77 (s, 2H); 7.21 (d, 2H, J = 6.1 Hz); 7.37 (t, 2H, J = 5.8 Hz); 7.81 (t, 2H, J = 7.3 Hz); 8.51 (bs, 2H).

2.8.12 Synthesis of $[Zn_2(CH_3L3)(CH_3COO)_2](PF_6)\cdot CH_3OH$

CH_3H**L3** (50 mg, 0.085 mmol) was dissolved in methanol (2 mL) and zinc(II) acetate dihydrate (0.37 mg, 0.17 mmol) in methanol (3 mL) was added dropwise at room temperature. The pale yellow solution was refluxed for 30 min, cooled slowly to room temperature and subsequently sodium hexafluorophosphate (43 mg, 0.26 mmol) was added. Colorless crystals emerged upon standing after 2 h which were filtered and dried in air (49 %, 41 mg).

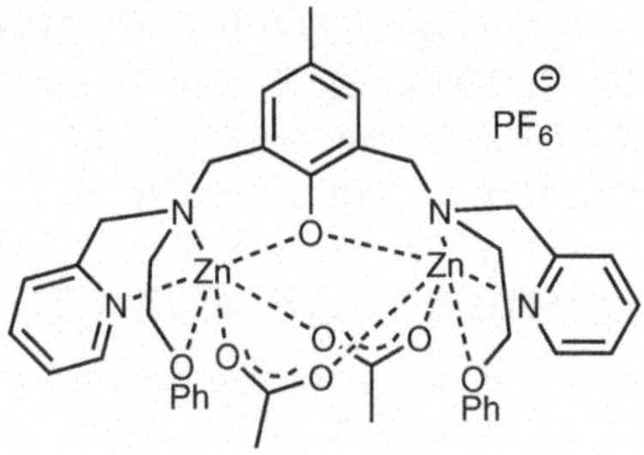

ESI mass spectrometry (methanol) m/z: 855.7 $[C_{41}H_{47}N_4O_8Zn_2]^+$; 825.7 $[C_{40}H_{47}N_4O_7Zn_2]^+$. **Microanalysis** Anal. Calc. for $C_{42}H_{48}Zn_2N_4O_8PF_6$ C 49.77, H 4.87, N 5.53 % found: C 49.50, H 4.60, N 5.55 %. **FT-IR spectroscopy** (ν, cm^{-1}): 2927 (w, C–H str), 1594 (m, C=O asym str, acetate); 1483 (m, C=O sym str, acetate); 1434 (m, C–H def); 1225 (w, C–O str), 833 (s P–F str), 760 (s, Ph–H); 694 (s, Ph–H); 555 (s, P–F str). **^{1}H NMR** (CD_3CN, 400.13 MHz); δ 1.72 (s, 6H, acetateCH_3); 2.24 (s, 3H, CH_3); 2.77 (m, 3H); 3.04 (m, 3H); 3.37 (d, 4H); 3.54 (m, 3H); 3.65 (d, 2H, J = 12.3 Hz); 3.77 (d, 2H, J = 11.8 Hz); 3.94 (m, 5H); 4.18 (m, 5H); 7.21 (m, 4H, arC*H*); 6.94 (m, 2H, arC*H*); 6.62 (d, 4H, J = 8.1 Hz, arC*H*); 7.41–7.50 (m, 4H, pyC*H*); 7.97 (td, 2H, J = 7.7, 1.4, pyC*H*); 8.71 (d, 2H, J = 5.1, pyC*H*). Second set of low intensity aromatic proton signals: 7.21 (m, 4H); 6.94 (m, 2H); 6.73 (d, 4H, J = 8.6 Hz); 7.41–7.50 (m, 4H); 7.93 (td, 2H, J = 7.7, 1.6 Hz); 8.75 (d, 2H, J = 5.3 Hz).

2.8.13 Synthesis of $[Zn_2(NO_2L2)(CH_3COO)_2](PF_6)$

NO_2H**L2** (50 mg, 0.1 mmol) was dissolved in methanol (2 mL) and was combined with zinc(II) acetate dihydrate (44 mg, 0.2 mmol) in methanol (2 mL) at room temperature. The pale yellow solution was refluxed for 30 min, cooled slowly to room temperature and subsequently sodium hexafluorophosphate (50 mg, 0.3 mmol) was added. Colorless needles emerged upon standing immediately which were filtered and dried in air (30 mg, 33 %).

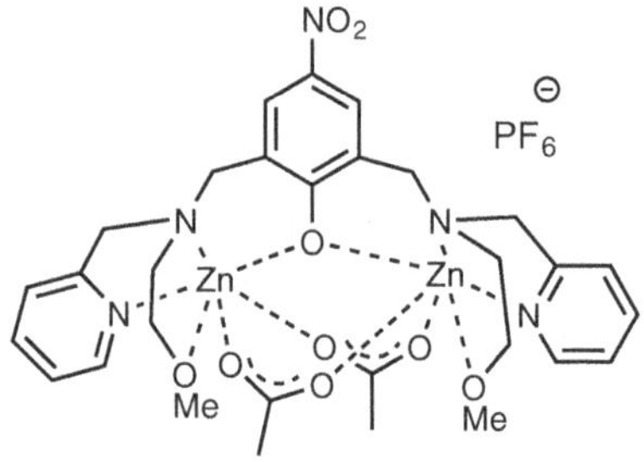

ESI mass spectrometry (methanol) m/z: 744.0 $[C_{30}H_{38}N_5O_9Zn_2]^+$. **Microanalysis** Anal. calc. for $C_{31}H_{41}Zn_2N_5O_{10}PF_6$ C 40.45, H 4.60, N 7.61 % found: C 39.36, H 4.27, N 7.67 %. **FT-IR spectroscopy** (ν, cm^{-1}) 2931 (w, C–H str); 1596 (m, C=O asym str, acetate); 1506 (w, NO_2 asym str); 1426 (m, C=O sym str, acetate); 1320 (m, NO_2 sym str); 1092 (m, C–O str); 833 (s P–F str); 755 (m, py C–H def); 659 (w, Ar–H def); 556 (s, P–F). **^{1}H NMR** (CD_3CN, 500.13 MHz); δ 1.91 (s, 6H, acetateCH_3); 2.57 (m, 4H, NCH_2); 2.74 (bm, 4H, CH_2CH_2O); 2.99 (s, 6H, OCH_3); 3.82 (d, 2H, arCH_2N, J = 12.3 Hz); 3.90 (d, 2H, NCH_2py, J = 14.8 Hz); 4.42 (d, 2H, NCH_2py, J = 15.1 Hz); 4.48 (d, 2H, arCH_2N, J = 12.7 Hz); 7.52 (m, 4H, pyC*H*); 8.01 (t, 2H, pyC*H*, J = 6.9 Hz); 8.2 (s, 2H. arC*H*); 8.72 (m, 2H, pyC*H*).

2.8.14 Synthesis of $[Zn_2(BrL2)(CH_3COO)_2](PF_6)$

BrH**L2** (100 mg, 0.2 mmol) was dissolved in methanol (8 mL) with zinc(II) acetate dihydrate (82 mg 0.4 mmol). The mixture was heated under reflux for 30 min, permitted to cool to room temperature and dry sodium hexafluorophosphate (95 mg, 0.6 mmol) added. After filtration the solution was left to stand at room temperature. Colorless crystals, suitable for X-ray crystallography, emerged after 2 h (62 mg, 33.6 %).

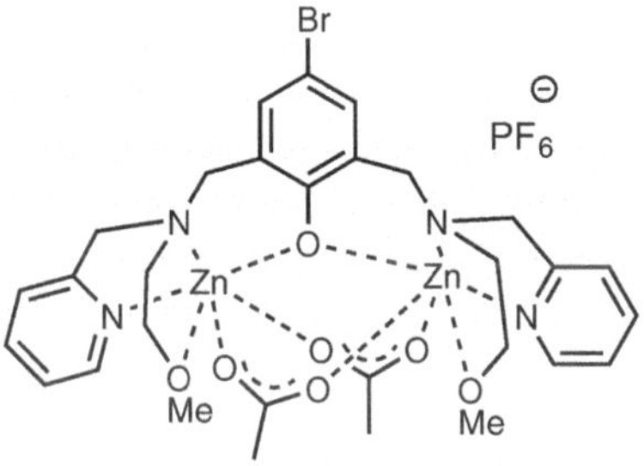

ESI mass spectrometry (methanol) m/z: 749.0 $[C_{29}H_{38}BrN_4O_6Zn_2]^+$, 707.3 $[C_{27}H_{36}BrN_4O_5Zn]^+$, 593.1 $[C_{26}H_{32}BrN_4O_3Zn]^+$. **Microanalysis** Anal calc. for $C_{30}H_{38}BrN_4O_7Zn_2PF_6$ C 39.07, H 4.15, N 6.07 % Found: C 39.33, H 4.13, N 6.14 % **FT-IR spectroscopy** (ν, cm^{-1}) 2930 (w, CH_2 str); 1597 (s, bridging acetate antisym. str); 1424 (s, bridging acetate sym. str); 829 (s, P–F str); 760, 658 (m, Ar–H def); 555 (m, P–F). **^{1}H NMR** (CD_3CN, 500.13 MHz); δ 1.99 (s, 6H, acetateCH_3); 2.54 (m, 4H, NCH_2); 2.72 (bm, 4H, CH_2CH_2O); 2.99 (s, 6H, OCH_3); 3.64 (d, 2H, arCH_2N, J = 12.1 Hz); 3.84 (d, 2H, NCH_2py, J = 14.7 Hz); 4.38 (d, 2H, NCH_2py, J = 14.4 Hz); 4.44 (d, 2H, arCH_2N, J = 11.3 Hz); 7.38 (s, 2H. arC*H*); 7.48–7.50 (m, 4H, pyC*H*, J = 5.4 Hz); 7.99 (t, 2H, pyC*H*, J = 7.1 Hz); 8.72 (m, 2H, pyC*H*).

2.8.15 Synthesis of $[Zn_4(BrL2)_2(PO_3F)_2(H_2O)_2](PF_6)_2$

BrH**L2** (60 mg, 0.1 mmol) was dissolved in methanol (6 mL) and was combined with zinc(II) acetate dihydrate (49 mg, 0.2 mmol). The complex solution was subsequently refluxed for 30 min, cooled to room temperature and sodium hexafluorophosphate (57 mg, 0.3 mmol) was added. After filtration the solution was left in a beaker to evaporate at room temperature. Colorless crystals were obtained after 2 days which were suitable for X-ray crystallography (40 mg, 22 %). As the sodium hexafluorophosphate employed was wet, the complex crystallized with PO_3F^{2-} instead of acetates as suggested by mass spectrometry, infrared spectroscopy and microanalysis.

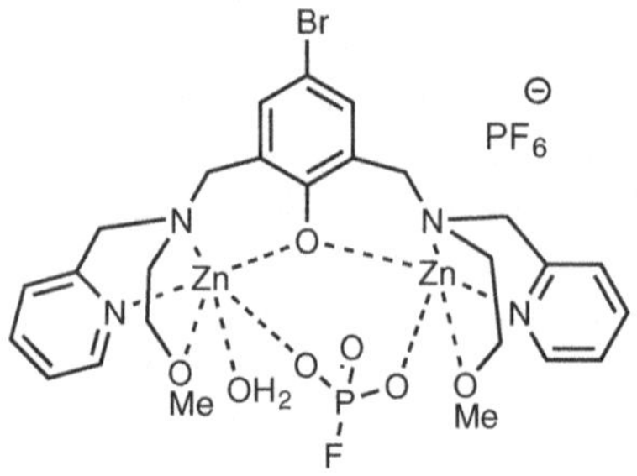

monomeric unit

ESI mass spectrometry (methanol) m/z: 754.99 $[C_{26}H_{32}BrN_4O_3ZnPFO_3]^+$, 593.12 $[C_{26}H_{32}BrN_4O_3Zn]^+$. **Microanalysis** Anal calc. for $C_{52}H_{68}Br_2N_8O_{14}Zn_4P_4F_{14}$ C 33.94, H 3.72, N 6.09; found: C 33.35, H 3.61, N 5.99 %. **FT-IR spectroscopy** (ν, cm^{-1}) 2854, 2929, 2893 (w, CH_2 str); 1460, 1445 (m, CH_2 def); 1183 (m, P = O, str); 1121 (s, C–O–C str); 831 (s, P–F str); 774 (m, Py–H def); 555 (m, P–F).

2.8.16 Synthesis of $[Zn_2(CH_3L4)(CH_3CO_2)_2](PF_6)$

CH_3H**L4** (80 mg, 0.2 mmol) and zinc(II) acetate dihydrate (86 mg, 0.4 mmol) were combined in methanol (5 mL) and refluxed for 30 min, after cooling to room temperature sodium hexafluorophosphate (65 mg, 0.4 mmol) was added and the pale yellow solution filtered and left on the bench to evaporate slowly. White crystals formed after 3 days that were dried in air (85 mg, 54.8 %).

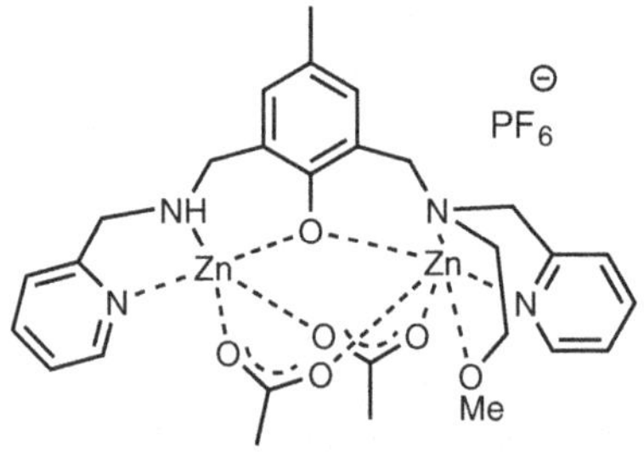

ESI mass spectrometry (methanol) m/z: 655.11 $[C_{28}H_{35}N_4O_6Zn_2]^+$, 595.06 $[C_{26}H_{31}N_4O_4Zn_2]^+$; (acetonitrile) m/z = 749.1, calc. m/z 749.17 (100 %), 748.17 (79.2 %), $[C_{60}H_{88}N_{10}O_{18}Zn_4]^{2+}$. **FT-IR spectroscopy** (ν, cm^{-1}) 3307 (w, N–H str); 2925 (w, C–H str); 1596 (s, bridging acetate antisym. str); 1422 (s, bridging acetate sym. str); 832 (s, P–F str); 768, 663 (m, pyH def); 556 (m, P–F). **Microanalysis** Anal calc. for $C_{28}H_{38}N_4O_6Zn_2PF_6$ C 41.29, H 4.24, N 7.13 %; found: C 41.21, H 4.25, N 6.72 %. **^{1}H NMR** (CD_3CN, 500.13 MHz); δ 1.97 (s, 6H, acetateCH_3); 2.55 (m, 2H, NCH_2CH$_2$); 2.75 (m, 1H, NCH$_2$CH_2); 2.82 (d, 1H NCH$_2$CH_2, J = 11.1 Hz); 3.04 (s, 3H, OCH_3); 3.65 (d, 1H, arCH_2N, J = 12.2 Hz); 3.79 (d, 1H, NCH_2py, J = 11.4 Hz); 3.85 (d, 1H, NCH_2py, J = 15.0 Hz); 3.99 (t, 1H, arCH_2N,

J = 12.8 Hz); 4.13 (t, 1H, NCH_2py, J = 11.6 Hz); 4.21 (dd, 1H, arCH_2N, J = 12.1, 5.1 Hz); 4.31 (d, 1H, NCH_2py, J = 15.2 Hz); 4.35 (d, 1H, arCH_2N, J = 12.3 Hz); 6.94 (s, 1H, ar*H*); 6.97 (s, 1H, ar*H*); 7.40–7.35 (m, 4H, py*H*); 7.98 (mc, 2H, py*H*, J = 7.7, 0.8 Hz); 8.60 (d, 1H, py*H*, J = 4.8 Hz); 8.77 (d, 1H, py*H*, J = 4.7 Hz). **^{13}C NMR** (CD_3CN, 100.62 MHz); 20.3 (CH_3); 24.3 (acetateCH_3); 51.5 (arCH2 N); 52.8 (NCH2py); 55.1 (*CH2*CH2); 59.1 (OCH_3); 61.9 (arCH_2N); 69.6 (CH2CH_2); 124.2 (py*C*H); 125.0 (py*C*H); 125.1 (py*C*H); 125.3 (py*C*H); 125.8 (ar*C*); 126.4 (arC); 126.4 (arCCH_3); 132.2 (ar*C*H); 132.8 (arCH); 140.8 (py*C*H); 141.0 (py*C*H); 148.6 (py*C*H); 148.7 (pyCH); 155.8 (py*C*); 161.1 (ar*C*O); 168.4 (CO_2^-).

2.8.17 Synthesis of $[Zn_2(CH_3L5)(CH_3CO_2)_2](PF_6)\cdot MeOH$

Work reported in this thesis regarding the CH_3H**L5** ligand and its Zn(II) complex was undertaken together with Miss Laurene Marty.

CH_3H**L5** (50 mg, 0.09 mmol) and zinc(II) acetate dihydrate (42 mg, 0.19 mmol) were combined in methanol (5 mL) and refluxed for 30 min, after cooling to room temperature sodium hexafluorophosphate (48 mg, 0.29 mmol) was added and the pale yellow solution filtered and left on the bench to evaporate slowly. A white powder was obtained after 3 days which was dried in air (29 mg, 37 %). Crystals suitable for X-ray structure analysis were obtained with the following method: CH_3H**L5** (30 mg, 0.05 mmol) and zinc(II) acetate dihydrate (25 mg, 0.11 mmol) were combined in methanol (3 mL) and refluxed for 30 min, after cooling to room temperature sodium tetraphenylborate (59 mg, 0.17 mmol) was added and the pale yellow solution filtered and left on the bench to evaporate slowly. The white powder was taken up in acetone/isopropanol and layered with hexane which yielded colorless crystals which desiccated readily upon removal from the solvent. X-ray structure analysis was thus conducted with copper radiation at 150 K. Further analysis was done with the hexafluorophosphate derivative.

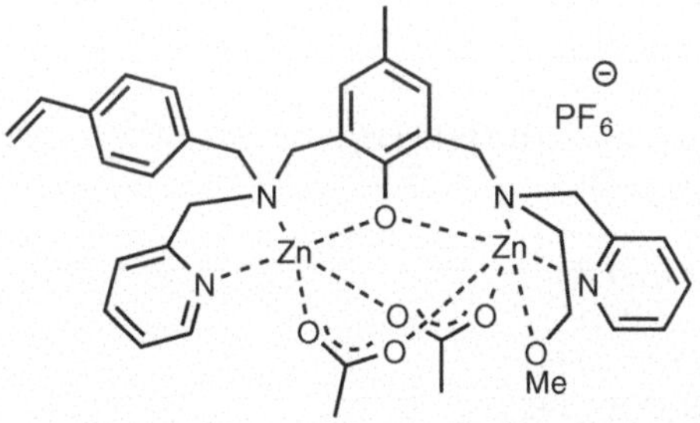

ESI mass spectrometry (methanol) m/z = 751.1 (100 %), 749.1 (95 %) calc. for $[C_{35}H_{47}N_4O_6Zn_2]^+$ m/z = 749.2 (100 %), 751.2 (96.1 %); (acetonitrile) m/z = 851.3, calc. 851.23 (100 %), 849.23 (85.5 %) $[C_{41}H_{49}N_6O_6Zn_2]^+$; m/z = 769.1, calc. 769.17 (100 %), 767.18 (87 %), $[C_{37}H_{43}N_4O_6Zn_2]^+$; m/z = 723.1, calc. 721.17 (100 %), 732.17 (96.6 %), $[C_{33}H_{43}N_4O_6Zn_2]^+$), m/z = 653.1, calc. m/z 653.11 (100 %), 655.11 (96.2 %), $[C_{28}H_{35}N_4O_6Zn_2]^+$. **FT-IR spectroscopy** (ν, cm^{-1}) 2927, 2856 (w, C–H str); 1598 (s, bridging acetate antisym. str); 1431 (s, bridging acetate sym. str); 832 (s, P–F str); 765, 729 (m, pyH def); 555 (m, P–F). **Microanalysis** Anal calc. for $C_{38}H_{48}N_4O_7Zn_2PF_6$ C 48.12, H 5.10, N 5.91 %; found: C 48.19, H 4.71, N 5.36 % ^{1}**H NMR** (CD_3CN, 500.13 MHz) δ 1.94 (s, 6H, $CH_3CO_2^-$); 2.14 (s, 3H, arCH_3); 2.50–2.56 (m, 2H, N$CH_2$$CH_2$); 2.84–2.90 (m, 2H, NCH_2CH_2); 3.03 (s, 3H, OCH_3); 3.57 (d, 1H, arCH_2N, J = 12.6 Hz); 3.64 (d, 1H, NCH_2py, J = 12.1 Hz); 3.71–3.91 (m, 5H, CH_2); 4.15 (d, 1H, arCH_2N, J = 12.5); 4.21–4.27 (m, 2H, CH_2); 5.20 (d, 1H, CH=CH_2, J_{cis} = 10.9 Hz); 5.25 (dd, 1H, CH=CH_2, J_{cis} = 11.0, 0.5 Hz, second isomer); 5.69 (d, 1H, CH=CH_2, J_{trans} = 17.6 Hz); 5.81 (d, 1H, CH=CH_2, J_{trans} = 17.6 Hz, second isomer); 6.62 (dd, 1H, *CH*=CH_2, J = 17.7, 10.9, 0.8 Hz); 6.71 (dd, 1H, *CH*=CH_2, J = 17.7, 10.9 Hz, second isomer); 6.81 (d, 2H, ar*CH*, J = 1.8 Hz); 6.89 (d, 2H, ar*CH*, J = 1.8 Hz); 6.95 (s, 2H, ar*CH*); 7.10–7.53 (m, 4H, py*CH*); 7.91 (m, 2H, py*CH*, J = 7.8, 1.6 Hz); 8.70 (m, 2H, py*CH*). ^{13}**C NMR** (CD_3CN, 100.62 MHz) δ 20.2 (arCCH_3); 23.1 ($CH_3CO_2^-$); 54.7 (CH_2); 58.6 (OCH_3); 59.1, 59.5, 60.8, 61.2, 61.3, 69.3 (CH_2); 114.9 (CH=CH_2); 115.2 (CH=CH_2, second isomer); 124.0, 124.3, 124.7, 125.0, 125.1 (py*C*H); 125.2, 125.3 (C_{quart}); 126.8, 126.9 (CH); 127.0, 127.3, 131.6 (C_{quart}), 132.8, 133.0 (CH); 133.2 (CH); 133.6 (CH); 133.7 (C_{quart}); 137.1 (*C*H=CH_2); 138.5 (arC); 140.8, 141.3, 141.5, 148.2, 148.5, 149.8 (CH); 155.8, 156.5, 160.8, 161.4 (C_{quart}), 179.0 (CH_3*C*O_2).

2.9 Crystal Structures Included in this Thesis

Complex	Chapter
$[Zn_2(CH_3\mathbf{L2})(CH_3COO)_2](PF_6)$	4
$[Zn_2(CH_3\mathbf{L3})(CH_3COO)_2](PF_6)$	4
$[Zn_2(Br\mathbf{L2})(CH_3COO)_2](PF_6)$	4
$[Zn_2(NO_2\mathbf{L2})(CH_3COO)_2](PF_6)$	4
$[Zn_4(Br\mathbf{L2})_2(PO_3F)_2(H_2O)_2](PF_6)_2$	4
$[Zn_4(CH_3\mathbf{L2})_2(NO_2C_6H_5OPO_3)_2(H_2O)_2](PF_6)_2$	4
$[Zn_2(CH_3\mathbf{L4})(CH_3COO)_2](PF_6)$	8
$[Zn_2(CH_3\mathbf{L5})(CH_3COO)_2](BPh_4)$	8
$[Co_2(CH_3\mathbf{L2})(CH_3COO)_2](PF_6)$	6
$[Co_2(Br\mathbf{L2})(CH_3COO)_2](PF_6)$	6
$[Co_2(CO_2Et\mathbf{L2})(CH_3COO)_2](PF_6)$	6
$[Co_2(CO_2EtH_2\mathbf{L1})(CH_3COO)_2](PF_6)$	6
$[Cd_4(CO_2EtH_2\mathbf{L1})_2(CH_3COO)_{3.75}Cl_{0.25}(H_2O)_2](PF_6)$	5

References

1. G. Ambrosi, M. Formica, V. Fusi, L. Giorgi, M. Micheloni, Coord. Chem. Rev. **252**, 1121–1152 (2008)
2. R. Robson, Aust. J. Chem. **23**, 2217–2224 (1970)
3. R. Robson, Inorg. Nucl. Chem. Lett. **6**, 125–128 (1970)
4. R.E. Mirams, S.J. Smith, K.S. Hadler, D.L. Ollis, G. Schenk, L.R. Gahan, J. Biol. Inorg. Chem. **13**, 1065–1072 (2008)
5. R.R. Buchholz, M.E. Etienne, A. Dorgelo, R.E. Mirams, S.J. Smith, S.Y. Chow, L.R. Hanton, G.B. Jameson, G. Schenk, L.R. Gahan, J. Chem. Soc., Dalton Trans. 6045–6054 (2008)
6. L.R. Gahan, S.J. Smith, A. Neves, G. Schenk, Eur. J. Inorg. Chem. **19**, 2745–2758 (2009)
7. S.J. Smith, A. Casellato, K.S. Hadler, N. Mitić, M.J. Riley, A.J. Bortoluzzi, B. Szpoganicsz, G. Schenk, A. Neves, L.R. Gahan, J. Biol. Inorg. Chem. **12**, 1207–1220 (2007)
8. H. Carlsson, M. Haukka, E. Nordlander, Inorg. Chem. **43**, 5681–5687 (2004)
9. H. Carlsson, E. Nordlander, M. Jarenmark, C. R. Chim. **10**, 433–462 (2007)
10. M. Jarenmark, E. Csapo, J. Singh, S. Wockel, E. Farkas, F. Meyer, M. Haukka, E. Nordlander, J. Chem. Soc., Dalton Trans. **39**, 8183–8194 (2010)
11. M. Jarenmark, S. Kappen, M. Haukka, E. Nordlander, J. Chem. Soc., Dalton Trans 993–996 (2008)
12. S. Albedyhl, Marie T. Averbuch-Pouchot, C. Belle, B. Krebs, Jean L. Pierre, E. Saint-Aman, S. Torelli, Eur. J. Inorg. Chem. **2001,** 1457–1464 (2001)
13. J.A. Larrabee, W.R. Johnson, A.S. Volwiler, Inorg. Chem. **48**, 8822–8829 (2009)
14. J.A. Larrabee, S.A. Chyun, A.S. Volwiler, Inorg. Chem. **47**, 10499–10508 (2008)
15. A. Mangalum, R.C. Smith, Tetrahedron **65**, 4298–4303 (2009)
16. X. Dong, X. Wang, Y. He, Z. Yu, M. Lin, C. Zhang, J. Wang, Y. Song, Y. Zhang, Z. Liu, Y. Li, Z. Guo, Chem. Eur. J. **16**, 14181–14189 (2010)
17. T. Routasalo, J. Helaja, J. Kavakka, A.M.P. Koskinen, Eur. J. Org. Chem. **2008**, 3190–3199 (2008)
18. T. Inomata, K. Shinozaki, Y. Hayashi, H. Arii, Y. Funahashi, T. Ozawa, H. Masuda, Chem. Commun. (Camb.) 392–394 (2008)
19. H. Sharghi, M.A. Nasseri, K. Niknam, J. Org. Chem. **66**, 7287 (2001)
20. A. Zinke, R. Ott, E. Leggewie, A. Hassanein, G. Zankl, Monatshefte für Chemie / Chemical Monthly **87**, 552–559 (1956)
21. V. Boehmer, F. Marschollek, L. Zetta, J. Org. Chem. **52**, 3200–3205 (1987)
22. R.T. Paine, Y.C. Tan, X.M. Gan, Inorg. Chem. **40**, 7009–7013 (2001)
23. J.d. Mendoza, P.M. Nieto, P. Prados and C. Sánchez, Tetrahedron, **46**, 671–682 (1990)
24. S. Striegler, M. Dittel, Inorg. Chem. **44**, 2728–2733 (2005)
25. E.F. Elslager, E.L. Benton, F.W. Short, F.H. Tendick, J. Am. Chem. Soc. **78**, 3453–3457 (1956)
26. Bruker, TopSpin 2.0 Software for NMR data processing
27. M. Hesse, H. Meier, B. Zeeh, *Spectroscopic Methods in Organic Chemistry* (G. Thieme, NY, 1997)
28. D.F. Evans, J. Chem. Soc., 1959 (2003)
29. E.M. Schubert, J. Chem. Educ. **69**, 62 (1992)
30. G.A. Bain, J.F. Berry, J. Chem. Educ. **85**, 532–536 (2008)
31. H. Sakiyama, Inorg. Chim. Acta **360**, 715–716 (2007)
32. Thermo-Scientific, Grams/AI 9.0 Software
33. M. J. Riley, VTVH 2.1.1 Program for the simulation and fitting variable temperature—variable field MCD spectra (2008)
34. H. Adamsky, AOMX Program (1996)
35. F. Ely, K.S. Hadler, N. Mitić, L.R. Gahan, D.L. Ollis, N.M. Plugis, M.T. Russo, J.A. Larrabee, G. Schenk, J. Biol. Inorg. Chem. **16**, 777–787 (2011)
36. G.M. Sheldrick, SHELXL97: Program for the refinement of crystal structures (1997)

37. L.J. Farrugia, J. Appl. Crystallogr. **32**, 837–838 (1999)
38. R.A. Peralta, A.J. Bortoluzzi, B. de Souza, R. Jovito, F.R. Xavier, R.A. A. Couto, A. Casellato, F. Nome, A. Dick, L.R. Gahan, G. Schenk, G.R. Hanson, F.C.S. de Paula, E.C. Pereira-Maia, S.d.P. Machado, P.C. Severino, C. Pich, T. Bortolotto, H. Terenzi, E.E. Castellano, A. Neves, M.J. Riley, Inorg. Chem. **49**, 11421–11438 (2010)
39. I.H. Segel, *Enzyme Kinetics: Behavior and Analysis of Rapid Equilibrium and Steady-State Enzyme Systems* (Wiley, NY, 1975)
40. N.V. Kaminskaia, B. Spingler, S.J. Lippard, J. Am. Chem. Soc. **122**, 6411–6422 (2000)
41. N.V. Kaminskaia, C. He, S.J. Lippard, Inorg. Chem. **39**, 3365–3373 (2000)
42. C.S. Ltd, CasaXPS: Processing Software for XPS, AES, SIMS and More (2009)
43. R.V. Jones, L. Godorhazy, N. Varga, D. Szalay, L. Urge, F. Darvas, J. Comb. Chem. **8**, 110–116 (2006)
44. K.S. Hadler, N. Mitic, S.H. Yip, L.R. Gahan, D.L. Ollis, G. Schenk, J.A. Larrabee, Inorg. Chem. **49**, 2727–2734 (2010)
45. C.J. Jackson, K.S. Hadler, P.D. Carr, A.J. Oakley, S. Yip, G. Schenk, D.L. Ollis, Acta crystallogr. F **64**, 681–685 (2008)
46. J.M. Walker, *The Proteomics Protocols Handbook* (Humana Press, Totowa, 2005)
47. V.M. D'Souza, B. Bennett, A.J. Copik, R.C. Holz, Biochemistry **39**, 3817–3826 (2000)
48. K.S. Hadler, E.A. Tanifum, S.H. Yip, N. Mitić, L.W. Guddat, C.J. Jackson, L.R. Gahan, K. Nguyen, P.D. Carr, D.L. Ollis, A.C. Hengge, J.A. Larrabee, G. Schenk, J. Am. Chem. Soc. **130**, 14129–14138 (2008)
49. H. Tsue, K. Enyo, K.-I. Hirao, Org. Lett. **2**, 3071–3074 (2000)
50. G.-C. Sun, Z.-H. He, Z.-J. Li, X.-D. Yuan, Z.-J. Yang, G.-X. Wang, L.-F. Wang, C.-R. Liu, Molecules **6**, 1001–1005 (2001)
51. T. Aoki, T. Kaneko, N. Maruyama, A. Sumi, M. Takahashi, T. Sato, M. Teraguchi, J. Am. Chem. Soc. **125**, 6346–6347 (2003)
52. W. Müller, P. Kipfer, D.A. Lowe, S. Urwyler, Helv. Chim. Acta **78**, 2026–2035 (1995)
53. A.S. Borovik, V. Papaefthymiou, L.F. Taylor, O.P. Anderson, L. Que, J. Am. Chem. Soc. **111**, 6183–6195 (1989)
54. E. Díez-Barra, J.M. Fraile, J.I. García, E. García-Verdugo, C.I. Herrerías, S.V. Luis, J.A. Mayoral, P. Sánchez-Verdú, J. Tolosa, Tetrahedron Asymmetry **14**, 773–778 (2003)
55. S. Carloni, V. Borzatta, L. Moroni, G. Tanzi, G. Sartori, R. Maggi, Catalysts Based on Metal Complexes for the Synthesis of Optically Active Chrysanthemic Acid (2005), p. 55 (Patent WO2005123254A1)
56. C.A. Bunton, S.J. Farber, J. Org. Chem. **34**, 767–772 (1969)
57. R.S. Molday, Canadian Patents and Development Limited (CA) Magnetic Iron-Dextran Microspheres (1984) (Patent 4452773)
58. B.-F. Pan, F. Gao, H.-C. Gu, J. Colloid Interface Sci. **284**, 1–6 (2005)

Chapter 3
Understanding the Overall Structure of GpdQ and Metal Binding

3.1 Introduction

3.1.1 Discovery of GpdQ and Its Overall Structure

The ability of the glycerophosphodiesterase GpdQ to degrade a variety of stable phosphate diesters first came to attention in the 1970s when Gerlt and co-workers purified GpdQ and investigated its function [1].[1] The natural substrate is glycerol-phosphoethanolamine [1] but GpdQ is reported to be also able to degrade other phosphodiester substrates along with phosphomonoester, phosphotriester and phosphorothiolate substrates (Fig. 3.1) [4, 5]. Its potential as bioremediator for nerve agents was noted in 2007 when Raushel reported the GpdQ catalyzed degradation of EA 2192, a degradation product of the nerve agent VX [2].

The crystal structure of GpdQ (Fig. 3.2) was first solved by Ollis and co-workers [3, 7]. A structure with higher resolution (1.9 Å) is, however, now available [6, 8].

The structure shows that GpdQ is a hexamer which consists of a trimer of dimers, having overall D_3 symmetry [6]. Figure 3.2 displays the hexamer (a) and the structure of the dimeric subunit (b). Each dimer is comprised of a cap domain (yellow), a dimerization domain shown in red and two catalytic domains (blue), each of them hosting a dinuclear active site.

In an attempt to evolve GpdQ towards a non-physiological substrate (BPNPP) a range of mutants were obtained by directed evolution [11]. These mutants exhibited increased activity towards BPNPP and the previously found hexameric structure was broken down in some mutants [11]. The mutation of a cysteine residue (C269) in the dimerization domain led to the formation of monomer and dimer forms of GpdQ. The smaller oligomeric (dimer) forms of GpdQ were shown to have enhanced activity towards BDNPP compared to the wild type form. For

---

[1] Parts of this Chapter have been reprinted from L.J. Daumann et al. "*Promiscuity Comes at a Price: Catalytic Versatility vs. Efficiency in Different Metal Ion Derivatives of the Potential Bioremediator GpdQ*" *BBA—Proteins and Proteomics* **2013**, 1834, 1, 425–432. with permission from Elsevier.

L. J. Daumann, *Spectroscopic and Mechanistic Studies of Dinuclear Metallohydrolases and Their Biomimetic Complexes*, Springer Theses, DOI: 10.1007/978-3-319-06629-5_3,

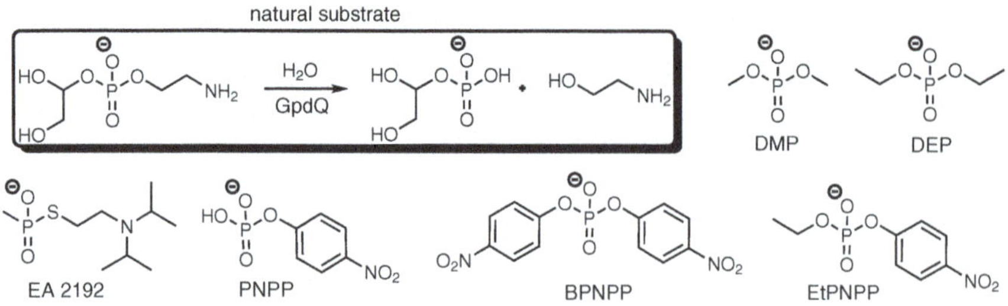

Fig. 3.1 Physiological substrate of GpdQ and other phosphoesters used in kinetic assays [1, 2, 4–6]

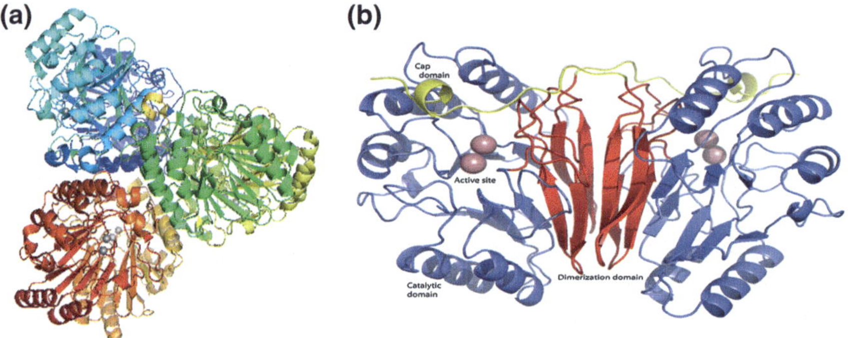

Fig. 3.2 PyMOL picture of hexameric (**a**) GpdQ (PDB number 3D03) [6] and the dimeric subunit (picture **b**) taken from Reference [9, 10]

example, the 8–3 mutant dimer (8th generation) showed an over 500-fold higher catalytic efficiency compared to the wild type [11]. This was attributed to the opening of the active site towards solvent and substrate molecules [11]. The more recent crystal structure at higher resolution (1.9 Å) illustrates an extensive hydrogen bonding network in the active site, which is comprised of two distinct metal binding sites (labeled as the α and β-sites). The α-site features four amino acid side chains, two aspartates (Asp50, Asp8) and two histidines (His10, His197), whereas the metal in the β-site is coordinated by two histidines (His156, His195), one aspartate (Asp50) and one asparagine (Asn80). The crystal structure also suggests the presence of a terminal H_2O bound to the α-metal and a bridging water/hydroxide molecule. The active site is surrounded by an extensive hydrogen bond network connecting it with the second coordination sphere [6] (Fig. 3.3).

The in vivo metal ion content in GpdQ has, to date, not been determined, although studies using anomalous scattering experiments suggest a Fe(II) bound in the α-site [8]. Catalytic activity can be reconstituted with a variety of divalent metal ions including Co(II), Zn(II), Fe(II), Mn(II) and Cd(II) [8, 12, 13]. No heterodinuclear derivatives of GpdQ have been reported in the literature although the native metal ion content in vivo is proposed to be of heterodinuclear

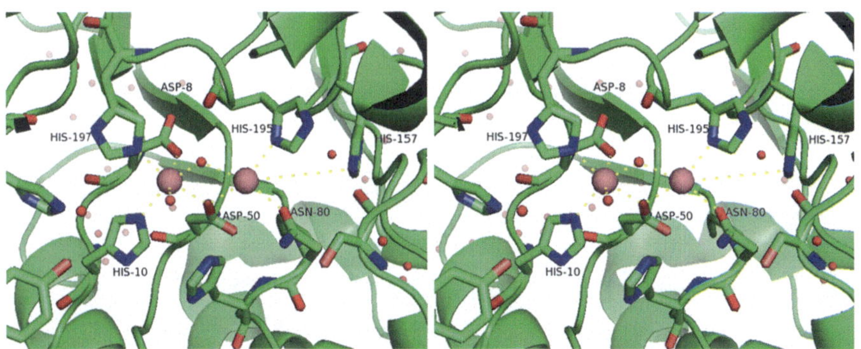

Fig. 3.3 Stereo view of active site of GpdQ. Picture generated with PyMOL [9]

composition [8]. Given the potential use of GpdQ in bioremediation much work has been undertaken to understand (i) the nature of its reaction mechanism (next section), and (ii) factors which influence substrate-binding.

3.1.2 Catalytic Mechanism

From magnetic circular dichroism (MCD) spectroscopy studies of Co(II)-substituted GpdQ it is known that GpdQ exists as a mononuclear resting state enzyme with only the α-site occupied by metal, as evidenced by the single band in the MCD spectrum at 495 nm (typical for 6-coordinate Co(II), a brief introduction about magnetic circular dichroism spectroscopy can be found in Sect. 6.1.2.) [8]. Upon coordination of substrate it is proposed that the active site ($E'_M.S$) undergoes rapid structural rearrangement facilitated by the hydrogen bond network that connects the first and second coordination sphere (Fig. 3.4) [12].

Hadler et al. [12] showed, using MCD, that the dinuclear Co(II)Co(II) metal center in GpdQ is only formed in the presence of phosphate, a substrate analog ($E_B.S$). The fact that bands from 5- and 6-coordinate Co(II) were observed in the MCD led to the proposal that the bond between Co(II) and Asn80 in the β-site is broken upon substrate coordination ($E^*_B.S$) [8, 12]. A terminal hydroxide then acts as nucleophile, product and β metal are released and the active site returns to its mononuclear resting state [12]. The formation of a dinuclear center in the presence of substrate was substantiated by electron paramagnetic resonance (EPR) spectroscopic studies of the Mn(II)Mn(II) derivative of GpdQ. The EPR data revealed K_d values for the metal ions in the α- and β-sites to be 29 and 344 μM, respectively, in the absence of a substrate analog [12]. In the presence of phosphate the affinity of the β-site increased significantly ($K_d = 56$ μM) [12]. It has also been shown that metal ion affinity of the β-site can be increased by changing the Asn80 residue to an aspartate, resulting in a fully occupied dinuclear center after the

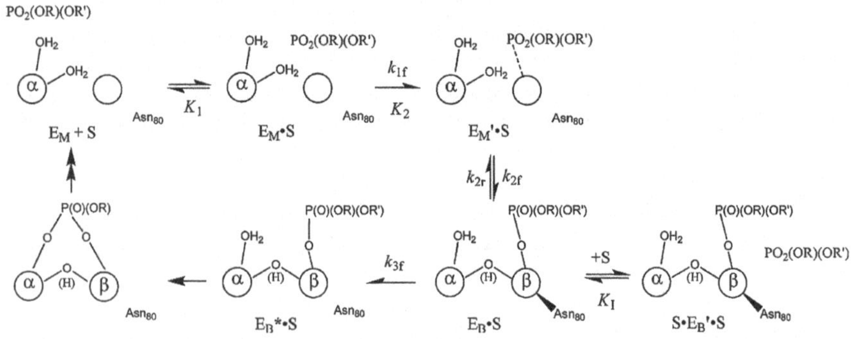

Fig. 3.4 Mechanism of hydrolysis of phosphate esters by GpdQ as determined in stopped-flow fluorescence studies [12]. Reprinted with permission from [12]. Copyright (2014) American Chemical Society

addition of two equivalents of Co(II) to the apoenzyme, even in the absence of substrate [8]. However the enzyme activity was impaired as shown by kinetic measurements, leading to the proposal that the coordination flexibility of Asn80 is required for high activity. The exchange of Asn80 by alanine led, as expected, to a dinuclear center only in the presence of substrate [8]. Also, this mutant showed increased reactivity while its affinity for substrate was greatly decreased. Further studies with GpdQ mutants have been conducted by Hadler et al. to elucidate the role of the surrounding hydrogen bond network [12]. Specifically, the role of His81 and His217 has been probed by replacing these residues by alanine in site directed mutagenesis experiments [12]. While replacement with alanine did not affect the pH dependence of the enzyme, an increased metal binding affinity resulted, being 2-(His81Ala) and 8-times (His217Ala) higher than in wt-GpdQ [12]. This shows that residues that are not directly interacting with the active site metal ions have an impact on the coordination environment through the hydrogen bond network mentioned above.

3.1.3 Aims of Chapter and Relevance

The metal ion coordination in the β-site and the reactivity of GpdQ is regulated by residue Asn80, which affords coordination flexibility in that site [12]. The metal ion in the α-site is bound more tightly than that in the β-site, and consequently GpdQ is predominantly present as a mononuclear enzyme in the resting state. However, in the presence of substrate, the metal ion affinity of the β-site is increased which leads to the formation of a catalytically competent dinuclear active site [12]. One aim of this chapter is to generate heterodinuclear forms of GpdQ such as Fe(II)Co(II), Fe(II)Mn(II), and Fe(II)Cd(II) as the metal ion content in vivo is proposed to be of heterodinuclear composition [8]. Metal binding affinities of the β-site for the heterodinuclear derivatives will be determined over a large pH range (pH 5–11). Also the catalytic properties of the heterodinuclear forms will be determined and compared with the purely homodinuclear counterparts reported previously [6, 12, 13]. Moreover, the role of hydrogen bonds within the second coordination sphere will be studied by site directed mutagenesis of the residues Ser127 (forming a hydrogen bond to Asn80) and Tyr19 (part of hydrogen bond network involving the terminal water molecule bound to the α-site). Kinetic assays, AAS and MCD spectroscopy will be employed to monitor the successful formation of Fe(II)Co(II) centers. Metal binding studies of Fe(II)Zn(II) wt-GpdQ and two histidine mutants (His217, His81) were conducted by B. McCarthy (B.Sc. (Hons)) but will be included in the results and discussion sections for completeness [14]. The site directed mutants utilized are shown in red below (Fig. 3.5). Moreover, the influence of residue Ser127 on Co(II) binding to the β-site (via Asn80) in the absence and presence of substrate will be probed by MCD. The data will be compared to previous studies with Co(II)Co(II) wt-GpdQ [6].

Whether the hexameric structure is essential for catalytic activity and stability will be investigated by studying the 8–3 mutant of GpdQ. This mutant contains the Ser127Ala, Tyr221His and Cys269Ala mutations. The latter residue is located in the dimerization domain and was proposed to cause hexameric GpdQ to break down to a dimeric form [11]. An evaluation whether this mutant, reported previously by Ollis and co-workers [11], is a good candidate for bioremedial applications will be part of this chapter. All experimental procedures such as enzyme expression and purification are described in Chap. 2 along with the conditions used in kinetics assays.

3.2 Results

3.2.1 Expression and Purification of wt-GpdQ and Mutants

3.2.1.1 wt-GpdQ

The expression and purification of GpdQ was conducted using a modification of a previously published protocol and involved three chromatographic steps. After a strong anion exchange column (HiPrep 16/10 DEAE FF) the proteins were subjected to an intermediate purification with a HiLoad 26/10 Phenyl Sepharose HP column (hydrophobic interaction) and finally a HiPrep 16/10 Sephacryl S-200 HR gel filtration column (size exclusion). AAS of purified GpdQ shows generally a super stoichiometric amount of iron with various amounts of other metal ions. In detail, the AAS of wild-type GpdQ expressed in *E. coli* indicated the presence of 1.35(5) equivalents Fe, 0.25(5) Zn, 0.30(5) Co and 0.15(4) Mn ions per active site, respectively.

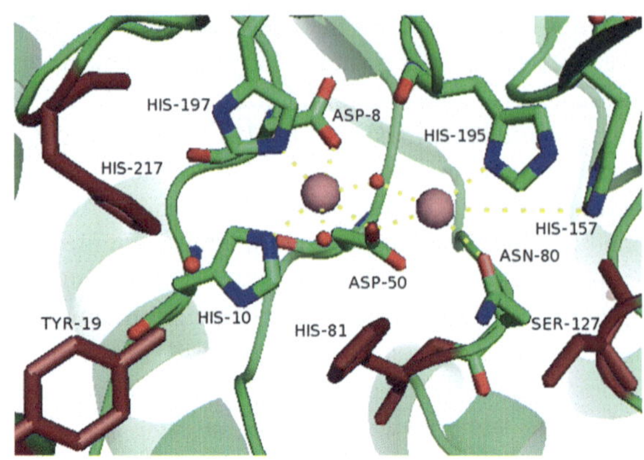

Fig. 3.5 Active site of GpdQ (PDB number 3D03) [6] with the second coordination sphere residues that were studied. Picture generated with PyMOL [9]

3.2.1.2 8–3-Mutant

The 8–3 mutant contains the Ser127Ala, His217Arg and Cys269Ala mutations (Fig. 3.6). The plasmid for this mutant was obtained from S. Yip [11]. The cells were grown using the same protocol as for wt-GpdQ. The purification involved two chromatographic steps (a detailed description can be found in Chap. 2).

After the first purification step involving a Q Sepharose (strong anion exchange) column, 8–3 GpdQ was precipitated with ammonium sulfate to give a 2.5 M final ammonium sulfate concentration. It should be noted that for wt-GpdQ the protocol dictates the opposite: other proteins are precipitated while GpdQ remains in the supernatant. It was noted that the 8–3 mutant GpdQ precipitated at much lower ammonium sulfate concentrations (below 0.2 M) as the wild type enzyme (above 1.3 M). The supernatant was discarded for the 8–3 mutant and the precipitated proteins resuspended in TRIS buffer and loaded onto a Superdex 200 gel filtration column. A size exclusion column functions on the principle that heavier/larger proteins are eluted first while smaller proteins remain longer in the small channels of the resin and are thus eluted after the larger proteins. Interestingly, the size exclusion column elution profile showed, in addition to hexameric GpdQ, the presence of both dimeric and monomeric forms (Fig. 3.7). The presence of monomer had not been reported for this mutant [11]. Kinetic assays were conducted to determine trends in activity and stability (Table 3.1). Whereas the specific activity (SA) in the assays with no metal ions added was similar in all fractions, the activity of the mainly monomer containing fractions 12 h after purification in the presence of 23 μM cobalt(II) sulfate was five times higher than that of the dimer and fifteen times higher than the SA of the mainly hexameric form of the 8–3 GpdQ mutant. To analyse the stabilities the specific activity was measured again after 24 h for the different oligomers. After 12 h the activity of the monomer was decreased sevenfold, whereas the specific activities of the dimer and hexamer remained the same.

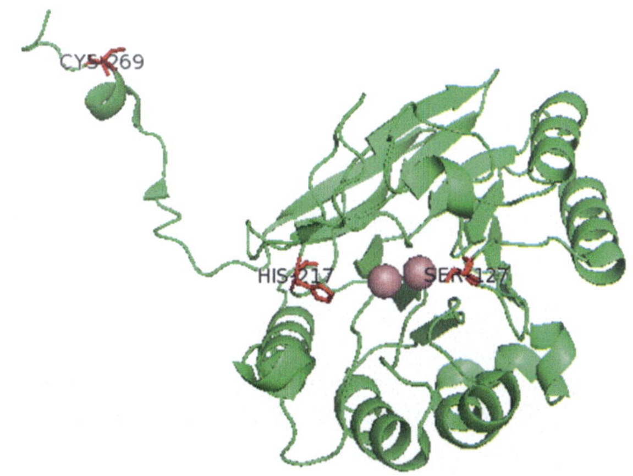

Fig. 3.6 Monomeric unit of GpdQ with the residues affected by mutation in red and the two metal ions in the active site as pink spheres

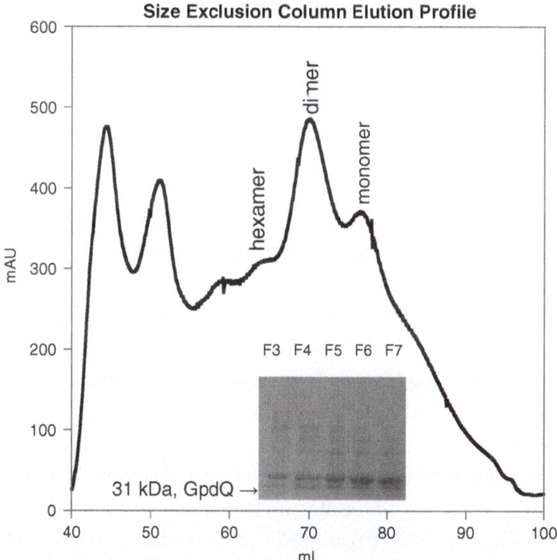

Fig. 3.7 Size exclusion column profile for the 8–3 mutant and a picture of the SDS-page gel of the GpdQ containing fractions

3.2.1.3 Ser127Ala and Tyr19Phe Mutants

The alanine and phenylalanine mutations to the Ser127 and Tyr19 residues were introduced by site directed mutagenesis. The expression and purification followed the same protocol as for wt-GpdQ.

Table 3.1 Specific activities [in μ mol min^{-1} mg^{-1}] of the oligomers of 8–3 Mutant GpdQ in the presence of different metals

Fraction Oligomer	3 hexamer	4 hexamer/dimer	5 dimer	6 dimer/monomer	7 monomer
Concentration [mg/mL]	1.28	1.59	2.07	2.00	1.67
SA no added metal (12 h)	0.27	0.14	0.21	0.19	0.20
SA Co(II) (12 h)	0.47	1.35	1.69	3.51	7.42
SA Mn(II) (12 h)	0.51	0.54	1.41	0.96	2.19
SA Co(II) (36 h)	0.65	–	1.54	–	1.06
SA Mn(II) (36 h)	0.20	–	0.72	–	0.52

3.2.2 Studies of Heterodinuclear GpdQ of the Fe(II)M(II) Type

The generation of a dinuclear center in GpdQ is not straightforward since one metal ion is bound tightly (to the α-site) whereas the binding of the second (to the β-site) requires the presence of a substrate (or substrate analog) [6]. Hence the strategy employed was to generate the half-apo enzyme first with Fe(II) (already present in the purified enzyme due to high abundance of this metal in the expression media) bound to the α-site, and then reconstitute a fully dinuclear active site by adding metal ions and either substrates (in activity assays) or the substrate analog phosphate (in MCD experiments).

3.2.2.1 Generation of Half-apo GpdQ

To generate a one-Fe(II) half-apo form of GpdQ the enzyme was incubated for one minute with a series of chelators to remove non-bound or weakly bound metal ions, followed by a desalting column to separate the enzyme from the chelators. A similar approach has been shown to be effective in removing metal ions from other dinuclear metallohydrolases such as OpdA [15] and some PAPs [16–19]. Moreover, all buffers and solutions were treated with Chelex prior to addition to the enzyme to avoid contamination by adventitious metals. However, none of the conditions tested promoted the quantitative removal of the metal ions. A prolonged time with the chelating solution caused the removal of Fe(II) from the α-site. This process is unfavorable since the external addition of the easily oxidized Fe(II) ion might result in Fe(III) bound the enzyme. In several preparations the Fe content as determined by AAS, varied from 1.0 to 1.3 and that of Zn, Mn and Co between 0 and 0.3. However, strongly impaired catalytic activity was observed when substrate was added without addition of extra metal ions to the assay solution ($\leq$4 % of full activity) which indicated that the metals are either bound to the α-site or not available to reconstitute the active (β)-site.

3.2.2.2 Magnetic Circular Dichroism of Fe(II)Co(II) wt-GpdQ

MCD protein samples of Fe(II)Co(II) wt-GpdQ were prepared by adding one equivalent Co(II) and ten equivalents of phosphate to the Fe(II) half-apo enzyme in buffered solution (pH 7). To assure the formation of a good glass, 60 % glycerol was added to the enzyme solution. The resultant spectrum (1.5 K, 7 T) was fitted to the minimum number of Gaussian peaks to achieve a satisfactory composite spectrum using the GRAMS AI software [20]. In previous work with GpdQ, MCD was used to demonstrate the binding of Co(II) to the six-coordinate α-site and Co(II) to the five-coordinate β-site in GpdQ; the former species has characteristic transitions around ~500 nm, the latter at ~570 nm [6, 21]. The MCD spectrum of half-apo GpdQ is shown in the inset of Fig. 3.8 (red). Only the relevant region between 480 nm to 600 nm is shown and compared to the corresponding spectrum of the half-apo enzyme with added Co(II) (blue). Virtually no six-coordinate Co(II) is present in half-apo GpdQ and only a small amount of five-coordinate Co(II), indicating that the majority of the sample contains mononuclear Fe(II) centers, consistent with the AAS measurement and the lack of kinetic activity. The full spectrum of Fe(II)Co(II) GpdQ in the presence of 10 equivalents of phosphate (to enhance the binding affinity of the β-metal ion site [21]) is shown in Fig. 3.8. The pseudo A-term feature between 360 and 440 nm is assigned to a Fe(II) metal-to-ligand charge transfer (MLCT) transition based on a comparison to related enzyme and model systems with a mixed N, O donor atom environment [22, 23].

Transitions arising from Fe(III) were not observed and it was thus concluded that the Fe(II) form of the enzyme was stable and not oxidized under ambient conditions. In contrast, as mentioned above, the bands at ~500 and ~570 nm are characteristic for six- and five-coordinate Co(II) species, respectively. Thus, although the addition of Co(II) led to the generation of heterodinuclear Fe(II)-Co(II) centers, homodinuclear Co(II)Co(II) centers are also present, suggesting that some Fe(II) ions in the α-site may have been replaced by Co(II) or had been vacant in the first place. Based on a comparison with the MCD spectrum of di-Co(II) GpdQ (300 mdeg/mM if 100 % occupied) it is estimated that at most 40 % of the α-sites contain Co(II) ions.

3.2.2.3 Determination of Metal Binding Affinity by Activity Measurements

The effect of the metal ion concentration on the activity of half-apo GpdQ was determined by varying the amount of added M(II), where M = $CoSO_4 \bullet 6H_2O$, $MnCl_2 \bullet H_2O$ or $CdCl_2 \bullet 2.5\ H_2O$, to the activity assay mixture (similar experiments were conducted by B. McCarthy with $ZnCl_2$, the data for the Fe(II)Zn(II) are included in subsequent discussions for comparison) [14]. All solutions (buffers, substrate and MilliQ water for making up the respective metal solutions) were treated with Chelex for 24 h prior to use to ensure complete removal of external metals that could influence the metal titration. The concentrations of the

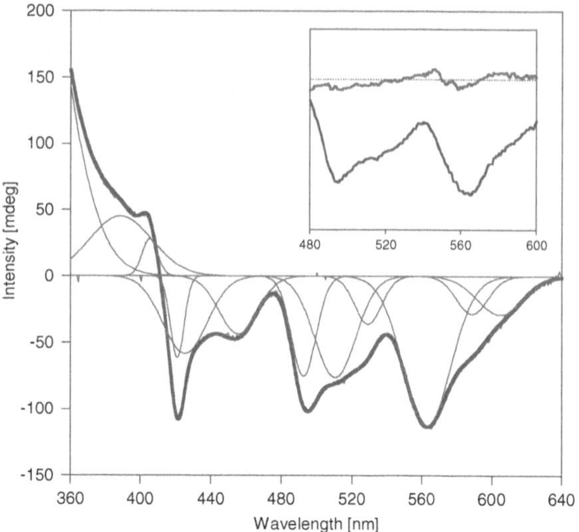

Fig. 3.8 MCD spectrum of Fe(II)Co(II)-GpdQ in presence of 10 eq. phosphate with Gaussian deconvolution Inset: *red* spectrum Fe(II) half apo, *blue* spectrum Fe(II)Co(II) plus 10 eq. phosphate. The spectra shown were acquired at a temperature of 1.5 K and a field of 7 T

half-apoenzyme (containing Fe(II) already bound to the α-site) and substrate (BPNPP) were kept constant at 200 and 10 mM, respectively, while that of M(II) increased from 0–50 μM. The addition of M(II) to half-apo of GpdQ at a constant pH and substrate concentration demonstrated saturation behavior similar to that described by the Michaelis-Menten equation (see Chap. 2, Eq. 2.5). The assays provided an estimate for the binding affinity of the metal ions to this site. The K_d, the dissociation constant for the β-site, is similar to the K_m value obtained in the Michaelis-Menten equation. An example of the data that were obtained is shown in Fig. 3.9.

The binding function (*r*), which is the molar ratio of the amount of metal ion bound to the total amount of enzyme active sites, was calculated according to Eq. (3.1), where *n* is the number of sites associated with K_d, the dissociation constant, and $[M]_{free}$ is the free metal ion concentration (Eq. 3.1) [24]. Plots of the data obtained for the titration of Co(II) into the Fe(II) half-apo wt-GpdQ and the Ser127Ala and Tyr19Phe mutant enzymes are shown in Fig. 3.9.

$$r = \frac{n[M]_{total}}{K_d + [M]_{total}} \tag{3.1}$$

Fitting the data for all metal derivatives yielded the number of available metal binding sites n for half-apo GpdQ (Table 3.2). The K_d values obtained at pH 7 are included (Table 3.2).

The inverse of the K_d values (referred herein as K_{Assoc}) obtained from measuring the enzyme activity as a function of added metal ion concentration were plotted as a function of pH. Bell-shaped profiles were obtained for the Co(II), Zn(II) and Mn(II) binding and a sigmoidal curve for Cd(II) binding to the β-site indicating that for the first three metals two deprotonation events influence the

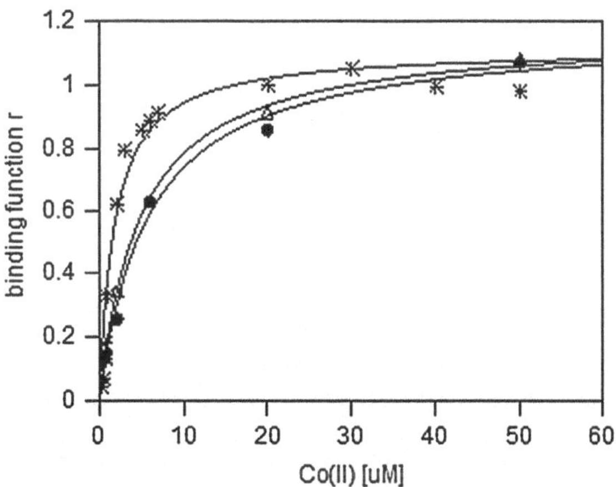

Fig. 3.9 Dependence of the catalytic activity of the Fe(II)Co(II) derivative of wt-GpdQ (*star*), Ser127Ala mutant (*open rectangle*) and Tyr19Phe mutant (*filled circle*) on the concentration of added Co(II). The data sets shown were measured at pH 7.0 and were fit to Eq. (3.1). Full activity was essentially reached when [Co(II)] ≥20 μM

Table 3.2 Summary of K_d and n values in different mixed metal ion derivatives of GpdQ at pH 7

Derivative	n	K_d [μM]
Fe(II)Zn(II) [14]	1.36 ± 0.14	1.26 ± 0.68
Fe(II)Mn(II)	1.35 ± 0.15	3.90 ± 1.49
Fe(II)Co(II)	1.11 ± 0.03	1.83 ± 0.29
Fe(II)Cd(II)	1.56 ± 0.17	0.81 ± 0.44
Fe(II)Co(II)$_{S127A}$	1.16 ± 0.05	4.96 ± 0.71
Fe(II)Co(II)$_{Y19F}$	1.16 ± 0.06	5.90 ± 1.20

metal binding whereas for Cd(II) only one deprotonation event affects the binding to the β-site (Fig. 3.10, Table 3.3).

Previously, McCarthy investigated whether the deprotonation of a histidine residue in the vicinity of the active site, His81 or His 217, can be attributed to pK_{a1} using the His81Ala and His217Ala mutants for the metal binding studies [14]. While for the His217Ala mutant pK_{a1} was still present when titrating Zn(II) into the Fe(II) half-apo enzyme, no apparent pH dependence was observed for the His81Ala mutant (Fig. 3.11, Table 3.3). Thus, pK_{a1} was attributed to this histidine residue [14].

To determine the origin of pK_{a2} (~10) for the Fe(II)Mn(II)/Co(II)/Zn(II) derivatives the GpdQ structure was searched for residues that may be deprotonated at this pH. Ser127 and Tyr19, two residues that are part of the extensive hydrogen bond network in the active site were initially considered as likely candidates. However, Co(II) binding to the Ser127Ala and Tyr19Phe mutants followed a similar pH dependence as for the wild type (Fig. 3.12). Deprotonation of these residues was thus ruled out to be the cause of decreasing metal binding affinity at pH values above 9.

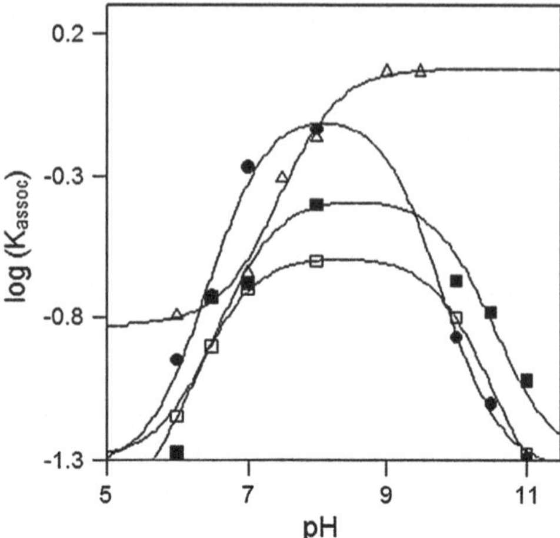

Fig. 3.10 pH dependence of K_{Assoc} for Fe(II)Co(II) (*filled circle*), Fe(II)Zn(II) (*open square*), [14] Fe(II)Mn(II) (*filled square*) and Fe(II)Cd(II) (*open rectangle*) wt GpdQ. K_{Assoc} was determined at each pH from the respective K_d values ($K_{Assoc} = 1/K_d$)

Table 3.3 pK_a values relevant for M(II) binding in wt- and mutant-GpdQ

GpdQ	Metal Ion Content	pK_{a1}	$pK_{a\ inter}$	pK_{a2}
wt [14]	Fe(II)Zn(II)	6.50 ± 0.40	–	10.50 ± 0.40
wt	Fe(II)Co(II)	6.40 ± 0.60	–	9.80 ± 0.50
His217Ala [14]	Fe(II)Zn(II)	6.59 ± 0.21	9.40 ± 0.04	10.70 ± 0.04
His81Ala [14]	Fe(II)Zn(II)	–	–	–
Tyr19Phe	Fe(II)Co(II)	7.57	–	10.65
Ser127Ala	Fe(II)Co(II)	7.21	–	10.11
wt	Fe(II)Cd(II)	7.46 ± 0.04	–	–
wt	Fe(II)Mn(II)	6.54 ± 0.07	–	10.59 ± 0.06

3.2.2.4 Kinetic Studies of the Fe(II)M(II) Derivatives

In addition to monitoring the activity as a function of metal ion concentration, the catalytic parameters k_{cat} and k_{cat}/K_m for each heterodinuclear derivative were measured at a range of pH values between pH 6.0 and 11 while keeping the metal concentration at a constant value (the optimal metal concentration was determined to be 30 μM of the respective metal in the experiments described above) and using the substrate BPNPP.

For the Fe(II)Co(II) derivative k_{cat} reaches a maximum at alkaline pH whereas the catalytic efficiency (k_{cat}/K_m) is highest under acidic conditions (Fig. 3.13). For comparative purposes corresponding data for the di-Co(II) derivative are included (Fig. 3.13) [12]. Reconstitution of catalytic activity for the remaining Fe(II)M(II) derivatives of GpdQ (where M = Zn, Mn or Cd) was monitored as described for the Fe(II)Co(II) derivative (Fig. 3.14). The catalytic properties of the Fe(II)Mn(II), Fe(II)Cd(II) and Fe(II)Zn(II) [14] derivative of GpdQ are shown in Fig. 3.14 and the catalytic parameters are summarized in Tables 3.4 and 3.5. The pH dependence of k_{cat} for the Fe(II)Cd(II) and Fe(II)Mn(II) derivatives follows a similar

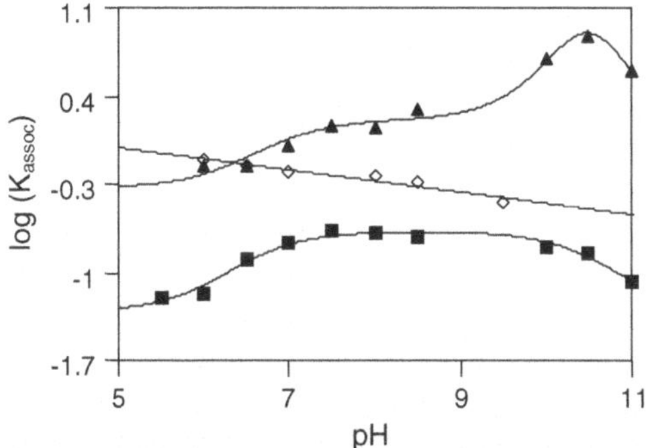

Fig. 3.11 pH dependence of the Zn(II) binding to the β-site in Fe(II)-half-apo GpdQ (data taken from B. McCarthy Honors Thesis) [14] Wild type (*filled square*), His81Ala mutant (*open diamond*) and His217Ala mutant (*filled rectangle*). K_{assoc} was determined at each pH from the respective K_d values ($K_{assoc} = 1/K_d$)

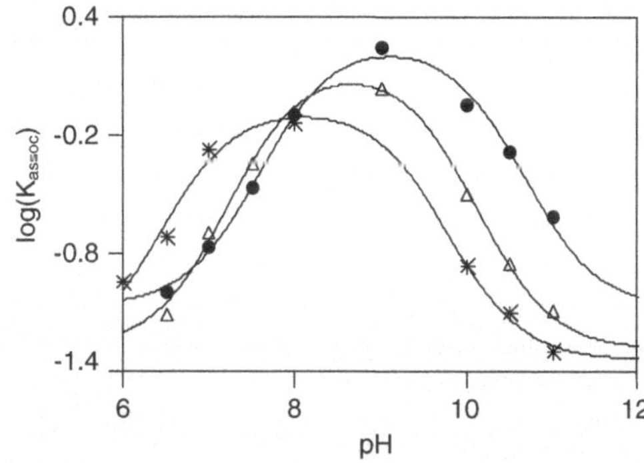

Fig. 3.12 pH dependence of Co(II) binding to the β-site in Fe(II)-half-apo GpdQ. Wild type (*star*), Tyr19Phe mutant (*filled circle*) and Ser127Ala mutant (*open rectangle*). K_{assoc} was determined at each pH from the respective K_d values

behavior to that found for the Fe(II)Co(II) and Co(II)Co(II) [12] derivatives whereas the data obtained for Fe(II)Zn(II) [14] analog showed no apparent pH dependence. The catalytic efficiencies (k_{cat}/K_m) of the Fe(II)Zn(II), Fe(II)Mn(II) and Fe(II)Cd(II) derivatives show very different pH-dependence behavior: while the catalytic efficiency of Fe(II)Zn(II) GpdQ is similar to the Fe(II)Co(II) derivative, optimal at acidic pH, the situation is reversed for the Fe(II)Mn(II) derivative. The Fe(II)Cd(II) derivative shows a more complex (bell-shaped) behavior with maximal catalytic efficiency at pH 8.

Table 3.4 shows a summary of the pK_e (pK_a value of the free enzyme) and pK_{es} (pK_a value of the enzyme substrate complex) values that were obtained by fitting the data shown in Figs. 3.13 and 3.14; data for homodinuclear derivatives reported previously in the literature are also included [6, 12, 13]. The pK_{es} values are between 9.0–10.7 for hetero- and homodinuclear derivatives and the pK_e values between 6.7 and 7.5 and apparent for all derivatives but Mn(II)Mn(II) and Cd(II)Cd(II). Fe(II)Cd(II) GpdQ displays an additional pK_e at 10.1 which is typical for a terminally Cd(II)-bound water molecule [25].

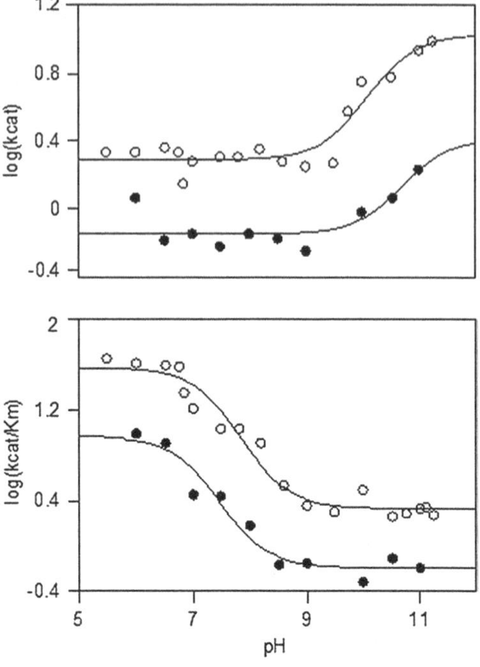

Fig. 3.13 pH dependence of the activity (k_{cat}) and catalytic efficiency (kcat/Km) of the hydrolysis of BPNPP by the Fe(II)Co(II) (*filled circle*) and Co(II)Co(II) [12] (*open circle*) wt GpdQ derivatives

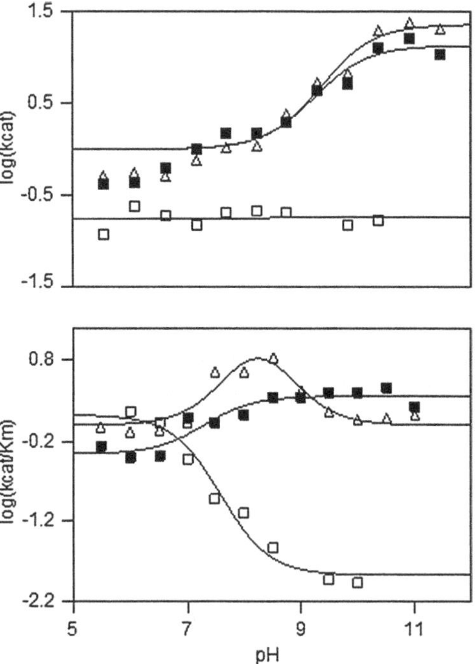

Fig. 3.14 pH dependence of the activity (k_{cat}) and catalytic efficiency (k_{cat}/K_m) of the hydrolysis of *BPNPP* by the Fe(II)Zn(II) (*open square*), [14] Fe(II)Mn(II) (*filled square*) and Fe(II)Cd(II) (*open rectangle*) wt GpdQ derivatives. Where a pH dependence was apparent the data were fit using equations derived for a mono- or diprotic system

Table 3.4 Summary of pK_e and pK_{es} values for the different metal ion derivatives of GpdQ

Derivative	pK_e	pK_{es}
Fe(II)–Co(II)	7.5	10.7
Co(II)–Co(II) [6, 12]	7.0	9.9
Fe(II)–Zn(II) [14]	7.7	–
Fe(II)–Mn(II)	7.3	9.0
Mn(II)–Mn(II) [12]	–	9.2
Fe(II)–Cd(II)	6.7, 10.1	9.1
Cd(II)–Cd(II) [13]	–	9.4

Table 3.5 Kinetic parameters for the hydrolysis of BPNPP by different metal ion derivatives of GpdQ at optimal pH (#: no data available)

Derivative	k_{cat} (s^{-1})	k_{cat}/K_m (mM^{-1} s^{-1})
Fe(II)–Co(II)	0.68 ± 0.00	0.68
Fe(II)–Co(II)$_{Y19F}$	2.32 ± 0.08	2.57
Fe(II)–Co(II)$_{S127A}$	1.18 ± 0.03	1.13
Co(II)–Co(II) [6]	1.62 ± 0.03	1.16
Fe(II)–Zn(II) [14]	0.21 ± 0.01	0.03
Zn(II)–Zn(II) [8]	0.11 ± 0.01	0.05
Fe(II)–Mn(II)	2.15 ± 0.00	2.38
Mn(II)–Mn(II) [12]	4.86 ± 0.33	1.67
Fe(II)-Cd(II)	2.35 ± 0.10	6.71
Cd(II)-Cd(II) [13]	15.0 ± 3.00	#

Table 3.5 shows the kinetic parameters obtained at optimal pH for all derivatives (homo- and heterodinuclear). The two most reactive forms of GpdQ are the previously reported homodinuclear Cd(II)Cd(II) and Mn(II)Mn(II) derivatives with k_{cat} values of 15 and 4.86 s^{-1}. However, the metal combination that led to the most efficient enzyme is Fe(II)Cd(II) with a k_{cat}/K_m of 6.71 mM^{-1} s^{-1}.

A detailed discussion of the ionizable residues that may influence catalytic activity and efficiency can be found in Sect. 3.3.4 along with an evaluation of contributions of the different metal ions to the mechanism and nucleophile generation.

3.2.3 *MCD of the Co(II)Co(II) Substituted Ser127Ala Mutant of GpdQ*

Previous studies by Hadler et al. have shown that the flexibility of active site residue Asn80 plays a crucial role in regulating Co(II) binding to the β-site [21]. To investigate whether metal binding can be influenced without directly altering Asn80, the serine residue, forming a hydrogen bond to Asn80, was exchanged for an alanine using site directed mutagenesis. By removing the hydrogen bond interaction between Ser127 and Asn80, the latter residue might be able to bind metals in the β-site more tightly without losing the flexibility necessary for efficient catalysis. As MCD data for Co(II)Co(II) wt-GpdQ in the presence and absence of phosphate (a substrate mimic) was readily available [6] it was decided to use the full apo form of the Ser127Ala mutant and to reconstitute it with Co(II) in both sites (α and β) as opposed to generating an Fe(II) half-apo enzyme. The resulting MCD spectrum is shown in Fig. 3.15 (as a blue line).

Mutant apo GpdQ was generated by adding a range of chelators (ethylenediaminetetraacetic acid tetrasodium salt, 1,10-phenanthroline, 2,6-pyridinedicarboxylic acid, 8-hydroxyquinoline-5-sulfonic acid and 2-mercaptoethanol) and incubating the enzyme for 24 h. Subsequently, the formerly yellow chelating solution had turned bright red, an indication that iron had been removed from the active site (formation of $[Fe(phen)_3]^{2+}$). After removal of chelators and excess metals with a desalting column two equivalents of Co(II) were added to the apo enzyme solution in buffer to which glycerol had been added as glassing agent. As can be seen in Fig. 3.15 in addition to the expected transition from Co(II) bound to the 6-coordinate α-site (~500 nm) a band of 5-coordinate Co(II) was also observed around 580 nm. That there is an noticeable increase in the population of the β-site in the absence of phosphate in the Ser127Ala mutant is more obvious when comparing it to the wt-GpdQ MCD spectrum where two equivalents of Co(II) had been added (red line, Fig. 3.15). In the presence of phosphate both spectra of mutant and wild type show a similar population of α- and β-site, respectively.

3.3 Discussion

3.3.1 Effects of Overall Structure on Activity: 8–3 Mutant

Ser127 is close to one of the ligands in the β-site (Asn80) in GpdQ. Removing the hydrogen bonding interaction by replacing Ser127 with an alanine leads to a more flexible β-site, which enables residue Asn80 to bind and release the metal ion during catalysis more easily [11]. The dissociation of the hexameric structure is suggested to be caused by the Cys269Ala mutation [11]. The presence of a monomer which had not been observed during previous purifications [11] is probably due to the fact that the pelleted cells were stored at −20 °C for 3 months prior to purification or the different expression protocol used at the University of Queensland. Previous studies by Yip et al. showed that the dimeric form of the 8–3 mutant had over 550 times higher catalytic efficiency compared to the wild type hexamer [11]. The specific activities measured in the experiments reported in this thesis showed that the activity of the mainly monomer-containing fractions 12 h after purification in the presence of 23 μM cobalt(II) sulfate was 5 times higher than that of the dimer and 15 times higher than the SA of the mainly hexameric form of the 8–3 GpdQ mutant (Table 3.1). This is likely to be due to the more

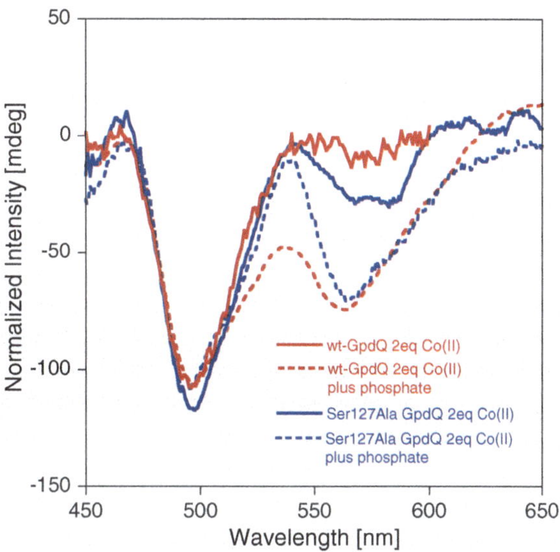

Fig. 3.15 MCD spectra of the Ser127Ala mutant after adding two equivalents Co(II) in the presence and absence of phosphate. For comparative reasons the normalized spectra of Co(II)Co(II) wt-GpdQ are also shown [6]. Reprinted with permission from L.J. Daumann et al. [28]. Copyright (2014) Elsevier

accessible active site in the monomer as opposed to the more enclosed active sites in dimer and hexamer. Hence, in the monomeric form of GpdQ substrate, metal and hydrolyzed substrate can be quickly exchanged. It was further observed that the activity of the monomer dropped dramatically within 24 h. In fact the monomer showed to be sevenfold less active after this time. Evidently, the mutations affect the stability to a large extent and denaturation of the enzyme occurs rapidly. Wt-GpdQ is an extremely stable enzyme and can be stored and handled at room temperature for months without loss of activity. As all forms of the 8–3 mutant are, however, very unstable further research on this mutant was not conducted. For applications on polluted water sources and soils it would imply that the enzyme has to be purified on site which would be an impracticable endeavor. These studies have, however, shown that changes to the oligomeric structure can improve the activity in GpdQ and moreover that differences in expression of GpdQ and in handling of the *E.coli* cells can cause the appearance of oligomeric structures (monomer) that had not been previously observed for this mutant.

3.3.2 *Effects of the Ser127 Mutation on Co(II) Binding to the β-site*

The MCD spectrum of Ser127Ala mutant GpdQ showed that the enzyme was able to assemble a dinuclear Co(II)Co(II) active site in the absence of substrate (to a certain extent, 40 % of the β-sites assembled). Ser127 forms a hydrogen bond to Asn80 which is responsible for the enzyme being a mononuclear enzyme in the resting state. Hadler et al. showed, using MCD, that when altering this residue (Asn80) to an aspartate, a residue with enhanced metal binding abilities, in fact the enzyme formed a dinuclear metal center in the resting state [21]. However the catalytic activity was greatly impaired suggesting that flexibility of Asn80 is required for an active enzyme. By not directly altering the metal binding residue Asn80 but Ser127, which is forming a hydrogen bond to this residue it was proposed that Asn80 should stay flexible but the metal binding affinity should be higher as electron density is no longer drawn towards Ser127. This proposal was supported by MCD measurements. Hence it was concluded that second coordination sphere residues like Ser127 have a strong impact on the assembly of a dinuclear catalytically competent center. While this mutation might not directly affect catalytic activity it might help the enzyme to use metals from a metal ion deprived environment more efficiently and thus help the enzyme adapt to environmental changes more rapidly.

3.3.3 Binding of Different Metals to the β-Site of Fe(II) Half-apo GpdQ

Half-apo GpdQ was obtained by subjecting the enzyme to a range of chelators and following removal of loosely bound metals and chelators with a desalting column. This yielded GpdQ with one Fe(II) bound to the α-site. This was verified by AAS, MCD and the fact that the enzyme was inactive without additional added metals. AAS measurements showed that the Fe content varied between 1–1.3 equivalents and that traces of other metals were present (0–0.3 equivalents of Zn, Mn and Co), the latter three are, however, not bound to the active site as the MCD spectrum of half apo Fe(II) GpdQ only showed a band (pseudo A-term) from high-spin Fe(II) (between 360–440 nm). As the hydrolytic activity of the enzyme was not distinguishable from autohydrolysis of the substrate BPNPP it was concluded that the traces of other metals are not available for the enzyme to reconstitute catalytic activity in the active site. Other binding sites other than the active site within the protein might be also a factor that has to be considered. The Fe(II) half-apo enzyme was stable and no oxidation to Fe(III) was observed over a prolonged time period (as shown by MCD).

The number of binding sites, n, estimated from the activity measurements in the metal titrations at pH 7 ranged from 1.11 (Fe(II)Co(II)) to 1.56 (Fe(II)Cd(II)). For Tyr19 it was suggested that an additional metal binding pocket could form upon deprotonation (pK_a of Tyr-OH: 9.8–10.5). The number of binding sites n at pH 11 for wt-GpdQ for the Fe(II)Co(II) derivative is 1.32 ± 0.05 whereas for the Tyr19Phe mutant (Fe(II)Co(II)) the number of binding sites is lower, with 1.12 ± 0.11. If compared at pH 7 when the Tyr19 residue should still be protonated, n is 1.11 for the wild type and similar (1.16) for the Y19F mutant (Fe(II)Co(II)). While the number of binding sites can only provide an estimate, the data indicate that multiple binding sites can be present in addition to the active site.

Metal ion association constants K_{assoc} ($K_{assoc} = 1/K_d$) in the various Fe(II)-M(II) derivatives of GpdQ (M = Co, Mn, Zn, Cd) were determined at various pH values. A bell-shaped pH dependence was observed for the Fe(II)Zn(II), Fe(II)-Co(II) and Fe(II)Mn(II) derivatives, which indicates that (at least) two ionisable residues (characterized by pK_{a1} and pK_{a2}) in the enzyme contribute to binding of metal ions. An overview of the pK_a values of amino acid residues near the active site in GpdQ is shown in Table 3.6.

The data were fitted to an equation derived for a diprotic system (Eq. 2.4, Chap. 2) [26]. For the Cd(II) derivative pK_{a2} is absent. pK_{a1} is in the range of 6.5–7.5 for the four (Fe(II)Cd(II), Fe(II)Mn(II), Fe(II)Co(II) and Fe(II)Zn(II)) derivatives and was assigned to a histidine residue whose deprotonation is anticipated to promote metal binding due to reduced electrostatic repulsion. Two histidine residues are in proximity to the active site, His81 and His217. Previously, it was shown that replacing these side chains by alanine does not affect the catalytic properties of the enzyme greatly. This observation is supported by the pH dependence of K_{assoc} for the Fe(II)Zn(II) derivatives of the His81Ala and His217Ala mutants of GpdQ [14]. In the His81Ala mutant K_{assoc} is largely pH independent (data above pH > 10 could not be measured reliably), an observation that suggests that pK_{a1} is associated with this residue. The pH dependence of the His217Ala mutant is more complex but indicates an increase in K_{assoc} between pH 6–7.5, a further increase between pH 8.5 and 10 before a sharp decrease at higher pH. This behavior is indicative of a minimum of three protonation equilibria (characterized by pK_{a1}, $pK_{a\text{-}inter}$ and pK_{a2}), and the data were fitted to an equation for a triprotic system (Chap. 2, Eq. 2.11). Since it is not possible to measure accurate data at pH values above 11.0 (due to enzyme lability) $pK_{a\text{-}inter}$, but especially the extrapolated $pK_{a\text{-}high}$ are only estimates.

pK_{a2} ($\geq$ 11) is associated with a residue whose deprotonation lowers the metal ion affinity. In close proximity to the active site there is one tyrosine, Tyr19, and one serine residue, Ser127. Like His81 and His217 both residues are part of the hydrogen bond network that links the α-metal ion with the β-metal ion via the substrate binding pocket. Tyr19 and Ser127 were replaced by a phenylalanine and an alanine, respectively, and the Fe(II)Co(II) derivatives of the two mutants were generated as described for the wild-type enzyme. While the effect of these mutations on the catalytic parameters of GpdQ was modest (Table 3.5), both k_{cat} and k_{cat}/K_m were enhanced. Furthermore, $pK_{a\text{-}high}$ is still present (Fig. 3.12). Thus, it appears unlikely that a residue in the immediate vicinity of the active site is associated with pK_{a2}. However for Cd(II) this pK_a is absent and this ion is considerably larger in size (1.01 Å) as the other divalent ions (0.82 for Zn(II), 0.89 for Mn(II) and 0.81 Å for Co(II)). It is thus possible that an alternative metal binding site is formed in the GpdQ hexamer at high pH (possibly upon deprotonation of a tyrosine or serine residue) and that only smaller ions (other than Cd(II)) can fit into this pocket. Since mutation did not alter the pH dependence of GpdQ in Ser127 and Tyr19, these residues were ruled out to be the main contributor for the decreased metal binding affinity at high pH.

Table 3.6 pK_a values of functional groups that may be present near the active site of GpdQ [26]

Group	pK_a
Phenolic OH (of Tyr)	9.8–10.5
Imidazolium (of His)	5.5–7.0
Guanidinium (of Arg)	11.6–12.6
Sulfhydryl (of Cys)	8.0–8.5
ε-amino (of Lys)	9.5–10.6

3.3.4 Catalytic Properties of Heterodinuclear GpdQ

It was shown in MCD measurements described in Sect. 3.2.2.2 that although a heterodinuclear Fe(II)Co(II) in GpdQ is formed upon the addition of Co(II) to the half-apo enzyme, it is not possible to generate a purely heterodinuclear derivative. Instead a mixture of Fe(II)Co(II) and homodinuclear Co(II)Co(II) centers is present. Addition of Co(II) to half-apo GpdQ is accompanied by an increase in catalytic activity as illustrated in Fig. 3.13, reaching a maximum value for k_{cat} of $0.7\ s^{-1}$ at pH 11. For comparison, at the same pH the homodinuclear Co(II) derivative has a k_{cat} of $1.6\ s^{-1}$. Taking into account that up to 40 % of the metal centers in the Fe(II)Co(II) derivatives are di-Co(II) the extrapolated activity of the heterodinuclear form alone is no larger than $0.1\ s^{-1}$, only ~5 % of that of its homodinuclear counterpart.

The most significant difference is observed in the magnitude of their catalytic efficiencies with the purely homodinuclear derivative having a more than twice larger k_{cat}/K_m ratio at optimal pH. The main contribution to this large difference is the high K_m value associated with the heterodinuclear form. Considering again that up to 40 % of the metal centers in this derivative are di-Co(II) this observation does not only indicate that the metal ion in the α-site plays an important role in substrate binding, but also that Fe(II) in the α-site is compromising interactions with the substrate significantly. Since, in contrast to Fe(II)Co(II)- GpdQ, it is less straightforward to estimate the proportion of homodinuclear centers formed upon the addition of the other metal ions, corresponding catalytic parameters, as far as available, for the homodinuclear Zn(II)Zn(II), Mn(II)Mn(II) and Cd(II)Cd(II) derivatives are also included for comparison (Table 3.5). For instance, the reported $k_{cat(max)}$ for the Mn(II)Mn(II) derivative of GpdQ is $4.9\ s^{-1}$ [6, 12, 13] while the corresponding value for the Fe(II)Mn(II) derivative is $2.2\ s^{-1}$. Although it is currently unknown what the proportion of homodinuclear centers in Fe(II)Mn(II) GpdQ is it is evident that the mixed metal ion derivative is less active, reaching a maximum turnover rate of no larger than ~45 % when compared to the di-Mn(II) form. Similarly, the Fe(II)Cd(II) derivative is, at best, ~15 % as active as the homodinuclear Cd(II) derivative. Only the replacement of Fe(II) in the α-site by Zn(II) results in a reduction of k_{cat}, leading to the following empirical series for the α-metal ion with respect to catalytic rate of GpdQ: Zn(II) $<$ - Fe(II) $< \ll$ Co(II) $<$ Mn(II) $<$ Cd(II). No relation to the softness/hardness according to the HSAB principle was found and neither did the size of the metal ion had an influence other than Cd(II) being the largest metal ion and also the one promoting the highest catalytic rates. The pH dependences of k_{cat} and k_{cat}/K_m for these three derivatives are shown in Fig. 3.14. While the Fe(II)Zn(II) derivative k_{cat} is largely independent on pH, the Fe(II)Mn(II) and Fe(II)Cd(II) derivatives reach maximum activity at pH values greater than 10, similar to the Fe(II)Co(II) and Co(II)Co(II) derivatives (Fig. 3.13), as well as the homodinuclear Mn(II)Mn(II) and Cd(II)Cd(II) derivatives [6, 12, 13] (note, that due to its low reactivity (Table 3.5) no pH profile for the Zn(II)Zn(II) derivative was determined). The corresponding pK_a values (pK_{es}) lie in the range between 9 and 10 (Table 3.4) and are likely to be due to a water molecule terminally coordinated to one of the divalent metal ions (Table 3.7).

Previously, this water was assigned to the metal ion located in the α-site and was proposed as the hydrolysis-initiating nucleophile [27]. Interestingly, in the Fe(II)Zn(II) derivative this protonation equilibrium is not relevant suggesting an alternative mechanistic strategy, possibly a different rate-limiting step (note that this derivative is considerably less active than most of the remaining ones; Table 3.5). The pH dependence of the k_{cat}/K_m ratio displays more diversity amongst the derivatives (Figs. 3.13 and 3.14). For the Fe(II)Zn(II) and Fe(II)Co(II) (also the Co(II)Co(II)) derivatives catalytic efficiency is maximal at pH values lower than 7, while the Fe(II)Mn(II) derivative operates most efficiently at high pH (see Table 3.4 for relevant pK_e values). In contrast, the catalytic efficiency of the Fe(II)Cd(II) derivative reaches a maximum at pH ~ 8.5, suggesting two distinct deprotonation events, one that increases catalytic efficiency (with a pK_e of ~7) and one (with a pK_e of ~10) that has the opposite effect. The pK_e of ~10, only observed in the Fe(II)Cd(II) derivative is likely to be due to a terminally Cd(II)-coordinated water molecule (Table 3.7). It is thus proposed that upon deprotonation the Cd(II) bound water/hydroxide is obstructing efficient substrate binding. The assignment of pK_e ~7, present in all heteronuclear derivatives, is not unambiguous as for some derivatives (Fe(II)Zn(II), Fe(II)Co(II) and also Co(II)-Co(II)) deprotonation of this residue leads to a decreased k_{cat}/K_M ratio while this has on the Fe(II)Mn(II) and Fe(II)Cd(II) derivatives the opposite effect. It is likely that pK_e corresponds to different ionisable residues. The fact that for the Mn(II)Mn(II) and Cd(II)Cd(II) derivatives this pK_e is not observed could indicate that pK_e in the Fe(II)Mn(II) and Fe(II)Cd(II) derivatives may be associated with a Fe(II) bound water molecule (the hydrolysis initiating nucleophile). This, however, implies that in the Mn(II)Mn(II) and Cd(II)Cd(II) derivatives this water molecule is only present in the enzyme-substrate complex (pK_{es}; Table 3.4), but not in the free enzyme. For the former metal ion derivatives (Fe(II)Zn(II), Fe(II)Co(II)) pK_e may be attributed to a terminally coordinated water molecule to the β metal ion. Deprotonation of this water would lead to slow substrate binding/exchange rates and thus a decreased catalytic efficiency at pH values above 7. In summary, it is evident that an unambiguous assignment of protonation equilibria relevant to catalytic performance was not possible, it was however shown that replacing metal

Table 3.7 pK_a values of water bound to divalent metal ions that were used in this study [25]

Group	pK_a
Fe(II)	3.3–9.5
Zn(II)	8.2–9.6
Co(II)	7.0–12.2
Cd(II)	9.4
Mn(II)	9.6–10.6

ions has a strong influence on the kinetic properties like pH optimum and catalytic rates. Depending on the metal ions used by GpdQ, the enzyme might employ different mechanisms as found previously for the related enzyme OpdA [15].

3.4 Conclusion

The site directed mutagenesis and studies with different metals ion show how subtle and complex the nature of interactions between the protein, the metal ions and the substrate in the active site of GpdQ is, which are modulated via the hydrogen bonding network that connects the metal ion in the β-site via its flexible ligand Asn80 and second coordination sphere residues Tyr19, His81, Ser127 and His217 to the metal ion in the α-site. Altering the oligomeric structure makes GpdQ a more efficient enzyme however with hugely compromised stability and altering the metal ion composition affects how the enzyme carries out hydrolysis, modulates catalytic efficiency and even alters the optimum pH range for the reaction. GpdQ was shown to be extremely versatile and flexible with the applied conditions (available metal ions, pH). However, the price the enzyme has to pay for this versatility is that its catalytic efficiency, on the enzymatic scale, is at best modest compared to other enzymes like OpdA. The advantage is that GpdQ is primed to adapt to new environmental challenges, making it an ideal candidate for bioremedial applications.

In the next chapter the development and characterization of Zn(II) model complexes for this enzyme will be discussed. A thorough mechanistic and structural analysis of the biomimetics will be conducted.

References

1. J.A. Gerlt, G.J.R. Whitman, J. Biol. Chem. **250**, 5053–5058 (1975)
2. E. Ghanem, Y. Li, C. Xu, F.M. Raushel, Biochemistry **46**, 9032–9040 (2007)
3. C.J. Jackson, P.D. Carr, J.W. Liu, S.J. Watt, J.L. Beck, D.L. Ollis, J. Mol. Biol. **367**, 1047–1062 (2007)
4. S.Y. McLoughlin, C. Jackson, J.W. Liu, D.L. Ollis, Appl. Environ. Microbiol. **70**, 404–412 (2004)

5. J.A. Gerlt, F.H. Westheimer, J. Am. Chem. Soc. **95**, 8166–8168 (1973)
6. K.S. Hadler, E.A. Tanifum, S.H. Yip, N. Mitić, L.W. Guddat, C.J. Jackson, L.R. Gahan, K. Nguyen, P.D. Carr, D.L. Ollis, A.C. Hengge, J.A. Larrabee, G. Schenk, J. Am. Chem. Soc. **130**, 14129–14138 (2008)
7. C.J. Jackson, P.D. Carr, H.K. Kim, J.W. Liu, D.L. Ollis, Acta Crystallogr. Sect. F: Struct. Biol. Cryst. Commun. **62**, 659–661 (2006)
8. C.J. Jackson, K.S. Hadler, P.D. Carr, A.J. Oakley, S. Yip, G. Schenk, D.L. Ollis, Acta crystallogr. F **64**, 681–685 (2008)
9. Schrödinger (2010) The PyMOL molecular graphics system, Version 1.3r1
10. D. Ollis, in *Power Point Presentation, Research School of Chemistry*, Australian National University, 2010
11. S.H.-C. Yip, J.-L. Foo, G. Schenk, L.R. Gahan, P.D. Carr, D.L. Ollis, Protein Eng. Des. Sel. **24**(12), 861–872 (2011)
12. K.S. Hadler, N. Mitić, F. Ely, G.R. Hanson, L.R. Gahan, J.A. Larrabee, D.L. Ollis, G. Schenk, J. Am. Chem. Soc. **131**(33), 11900–11908 (2009)
13. R.E. Mirams, S.J. Smith, K.S. Hadler, D.L. Ollis, G. Schenk, L.R. Gahan, J. Biol. Inorg. Chem. **13**, 1065–1072 (2008)
14. B.Y. McCarthy, Honours Thesis, University of Queensland, School of Chemistry and Molecular Biosciences, 2010
15. F. Ely, K.S. Hadler, L.R. Gahan, L.W. Guddat, D.L. Ollis, G. Schenk, Biochem. J. **432**, 565–573 (2010)
16. N. Mitić, K.S. Hadler, L.R. Gahan, A.C. Hengge, G. Schenk, J. Am. Chem. Soc. **132**, 7049–7054 (2010)
17. N. Mitić, C.J. Noble, L.R. Gahan, G.R. Hanson, G. Schenk, J. Am. Chem. Soc. **131**, 8173–8179 (2009)
18. G. Schenk, R.A. Peralta, S.C. Batista, A.J. Bortoluzzi, B. Szpoganicz, A.K. Dick, P. Herrald, G.R. Hanson, R.K. Szilagyi, M.J. Riley, L.R. Gahan, A. Neves, J. Biol. Inorg. Chem. **13**, 139–155 (2008)
19. S.J. Smith, A. Casellato, K.S. Hadler, N. Mitić, M.J. Riley, A.J. Bortoluzzi, B. Szpoganicz, G. Schenk, A. Neves, L.R. Gahan, J. Biol. Inorg. Chem. **12**, 1207–1220 (2007)
20. Thermo-Scientific, Grams/AI 9.0 Software
21. K.S. Hadler, N. Mitic, S.H. Yip, L.R. Gahan, D.L. Ollis, G. Schenk, J.A. Larrabee, Inorg. Chem. **49**, 2727–2734 (2010)
22. E.I. Solomon, T.C. Brunold, M.I. Davis, J.N. Kemsley, S.-K. Lee, N. Lehnert, F. Neese, A.J. Skulan, Y.-S. Yang, J. Zhou, Chem. Rev. **100**, 235–349 (2000) (Washington)
23. E.G. Pavel, M. Gunsior, C.A. Townsend, E.I. Solomon, R.W. Busby, J. Zhou, J. Am. Chem. Soc. **120**, 743–753 (1998)
24. V.M. D'Souza, B. Bennett, A.J. Copik, R.C. Holz, Biochemistry **39**, 3817–3826 (2000)
25. J. Burgess, *Metal Ions in Solution* (Halsted Press, Chichester, 1978)
26. I.H. Segel, *Enzyme Kinetics: Behavior and analysis of Rapid Equilibrium and Steady-State Enzyme Systems* (Wiley-Interscience, NY, 1975)
27. K.S. Hadler, L.R. Gahan, D.L. Ollis, G. Schenk, J. Inorg. Biochem. **104**, 211–213 (2010)
28. L.J. Daumann et al., Immobilization of the enzyme GpdQ on magnetite nanoparticles for organophosphate pesticide bioremediation. J. Inorg. Biochem. **131**, 1–7 (2014)

Chapter 4
Structural and Mechanistic Studies of Zn(II) Complexes as Phosphoesterase Models

4.1 Introduction

4.1.1 Zinc: An Essential Metal

Zn(II) is a common metal in all forms of life; it is not only the second most abundant metal in biological systems after iron, but also occurs in the active site of over 200, mostly hydrolytic, enzymes [1, 2]. It is suggested that the high occurrence of Zn(II) in biological systems is due to the abundance of soluble forms of this element in the environment [2]. The advantages of having Zn(II) in an active site of an enzyme include its high Lewis acidity which is mostly influenced by the ionic potential Z_{eff}/r (Z_{eff} = effective electrical charge, r = effective radius of the ion) [2]. When plotting Z_{eff}/r of an hexa-aqua coordinated M(II) ion against logK_a (from the corresponding deprotonation of one bound water molecule) Ochiai found a strong linear correlation between K_a and the ionic potential, and that Cu(II) and Zn(II) were in fact the most effectively polarizing divalent cations [2]. Another advantage of Zn(II) is the coordination flexibility arising from its d^{10} configuration. However, the closed shell configuration causes problems for the (bio)chemist studying Zn(II) systems as this configuration is spectroscopically silent. The problem is often solved by replacing in vitro Zn(II) with Co(II) or Cu(II) and utilizing the excellent spectroscopic properties of these ions [3], or with Cd(II) and using ^{113}Cd-NMR spectroscopy as a probe [4]. Besides the dinuclear Zn(II) hydrolases which are focus of this chapter, mono- and trinuclear enzymes are also common in nature [5]. Among the natural occurring Zn_2/Zn_3-enzymes are the phosphotriesterase from *Pseudomonas diminuta* [6], the alkaline phosphatase [7], phospholipase C [8], P1 nuclease [9], the metallo-β-lactamases [10–12] and also some amino peptidases [13–16]. A few examples of the active sites in these enzymes are given in Fig. 4.1.[1] The structurally

---

[1] Parts of this chapter have appeared in [L.J. Daumann et al., "*Spectroscopic and mechanistic studies of dinuclear metallohydrolases and their biomimetic complexes*" Dalton Trans. **2014**, 43, 910–928. and L.J. Daumann et al., "*The Role of Zn-OR and Zn-OH Nucleophiles and the Influence of p-Substituents in the Reactions of Binuclear Phosphatase Mimetics*" Dalton Trans. **2012**, 41, 1695–1708.]—Reproduced by permission of The Royal Society of Chemistry.

L. J. Daumann, *Spectroscopic and Mechanistic Studies of Dinuclear Metallohydrolases and Their Biomimetic Complexes*, Springer Theses, DOI: 10.1007/978-3-319-06629-5_4,

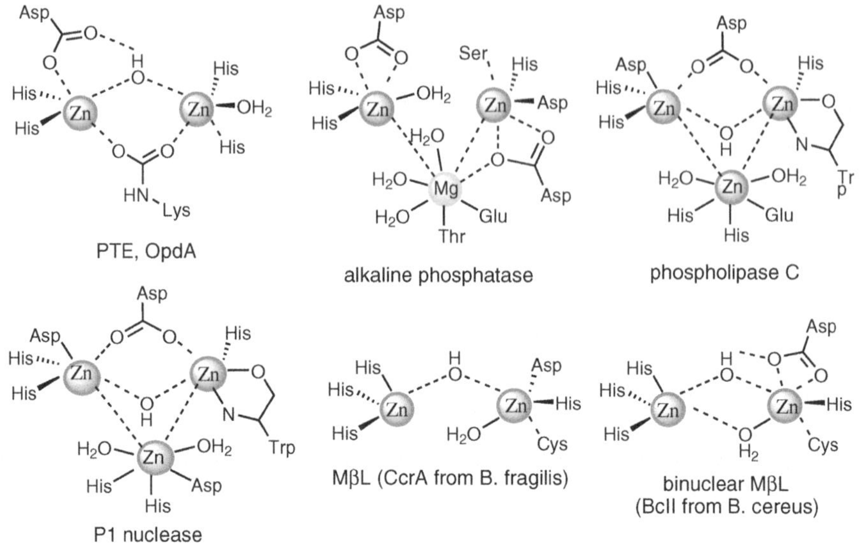

PTE, OpdA

alkaline phosphatase

phospholipase C

P1 nuclease

MβL (CcrA from B. fragilis)

binuclear MβL (BcII from B. cereus)

Fig. 4.1 Commonly found zinc enzymes in nature [5–11, 13]

related, trinuclear representatives alkaline phosphatase, phospholipase C and P1 nuclease all possess an adjacent M(II) ion in close proximity to the dinuclear Zn(II) site [5]. Phospholipase C and P1 nuclease catalyze the cleavage of phosphodiester bonds while the alkaline phosphatase hydrolyzes monoesters at both alkaline and acidic pH [5]. The active nucleophile for alkaline phosphatases is believed to be the serine residue activated by the second Zn(II) ion [5, 7, 17]. The metallo-β-lactamases (MβLs) contain at least one Zn(II) ion in the active site and a debate about the necessity of the second Zn(II) is currently ongoing [12]. MβLs are part of the defense mechanism that has evolved in bacteria to protect them against lactam-based antibiotics [10, 18]. The mode of action of some MβLs has been investigated thoroughly, using NMR, X-ray crystallography, stopped flow kinetic methods and DFT calculations. However, new classes of MβLs are found regularly, probably evolving due to environmental pressure (increasing use of antibiotics) [10, 19–26].

With the help of small Zn(II)Zn(II) model systems, where a facile characterization of the mechanism of action is possible using, for example, NMR spectroscopy and mass spectrometry techniques, the study of natural Zn(II)Zn(II) systems can be simplified [5, 27–32].

4.1.2 Dinuclear Zinc Hydrolase Mimics

Numerous research groups have reported Zn(II)Zn(II) hydrolase mimics [5, 24, 28–31, 33–39]. The strategies for ligand and complex design are diverse and have been recently reviewed [5, 32, 40]. Thus, only a brief overview of different ligands

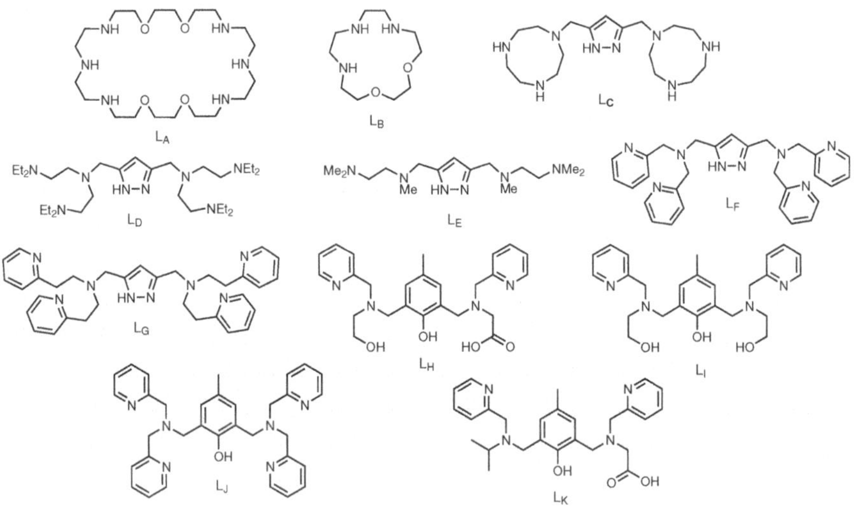

Fig. 4.2 Overview of ligands for dinuclear phosphatase mimics

will be given (Fig. 4.2; Table 4.1). Bringing two Zn(II) ions in close proximity can be achieved with the help of bridging ligands, like phenolate [33, 41, 42], carboxylate [43] and pyrazolyl [44], or by incorporating them into a macrocycle (i.e. crown ether), or, finally, by a simple aliphatic linker between two ligands, each of them coordinating one Zn(II) ion. The rate enhancement in phosphate ester hydrolysis by two Zn(II) ions (opposed to catalysis by just one Zn(II) ion) was shown in a study by Bazzicalupi et al. [45, 46]. Employing the macrocyclic ligands $\mathbf{L_A}$ and $\mathbf{L_B}$ (Fig. 4.2), it was demonstrated that the rate of hydrolysis of BPNPP underwent a 10-fold increase by the dinuclear $[Zn_2\mathbf{L_A}(OH)_2]^{2+}$ opposed to the mononuclear $\mathbf{L_B}Zn\text{-}OH^+$ complex (Table 4.1) [45]. In an attempt to crystallize $[Zn_2\mathbf{L_A}(OH)_2]^{2+}$ with diphenyl phosphate a bridging coordination of the phosphate ester to the two Zn(II) ions was found, further substantiating the necessity of two Zn(II) ions during hydrolysis [45]. The bidentate-bridging binding mode of the diphenyl phosphate was also proposed in solution using ^{31}P-NMR measurements because of the rather large shifts observed for the phosphate ester signal [45].

The separation of the two zinc centers in the biomimetics has been shown to be important. Meyer et al. correlated the hydrolytic activity of dizinc phosphodiesterase models (Table 4.1, Fig. 4.2, $\mathbf{L_G}$ and $\mathbf{L_F}$) with their Zn···Zn distances [38]. Here, the length of the ligand side chains determines the Zn···Zn separation. The distance of the two Zn(II) ions in the complex with $\mathbf{L_G}$ is rather short with 3.4 Å, while the metal ions are separated by more than 4.1 Å with ligand $\mathbf{L_F}$. The authors proposed that the separation of the Zn(II) centers is significant in terms of the ability of the metal complexes to efficiently catalyze the hydrolysis of the substrate, in this case BPNPP [38]. The k_{cat} for the Zn(II) complex $[Zn_2(\mathbf{L_F}H_{-1})(MeOH)(OH)](ClO_4)_2$, where the Zn(II) ions have a greater

Table 4.1 Kinetic properties of Zn(II) phosphoesterase mimics

Complex	k_{cat} [s^{-1}]	K_M [mM]	Substrate	pH opti-mum	T [K]	Solvent system	Kinetic pK_a	Potentio-metric pK_a
$Zn_2\mathbf{L}_{\mathbf{A}\text{-}2H}(OH)_2]^{2+}$ [45, 46]	1.15×10^{-4a}	–	BPNPP	10.5	308	water	~9.5	7.6, 9.2
$[Zn\mathbf{L}_{\mathbf{B}}(OH)_2]^{2+}$ [45, 46]	1.31×10^{-5a}	–	BPNPP	10	308	water	~8.9	8.8
$[Zn_2(\mathbf{L}_{\mathbf{C}}H_{-2})]$ [47]	2.24×10^{-6}	12	BPNPP	9.73	308	water	–	8.15
$[Zn_2(\mathbf{L}_{\mathbf{D}}H_{-1})(MeOH)(OH)](ClO_4)_2$ [48]	1.9×10^{-6}	42	BPNPP	8.28	323	DMSO/water 1:1	7.57	7.57
$[Zn_2(\mathbf{L}_{\mathbf{E}}H{-}1)]$ [48]	4.2×10^{-5}	55	BPNPP	8.28	323	DMSO/ water 1:1	7.66	6.53, 6.58, 7.66
$[Zn_2(\mathbf{L}_{\mathbf{F}}H_{-1})(MeOH)(OH)](ClO_4)_2$ [38]	2.3×10^{-5}	56	BPNPP	8.28	323	DMSO/ water 1:1	7.60	3.44, 4.36, 7.60
$[Zn_2(\mathbf{L}_{\mathbf{G}}H_{-1})(OH)](ClO_4)_2$ [38]	4.9×10^{-6}	51	BPNPP	8.28	323	DMSO/ water 1:1	7.96	5.27, 7.96
$[Zn_2(\mathbf{L}_{\mathbf{H}}H_{-2})(OAc)](PF_6)$ [33]	1.26×10^{-6}	1.96	BPNPP	9.0	323	CH_3CN/ water 1:1	7.87	–
$[Zn_2(\mathbf{L}_{\mathbf{I}}H_{-2})(OAc)(H_2O)](PF_6)$ [30]	4.6×10^{-6}	61.5	BDNPP	7.2	323	DMSO/ water 3:7	7.31	7.20, 8.15
$[Zn_2(\mathbf{L}_{\mathbf{J}}H_{-1})(OH)]^{2+}$ [35]	6.4×10^{-4}	13.5	HPNP	–	298	DMSO/ water 3:7	7.4	7.60
$[(Zn_2(\mathbf{L}_{\mathbf{K}}H_{-2})(OAc))_2](PF_6)_2$ [29]	1.2×10^{-4}	3.6	HPNP	8.5	298	CH_3CN/ water 1:1	~8.5	–
$[(Zn_2(\mathbf{L}_{\mathbf{K}}H_{-2})(OAc))_2](PF_6)_2$ [29]	6.4×10^{-4}	16	BDNPP	8.5	298	CH_3CN/ water 1:1	6.63	–

[a] second order rate constants in [M^{-1} s^{-1}]

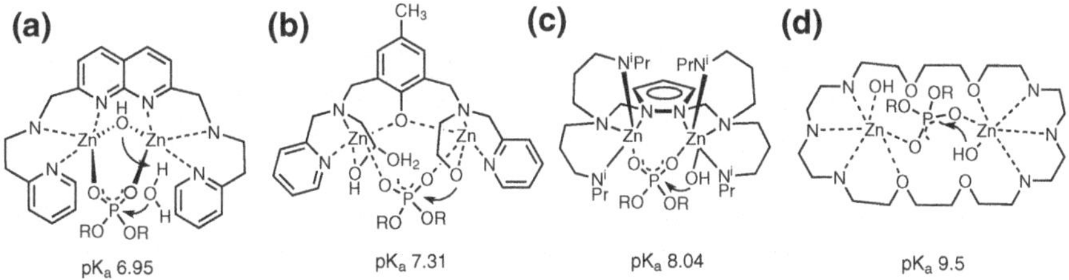

Fig. 4.3 Different hydrolysis initiation nucleophiles found in dinuclear Zn(II) complexes

separation, is 2.3×10^{-5} s^{-1}, greater than for the derivative with short intramolecular distances $[Zn_2(\mathbf{L_G}H_{-1})(OH)](ClO_4)_2$, k_{cat} 4.9×10^{-6} s^{-1} (Table 4.1) [38]. A study by Nordlander and co-workers employed an investigation of the necessity for an asymmetric coordination environment for phosphotriesterase activity [27, 29]. Kinetic studies of the hydrolysis of BDNPP and the transesterification with the substrate 2-hydroxypropyl-p-nitrophenyl-phosphate (HPNP) utilizing Zn(II) complexes of the asymmetric ligand $\mathbf{L_K}$ (Table 4.1, Fig. 4.2) and a similar symmetric ligand showed that the more open coordination sphere in the $\mathbf{L_K}$ ligand led to a fivefold enhancement of k_{cat}.

The nucleophile in zinc complex-catalyzed hydrolysis reactions is in most cases proposed to be a terminal water molecule ($pK_a = 6.60$–9.5 (c) and (d) in Fig. 4.3) [33, 49]. If the complex contains an additional alcohol moiety the assignment of the hydrolysis initiating nucleophile can be ambiguous [30, 49]. In monomeric Zn(II) systems it was proposed, based on DFT calculations and reactivity studies, that a metal bound alkoxide is more reactive than a water molecule [17, 50, 51]. Deprotonation constants of water bound to free Zn(II) have been observed to range from 8.2–9.2, [52] whereas the pK_a of a zinc-bound alcohol was shown to be around 7.2 [30, 49]. A bridging hydroxide was initially proposed in the zinc complex (Fig. 4.3a). The authors, however, proposed that the bridging hydroxide acts as a general base to activate an external water molecule [53]. The pK_a of 7.31 was assigned to the zinc-bound alcohol arm in complex (Fig. 4.3b). However the authors did not indicate how the catalyst is recovered after nucleophilic attack of the ligand arm [30].

4.1.3 Target Ligands and Aims of this Chapter

In this work five symmetric ligands were prepared in order to create functional mimics for phosphoesterase enzymes. The ligands were prepared as described in Chap. 2. They contain either a methyl- (**L2** ligand) or phenyl- (**L3** ligand) ether, replacing the potentially nucleophilic alcohol moiety in a previously reported ligand, CH_3H_3**L1** (Fig. 4.4) [30].

R' = CH_3, R" = H	CH_3H_3**L1**
R' = CH_3, R" = CH_3	CH_3H**L2**
R' = CH_3, R" = Ph	CH_3H**L3**
R' = NO_2, R" = CH_3	NO_2H**L2**
R' = Br, R" = CH_3	BrH**L2**

Fig. 4.4 Ligands employed to mimic the active site residues of phosphoesterases

CH_3H_3**L1** was prepared in order to compare the properties of its Zn(II) complex with those of the various Zn(II) complexes reported in this chapter. Moreover, in order to compare the influence of the substituent in the *para*-position to the bridging phenol, ligands were employed with methyl-, bromo- and nitro-groups. To investigate the steric effects of pendant ligand arms during phosphate ester hydrolysis, CH_3H**L2** with methyl ether arms [54] and CH_3H**L3** the (more bulky) phenyl ether derivative were synthesized. Mass spectrometry in the presence and absence of substrates, ^{1}H, ^{13}C and ^{31}P NMR including ^{18}O-labeling studies, X-ray crystallography and kinetic assays were employed to address questions like substrate binding mode, nucleophiles in OP hydrolysis and in-solution structure of the Zn(II) complexes. Experimental procedures including ligand and complex syntheses can be found in Chap. 2.

4.2 Results

4.2.1 Syntheses of the Dinuclear Zn(II) Complexes

The Zn(II) complexes were typically synthesized by adding two equivalents of zinc(II) acetate dihydrate dissolved in a small amount of methanol to a ligand solution in methanol and refluxing the mixture for 30 min to ensure complete binding of two equivalents of Zn(II) to the ligands. Sodium hexafluorophosphate was employed as counter ion as it proved to be the most successful anion for obtaining crystals suitable for X-ray structure determination (Fig. 4.5).

4.2.2 Crystal Structures of the Zn(II) Complexes

The complex $[Zn_2(CH_3$**L2**$)(CH_3COO)_2](PF_6)$ [54] crystallized in the monoclinic P21/c space group, $[Zn_2(CH_3$**L3**$)(CH_3COO)_2](PF_6)\cdot CH_3OH$ and $[Zn_2(Br$**L2**$)(CH_3COO)_2](PF_6)$ in the triclinic P-1 space group, and $[Zn_2(NO_2$**L2**$)(CH_3COO)_2](PF_6)$ in the monoclinic P21/n space group. All structures were composed of one ligand with two Zn(II) bridged by two acetates and hexafluorophosphate as counter ion, which was in some cases resolved into two disordered octahedral anions

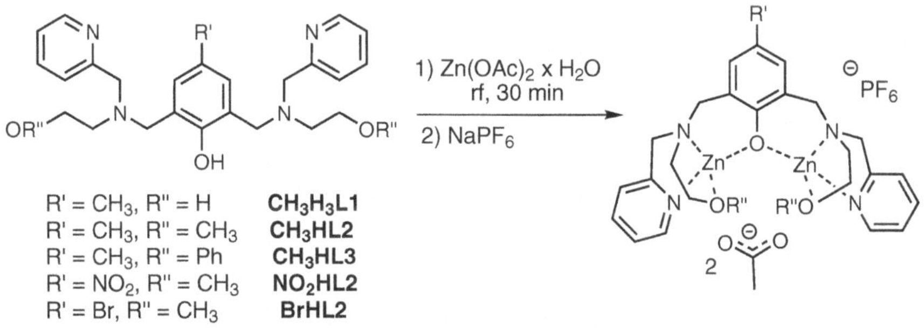

1) $Zn(OAc)_2 \times H_2O$ rf, 30 min
2) $NaPF_6$

R' = CH_3, R'' = H **CH_3H_3L1**
R' = CH_3, R'' = CH_3 **CH_3HL2**
R' = CH_3, R'' = Ph **CH_3HL3**
R' = NO_2, R'' = CH_3 **NO_2HL2**
R' = Br, R'' = CH_3 **BrHL2**

Fig. 4.5 Overview of the complexes synthesized

sharing the same phosphorus central atom. For the complex with CH_3**L3** the structure was refined with a methanol solvate. The crystallographic data obtained for the complexes are displayed in Table 4.2, selected bond lengths and angles are shown in Table 4.3. The structures show that the Zn(II) ions are triple-bridged by the deprotonated phenol moiety of the ligand and by two acetate moieties. The hexa-coordinate coordination sphere is completed by a tertiary and a pyridine nitrogen, as well as an oxygen atom from the ligands' ether arms. In the case of $[Zn_2(CH_3$**L3**$)(CH_3COO)_2]$-(PF_6).CH_3OH the (phenol)ether-zinc distances are at the long end of the range seen previously for similar Zn(II) complexes [42, 55, 56]. In general, the ether arms present weaker ligands than alcoholic moieties for Zn(II). For all complexes the ether arms are coordinated *trans* to each other from each side of the molecule giving the molecule nearly perfect C_2 symmetry. No water molecules are present, in contrast to previously reported similar systems [30]. Both Zn(II) centers are thus identical and hexacoordinate in all complexes.

Para-substituents to the bridging phenolic oxygen influence the Zn…Zn distance only to a very small extent. The Zn…Zn distance in the most strongly electron withdrawing NO_2**L2** ligand is slightly elongated (3.341 Å) compared to the Br**L2** (3.304 Å) and CH_3**L2** (3.296 Å) ligands, and while the Zn-O-Zn angle of the **L2** ligands is similar (109.70°, 109.73°, 110.21°) and close to a perfect tetrahedral angle, the **L3** ligand with the bulky phenyl ether arms enforces a larger deviation from 109° (107.49°). The ORTEP [57] plots for the complexes are shown in Fig. 4.6.

4.2.3 Crystal Structures of Zn(II) Complexes with Substrate Mimics

To further investigate the nature of substrate binding the complex $[Zn_2(CH_3$**L2**$)(CH_3COO)_2](PF_6)$ was crystallized from acetonitrile/water 1:1 with 25 equivalents of BPNPP. Colorless crystals suitable for X-ray crystallography

Table 4.2 Crystallographic Data for $[Zn_2(CH_3\mathbf{L2})(CH_3COO)_2](PF_6)$, $[Zn_2(CH_3\mathbf{L3})(CH_3COO)_2](PF_6)$, $[Zn_2(NO_2\mathbf{L2})(CH_3COO)_2](PF_6)$ and $[Zn_2(Br\mathbf{L2})(CH_3COO)_2](PF_6)$

	$[Zn_2(CH_3\mathbf{L2})(CH_3COO)_2](PF_6)$ [54]	$[Zn_2(CH_3\mathbf{L3})(CH_3COO)_2](PF_6)\cdot CH_3OH$	$[Zn_2(NO_2\mathbf{L2})(CH_3COO)_2](PF_6)$	$[Zn_2(Br\mathbf{L2})(CH_3COO)_2](PF_6)$ twinned
Empirical formula	$C_{31}H_{41}F_6N_4O_7PZn_2$	$C_{41}H_{45}F_6N_4O_7PZn_2$	$C_{30}H_{38}F_6N_5O_9PZn_2$	$C_{30}H_{38}BrF_6N_4O_7PZn_2$
Formula weight	857.39	981.52	888.36	922.26
Temperature (K)	293(2)	293(2)	293(2)	293(2)
Wavelength (Å)	0.71073 (MoK_α)	0.71073 (MoK_α)	1.5418 (CuK_α)	1.5418 (CuK_α)
Crystal system	Monoclinic	Triclinic	Monoclinic	Triclinic
Space group	P21/c	P-1	P21/n	P-1
a (Å)	10.7745(5)	10.4535(7)	8.9260(10)	8.9022(5)
b (Å)	14.8184(8)	13.1726(8)	31.571(2)	13.1115(7)
c (Å)	23.1270(10)	16.4175(8)	13.3124(9)	16.6071(9)
α (°)		83.319(5)		99.071(5)
β (°)	102.967(5)	85.597(5)	103.264(9)	91.761(5)
γ (°)		84.177(5)		102.970(5)
Vol ($Å^3$)	3598.3(3)	2228.9(2)	3651.4(5)	1860.92(18)
Z	4	2	4	2
μ (mm^{-1})	1.459	1.188	2.823	3.954
F(000)	1760	1008	1816	932
ρ (Mg/m^3)	1.583	1.462	1.616	1.646
Reflns col.	15336	15104	20343	19461
Ind. Reflns	6333 (0.0456)	7839 (0.0565)	5800 (0.0621)	19461
θ range (°)	2.93–25	2.84–25	3.69–62.49	3.51–60.57
GOOF on F^2	0.79	0.704	0.93	1.042
final R indices [$I > 2\sigma(I)$]	R1 = 0.0399, wR2 = 0.0735	R1 = 0.0483, wR2 = 0.0713	R1 = 0.0476 wR2 = 0.1149	R1 = 0.07, wR2 = 0.1872
R indices (all data)	R1 = 0.0913, wR2 = 0.0791	R1 = 0.1673, wR2 = 0.0855	R1 = 0.0787 wR2 = 0.1257	R1 = 0.0924, wR2 = 0.2067

Table 4.3 Bond lengths (Å) and angles (°) for the first coordination sphere of [Zn_2(CH_3**L2**)(CH_3COO)$_2$](PF_6), [Zn_2(CH_3**L3**)(CH_3COO)$_2$](PF_6)·CH_3OH, [Zn_2(NO_2**L2**)(CH_3COO)$_2$] and [Zn_2(Br**L2**)(CH_3COO)$_2$](PF_6)

	[Zn_2(CH_3**L2**) (CH_3COO)$_2$] (PF_6)	[Zn_2(CH_3**L3**) (CH_3COO)$_2$] (PF_6)·CH_3OH	[Zn_2(NO_2**L2**) (CH_3COO)$_2$] (PF_6)	[Zn_2(Br**L2**) (CH_3COO)$_2$] (PF_6)
Zn(1)-O(1)	2.006(2)	2.003(3)	2.043(3)	2.007(3)
Zn(1)-O(2)	2.049(3)	1.979(3)	1.992(3)	2.000(4)
Zn(1)-O(5)	2.397(3)		2.380(4)	2.433(4)
Zn(1)-O(6)	1.997(3)	2.044(3)	2.055(3)	2.067(3)
Zn(1)-N(3)	2.184(3)	2.200(4)	2.193(3)	2.196(4)
Zn(1)-N(4)	2.123(3)	2.108(4)	2.134(4)	2.114(4)
Zn(2)-O(1)	2.024(3)	2.024(3)	2.041(3)	2.020(3)
Zn(2)-O(3)	1.992(2)	2.021(3)	2.076(3)	2.068(4)
Zn(2)-O(4)	2.467(3)		2.323(3)	2.368(4)
Zn(2)-O(7)	2.076(3)	1.982(4)	1.997(3)	1.991(3)
Zn(2)-N(1)	2.188(3)	2.187(4)	2.197(3)	2.206(4)
Zn(2)-N(2)	2.148(3)	2.123(4)	2.129(4)	2.127(4)
Zn(1)…Zn(2)	3.296	3.247	3.341	3.304
O(1)-Zn(1)-O(2)	91.69(11)	95.65(14)	98.64(13)	97.86(14)
O(1)-Zn(1)-O(5)	89.46(10)		91.72(13)	91.14(15)
O(1)-Zn(1)-O(6)	99.21(11)	95.01(13)	91.30(13)	92.35(14)
O(1)-Zn(1)-N(3)	91.24(11)	89.34(16)	89.11(11)	89.66(13)
O(1)-Zn(1)-N(4)	167.50(13)	167.69(17)	167.04(13)	167.80(16)
N(3)-Zn(1)-N(4)	77.71(13)	78.48(19)	78.24(13)	78.25(15)
O(1)-Zn(2)-O(3)	100.72(11)	93.12(12)	92.03(13)	91.79(14)

(continued)

Table 4.3 (continued)

	[Zn_2(CH_3**L2**) ($CH_3COO)_2$] (PF_6)	[Zn_2(CH_3**L3**) ($CH_3COO)_2$] (PF_6)·CH_3OH	[Zn_2(NO_2**L2**) ($CH_3COO)_2$] (PF_6)	[Zn_2(Br**L2**) ($CH_3COO)_2$] (PF_6)
O(1)-Zn(2)-O(4)	88.43(10)		93.56(12)	95.33(14)
O(1)-Zn(2)-O(7)	92.69(11)	97.80(14)	98.56(12)	99.09(13)
O(1)-Zn(2)-N(1)	88.93(10)	89.24(15)	88.77(11)	89.33(13)
O(1)-Zn(2)-N(2)	167.01(11)	167.12(17)	166.63(13)	166.91(15)
N(1)-Zn(2)-N(2)	78.08(12)	78.02(18)	77.88(14)	77.66(16)
Zn(1)-O(1)-Zn(2)	109.70(11)	107.49(16)	109.73(13)	110.21(14)

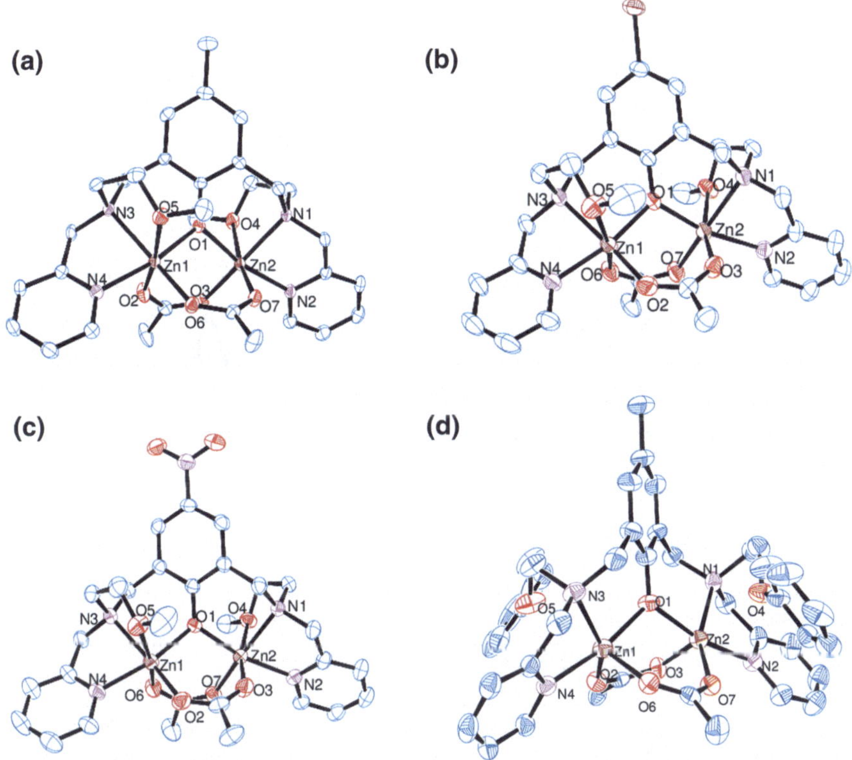

Fig. 4.6 ORTEP plots of **a** $[Zn_2(CH_3\mathbf{L2})(CH_3COO)_2](PF_6)$ **b** $[Zn_2(Br\mathbf{L2})(CH_3COO)_2](PF_6)$ **c** $[Zn_2(NO_2\mathbf{L2})(CH_3COO)_2](PF_6)$ and **d** $[Zn_2(CH_3\mathbf{L3})(CH_3COO)_2](PF_6).CH_3OH$ with 25 % probability level of thermal ellipsoids. Hydrogen atoms, counter ions and non-coordinated solvent molecules have been omitted for clarity, as have atom labels of non-coordinating residues

formed after 5 days. The structure revealed that the complex contained the hydrolysis product of BPNPP, 4-nitrophenylphosphate (PNPP) and crystallized as the tetramer (Fig. 4.7, Table 4.4): $[Zn_4(CH_3\mathbf{L2})_2(NO_2C_6H_5OPO_3)_2(H_2O)_2](PF_6)_2$, where two complex units are μ^3,η^3-bridged by two PNPP ions. One of the $[Zn_2(CH_3\mathbf{L2})]^+$ dimers is hereby coordinated by the $NO_2C_6H_5OPO_3^-$ in a bidentate fashion while the other fragment has a monodentate coordination to PNPP-oxygen. The hexacoordinated sphere is completed with two water molecules coordinated to each Zn(II) ion of the second fragment forming hydrogen bonds to the nearby oxygen atoms of PNPP (Fig. 4.7).

A similar tetranuclear structure had been observed with a zinc complex of a pyrazolate based ligand [48]. PNPP was found to form a bridge between two Zn(II) ions of the same ligand and one to a neighboring unit. The Zn…Zn distance in the same ligand unit is 4.257 Å, whereas the distances to the closest zinc ion of the neighboring unit are 3.146 and 3.545 Å, respectively [48]. The Zn(II) centers in $[Zn_4(CH_3\mathbf{L2})_2(NO_2C_6H_5OPO_3)_2(H_2O)_2](PF_6)_2$ are separated by 3.747 Å (Zn1/

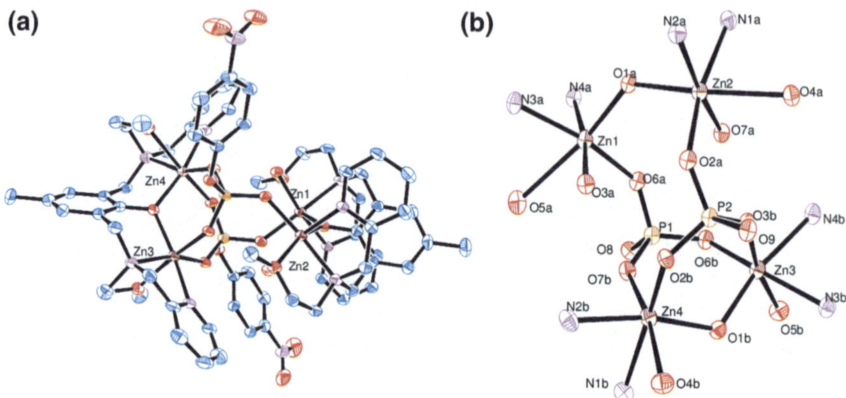

Fig. 4.7 **a** Structure of $[Zn_4(CH_3\mathbf{L2})_2(NO_2C_6H_5OPO_3)_2(H_2O)_2](PF_6)_2$ with 10 % probability level of thermal ellipsoids. **b** First coordination sphere of the Zn(II) ions including the phosphate cores. Hydrogen atoms, counter ions and non-coordinated solvent molecules have been omitted for clarity

Table 4.4 Data for $[Zn_4(CH_3\mathbf{L2})_2(NO_2C_6H_5OPO_3)_2(H_2O)_2](PF_6)_2$ and $[Zn_4(Br\mathbf{L2})_2(PO_3F)_2(H_2O)_2](PF_6)_2$

	$[Zn_4(CH_3\mathbf{L2})_2(NO_2C_6H_5OPO_3)_2(H_2O)_2](PF_6)_2$	$[Zn_4(Br\mathbf{L2})_2(PO_3F)_2(H_2O)_2](PF_6)_2$
Empirical formula	$C_{66}H_{80}F_{12}N_{10}O_{20}P_4Zn_4$	$C_{52}H_{68}Br_2F_{14}N_8O_{14}P_4Zn_4$
Formula weight	1946.76	1840.32
Temperature (K)	293(2)	150(2)
Wavelength (Å)	1.54180 (CuK_α)	1.5418 (CuK_α)
Crystal system	Monoclinic	Triclinic
Space group	P21/c	P-1
a (Å)	26.602(5)	13.3002(6)
b (Å)	14.3664(7)	15.8922(6)
c (Å)	23.0070(10)	19.6397(7)
α (°)		94.791(3)
β (°)	107.464(6)	109.511(4)
γ (°)		112.586(4)
Vol ($Å^3$)	8387.4(8)	3504.1(2)
Z	4	2
μ (mm^{-1})	2.883	4.658
F(000)	3976	1848
ρ (Mg/m^3)	1.542	1.744
Reflns collected	42499	36029
Ind reflections (R_{int})	13255 (0.0592)	11083 (0.0402)
θ range (°)	3.48–62.5	3.11–62.41
GOOF on F^2	0.816	1.037
final R indices [I > 2σ(I)]	R1 = 0.05, wR2 = 0.1248	R1 = 0.053, wR2 = 0.1522
R indices (all data)	R1 = 0.1089, wR2 = 0.1415	R1 = 0.0778, wR2 = 0.1675

Zn2) and 3.521 Å (Zn3/Zn4), respectively. Relevant bond distances and angles for $[Zn_4(CH_3\mathbf{L2})_2(NO_2C_6H_5OPO_3)_2(H_2O)_2](PF_6)_2$ are reported in Table 4.5. A similar tetranuclear structure with bridging PO_3F^{2-} was observed in an attempt to make $[Zn_2(Br\mathbf{L2})(CH_3COO)_2](PF_6)$ with wet sodium hexafluorophosphate. The sodium hexafluorophosphate had partially hydrolyzed to PO_3F^{2-} and other phosphate species as confirmed by ^{31}P-NMR. In the complex obtained, the bridging acetates were replaced by PO_3F^{2-} and as seen in the complex above in one fragment by two water molecules forming $[Zn_4(Br\mathbf{L2})_2(PO_3F)_2(H_2O)_2](PF_6)_2$. The incorporation of partial hydrolysis products of $NaPF_6$ into zinc complexes has been observed previously by Buchholz et al. [33]. The structure of $[Zn_4(Br\mathbf{L2})_2(PO_3F)_2(H_2O)_2](PF_6)_2$ is shown in Fig. 4.8 and Table 4.5 lists relevant angles and bond lengths for this structure.

The Zn-O-Zn angles $[Zn_4(Br\mathbf{L2})_2(PO_3F)_2(H_2O)_2](PF_6)_2$ and $[Zn_4(CH_3\mathbf{L2})_2(NO_2C_6H_5OPO_3)_2(H_2O)_2](PF_6)_2$ were found to be at least 10° larger than in the respective complexes with bridging acetate. Mass spectral analysis, IR and microanalysis indicated that the structure was representative of the whole sample obtained in the synthesis with wet sodium hexafluorophosphate.

4.2.4 Infrared Spectroscopy

The infrared spectra of the complexes showed a shift of the asymmetric and symmetric acetate stretches to higher frequencies (with respect to free acetate); [58] the separation $\Delta\nu_{asym-sym}$ confirming a symmetric bidentate coordination for all Zn(II) complexes in the solid state [58]. The hexafluorophosphate stretch and deformation bands were generally found in all spectra around 830 and 555 cm^{-1}. An example infrared fingerprint region of $[Zn_2(Br\mathbf{L2})(CH_3COO)_2](PF_6)$ is shown in Fig. 4.9a The fingerprint region of $[Zn_4(CH_3\mathbf{L2})_2(NO_2C_6H_5OPO_3)_2(H_2O)_2](PF_6)_2$ is shown below (Fig. 4.9b). Beside the characteristic bands from hexafluorophosphate, the asymmetric and symmetric stretches from bidentate bound PNPP are found [58].

4.2.5 Mass Spectrometry of the Zn(II) Complexes

Mass spectral data, obtained in both methanol and under the conditions employed for kinetic measurements (water/acetonitrile 1:1), showed the distinctive isotopic pattern of dinuclear Zn(II) complexes [59]. For the Zn(II) complex with $NO_2H\mathbf{L2}$ the highest mass peak was at *m/z* 744.0, corresponding to a $[Zn_2(NO_2\mathbf{L2})(CH_3COO)_2]^+$ species (calc. *m/z* 742.1 (100 %), 744.1 (97 %)). The isotopic pattern for $[Zn_2(Br\mathbf{L2})(CH_3COO)_2]^+$ was more complex due to the contribution of the isotopes of bromine. Interestingly a species with only one acetate bound was observed under these conditions: $[Zn_2(Br\mathbf{L2})(CH_3COO)(OCH_3)]^+$ (found *m/z*

Table 4.5 Selected bond lengths (Å) and angles (°) for $[Zn_4(Br\mathbf{L2})_2(PO_3F)_2(H_2O)_2](PF_6)_2$ and $[Zn_4(CH_3\mathbf{L2})_2(NO_2C_6H_5OPO_3)_2\ (H_2O)_2](PF_6)_2$

	$[Zn_4(Br\mathbf{L2})_2(PO_3F)_2(H_2O)_2](PF_6)_2$	$[Zn_4(CH_3\mathbf{L2})_2(NO_2C_6H_5OPO_3)_2(H_2O)_2](PF_6)_2$
Zn(1)-O(6A)	2.086(4)	2.001(4)
Zn(1)-O(1A)	2.028(4)	2.065(4)
Zn(1)-O(3A)	2.015(4)	2.080(3)
Zn(1)-N(4A)	2.118(5)	2.111(4)
Zn(1)-N(3A)	2.164(5)	2.154(4)
Zn(2)-O(2A)	2.083(4)	2.011(4)
Zn(2)-O(1A)	2.159(4)	2.096(3)
Zn(2)-O(7A)	2.024(4)	2.128(3)
Zn(2)-N(2A)	2.114(4)	2.129(5)
Zn(2)-N(1A)	2.156(4)	2.164(5)
Zn(2)-O(4A)	2.296(4)	2.374(4)
Zn(3)-O(6B)	1.962(4)	1.982(4)
Zn(3)-O(1B)	2.057(4)	2.050(4)
Zn(3)-O(3B)	2.099(4)	2.082(3)
Zn(3)-N(3B)	2.165(6)	2.170(5)
Zn(3)-N(4B)	2.173(5)	2.184(6)
Zn(3)-O(5B)	2.269(4)	2.262(4)
Zn(4)-O(2B)	1.999(4)	1.993(3)
Zn(4)-O(1B)	2.054(4)	2.013(4)
Zn(4)-O(7B)	2.078(5)	2.065(4)
Zn(4)-N(2B)	2.159(6)	2.126(6)
Zn(4)-N(1B)	2.182(5)	2.202(5)
Zn(4)-O(4B)	2.285(5)	2.389(4)
O(1A)-Zn(1)-N(4A)	108.47(17)	93.61(16)
O(1A)-Zn(1)-N(3A)	93.55(17)	93.81(18)
N(4A)-Zn(1)-N(3A)	79.95(19)	80.01(18)
N(2A)-Zn(2)-N(1A)	80.38(17)	80.2(2)
N(2A)-Zn(2)-O(1A)	87.13(16)	90.71(16)
N(1A)-Zn(2)-O(1A)	92.02(17)	93.23(17)
O(1B)-Zn(3)-N(3B)	90.11(18)	89.64(19)
O(1B)-Zn(3)-N(4B)	164.7(2)	165.5(2)
N(3B)-Zn(3)-N(4B)	75.6(2)	77.3(2)
O(1B)-Zn(4)-N(2B)	166.47(18)	167.4(2)
O(1B)-Zn(4)-N(1B)	90.36(17)	89.8(2)
N(2B)-Zn(4)-N(1B)	76.8(2)	77.6(2)
Zn(1)-O(1A)-Zn(2)	125.56(19)	128.4(2)
Zn(4)-O(1B)-Zn(3)	120.67(19)	120.1(2)

749.0, calc. *m/z* 749.1 (100 %), 747.1 (95 %), Fig. 4.10) and $[Zn_2(Br\mathbf{L2})(OH)(OCH_3)]^+$ (found *m/z* 707.3, calc. *m/z* 707.0 (100 %), 705.0 (95 %)).

Only labile coordination of the phenyl ether arms of the $CH_3H\mathbf{L3}$ ligand to the Zn(II) ions is proposed (due to steric demand of the phenyl groups). Therefore,

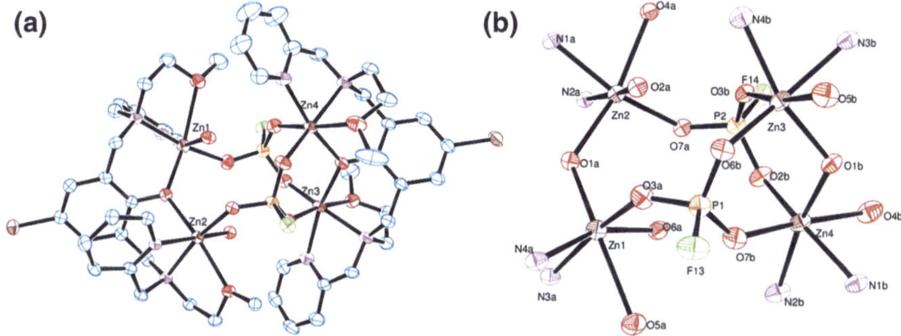

Fig. 4.8 ORTEP plot of **a** $[Zn_4(Br\mathbf{L2})_2(PO_3F)_2(H_2O)_2](PF_6)_2$ with 25 % probability level of thermal ellipsoids. **b** First coordination sphere of the Zn(II) ions including the phosphate cores. Hydrogen atoms, and counter ions have been omitted for clarity, as have carbon atom labels

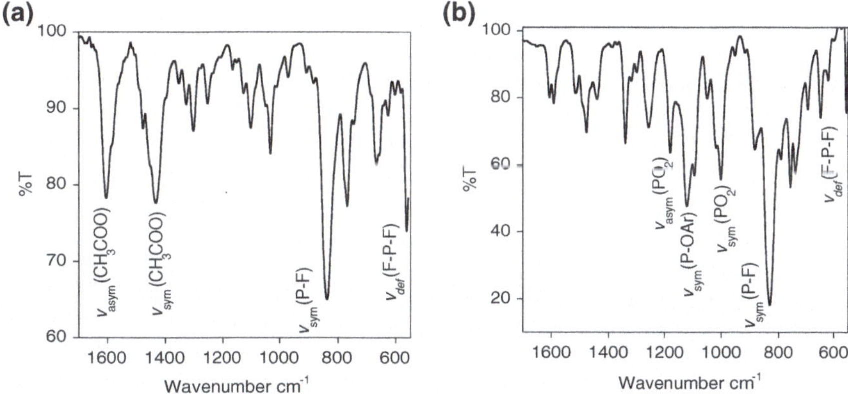

Fig. 4.9 **a** Fingerprint region of $[Zn_2(Br\mathbf{L2})(CH_3COO)_2](PF_6)$ and **b** Fingerprint IR region of $[Zn_4(CH_3\mathbf{L2})_2(NO_2C_6H_5OPO_3)_2(H_2O)_2](PF_6)_2$

they can be easily replaced by other ligands such as water or solvent molecules. The mass spectrum supports this proposal with the major peak being at *m/z* 855.7 arising from a $[Zn_2(CH_3\mathbf{L3})(CH_3COO)_2(H_2O)]^+$ species (calc. *m/z* 853.2 (100 %), 855.2 (97 %)). With the assumption that the acetates are bound in a bidentate fashion the additional coordination position for the water molecule could stem from a phenyl ether being displaced. For $[Zn_2(CH_3\mathbf{L2})(CH_3COO)_2](PF_6)$ the major peak was observed at *m/z* 711.2, which can be attributed to a $[Zn_2(CH_3\mathbf{L2})(CH_3COO)_2]^+$ species (Fig. 4.11, calc. *m/z* 711.2 (100 %), 713.2 (97 %)).

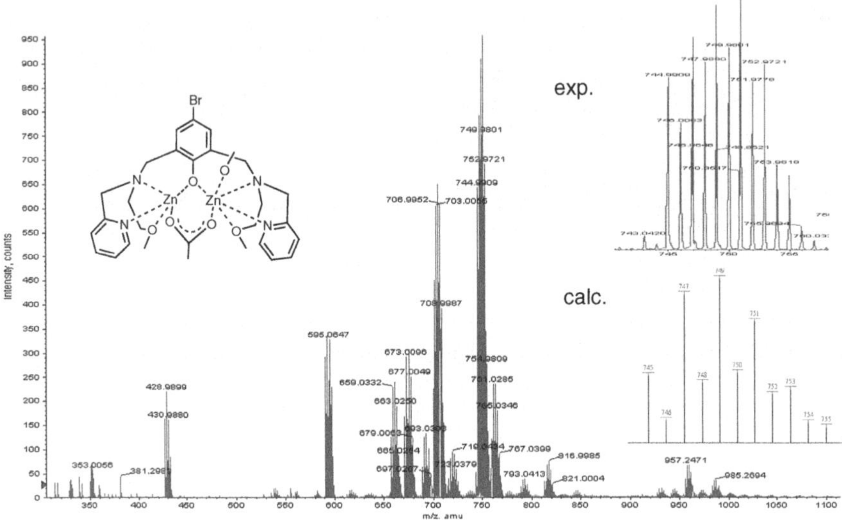

Fig. 4.10 Mass spectrum of $[Zn_2(Br\mathbf{L2})(CH_3COO)_2]^+$ in methanol. Inset recorded splitting pattern (*blue*) and calculated (*black*)

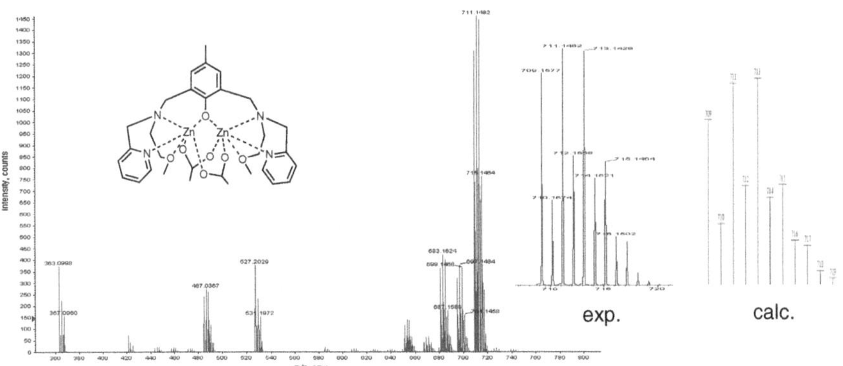

Fig. 4.11 Mass spectrum of $[Zn_2(CH_3\mathbf{L2})(CH_3COO)_2]^+$ in methanol. Inset: recorded splitting pattern (*blue*) and calculated (*black*)

4.2.6 Mass Spectrometry of Complex $[Zn_2(CH_3L2)(CH_3COO)_2]^+$ in the presence of Substrates

Under conditions similar to the kinetic studies (water/acetonitrile 1:1) the complex $[Zn_2(CH_3\mathbf{L2})(CH_3COO)_2]PF_6$ was mixed with 25 equivalents BPNPP (a slow substrate employed for enzyme kinetic studies). The mass spectrum (Fig. 4.12) measured one hour after mixing showed the presence of $[Zn_2(CH_3\mathbf{L2})$-(BPNPP)-

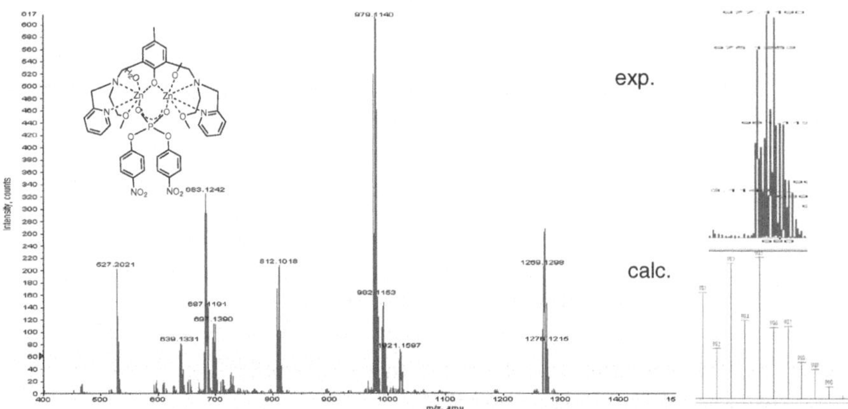

Fig. 4.12 Mass spectrum of $[Zn_2(CH_3\mathbf{L2})(CH_3COO)_2]PF_6$ measured in acetonitrile/water 1:1 in the presence of 25 equivalents BPNPP, magnified pattern of the major peak (*blue*) calculated splitting pattern (*black*) and proposed structure

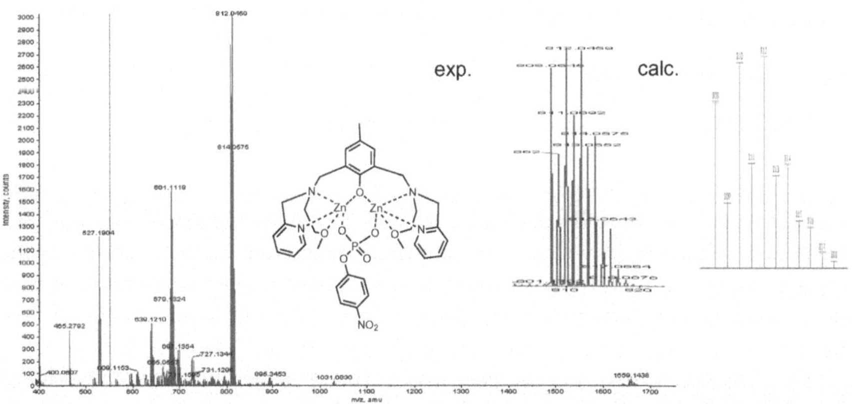

Fig. 4.13 Mass spectrum of $[Zn_2(CH_3\mathbf{L2})(CH_3COO)_2]PF_6$ measured in acetonitrile/water 1:1 in the presence of excess PNPP, magnified pattern of the major peak (*blue*) calculated splitting pattern (*black*) and proposed structure

$(H_2O)(OCH_3)]^+$ (found *m/z* 979.1, calc. *m/z* 983.1 (100 %), 981.1 (97 %)) and in addition PNPP, the hydrolysis product of BPNPP, bound to the complex $[Zn_2(CH_3\mathbf{L2})(PNPP)]^+$ (found *m/z* 811.4, calc. *m/z* 812.1 (100 %), 810.1 (97 %)).

When the mass spectrum $[Zn_2(CH_3\mathbf{L2})(CH_3COO)_2]PF_6$ was recorded in the presence of an excess of the hydrolysis product PNPP the formation of both a tetrameric species $[Zn_4(CH_3\mathbf{L2})_2(PNPP)_2(OH)(H_2O)]^+$ (found *m/z* 1659.1, calc. *m/z* 1657.2 (100 %), 1659.2 (96 %)), and a dimeric species $[Zn_2(CH_3\mathbf{L2})(PNPP)_2 + 2H]^+$ (found *m/z* 1031.1, calc. *m/z* 1031.1 (100 %)) was observed (Fig. 4.13).

$[Zn_2(CH_3\mathbf{L2})(PNPP)]^+$ (found *m/z* 811.4, calc. *m/z* 812.1 (100 %), 810.1 (97 %)) was present during the above mentioned experiment as well as formate species (present in the operating systems of the mass spectrometer) such as $[Zn_2(CH_3\mathbf{L2})(HCOO)\text{-}H]^+$ (found *m/z* 639.4, calc. *m/z* 639.1 (100 %), 637.1 (98 %)) and $[Zn_2(CH_3\mathbf{L2})(HCOO)_2]^+$ (found *m/z* 684.4, calc. *m/z* 685.1 (100 %), 683.1 (98 %)).

4.2.7 NMR Spectroscopy of the Complexes

4.2.7.1 ^1H NMR-titration and Temperature Experiment of $[Zn_2(CH_3L2)(CH_3COO)_2]PF_6$ and $[Zn_2(CH_3L3)(CH_3COO)_2]PF_6$

The coordination of ligands to Zn(II) ions in solution can be studied by ^1H NMR-titration experiments. Successively, one and two equivalents of zinc(II) acetate dihydrate were added to a solution of $CH_3H\mathbf{L2}$ in deuterated acetonitrile at room temperature. Upon the addition of one equivalent zinc(II) acetate the aromatic region of the ^1H NMR spectrum of $CH_3H\mathbf{L2}$ exhibited a set of downfield shifted signals, in addition to those of the free ligand. The former signals did not change upon the addition of the second equivalent of zinc acetate (Fig. 4.14). It is proposed that the spectrum with one equivalent of zinc(II) acetate shows only free ligand and complex with two Zn(II) ions bound. Interestingly, in the spectrum with two equivalents of zinc(II) acetate a second set of low intensity signals was observed at higher field which is suggested to arise from another isomer of the $[Zn_2(CH_3\mathbf{L2})(CH_3COO)_2]^+$ complex (Fig. 4.14, also present in the spectrum with one equivalent but partially overlapping with free ligand signals and thus not characterized).

The NMR spectrum of the complex was recorded at four different temperatures (298, 316, 333 and 343 K) to investigate the effect of temperature on this (proposed) isomer equilibrium. The intensities of the signals were temperature dependent; the less intense signals gained intensity upon increase in temperature (Figs. 4.15 and 4.16). The effect was shown to be reversible upon cooling of the sample.

To test whether the second set of low intensity signals arose from a species where the ether arms are not coordinated to the Zn(II) centers, the ^1H-NMR spectrum of $[Zn_2(CH_3\mathbf{L3})(CH_3COO)_2](PF_6)$ was recorded in acetonitrile. In this complex the methyl ethers of $CH_3H\mathbf{L2}$ are replaced by phenyl ether groups, which are expected to favor the ‘ether arms-off’ form rather than the ‘on’ form. The comparison of one aromatic signal of this complex with that for the complex with $CH_3H\mathbf{L2}$ is displayed in Fig. 4.17 and it is observed that the intensity distribution is reversed. The higher field signals increase in intensity as the temperature is increased similar as found in $[Zn_2(CH_3\mathbf{L2})(CH_3COO)_2](PF_6)$, supporting the assignment to a form of complex where the ether donors are not coordinated.

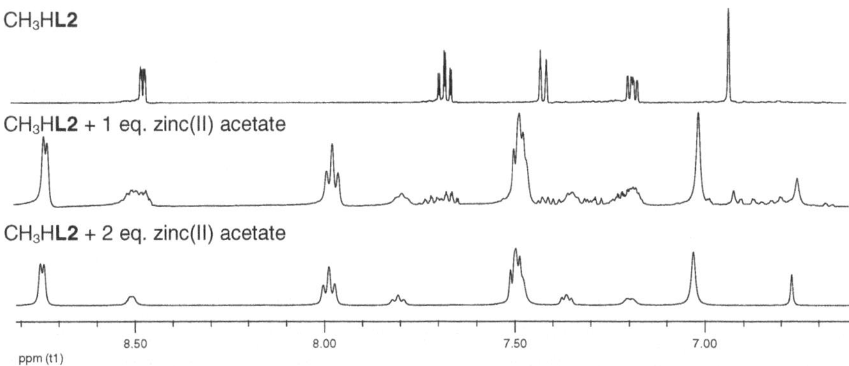

Fig. 4.14 Aromatic region of the NMR-titration of one and two equivalents of zinc(II) acetate into CH_3H**L2** (CD_3CN)

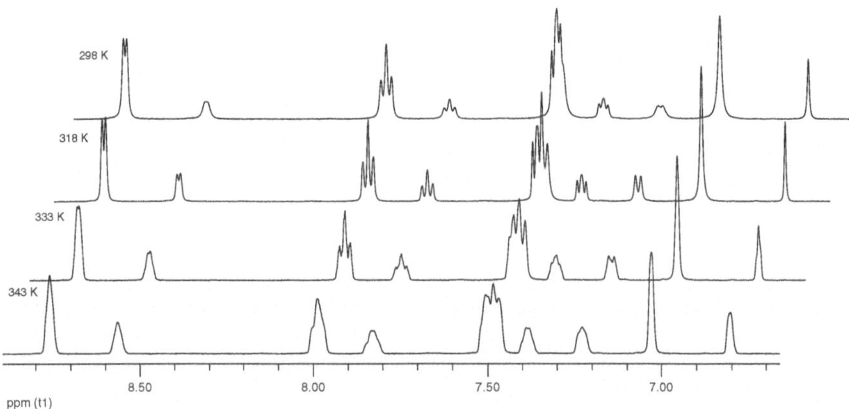

Fig. 4.15 Aromatic region of $[Zn_2(CH_3$**L2**$)(CH_3COO)_2](PF_6)$ in CD_3CN at different temperatures

Another experiment was conducted with $[Zn_2(CH_3$**L2**$)(CH_3COO)_2](PF_6)$. Here, the ^{1}H-NMR spectrum was recorded in the more strongly coordinating solvent d^6-$(CD_3)_2SO$ to shift the equilibrium between the isomers to the 'ether arms-off' isomer. The change in intensities can be seen in Fig. 4.17. The integral ratio has changed substantially.

4.2.7.2 ^{31}P NMR and ^{18}O-Labeling Studies for BDNPP Hydrolysis

In order to explore the mechanism employed by the Zn(II) complexes, the incorporation of ^{18}O into the reaction product was monitored by ^{31}P NMR. In addition to the Zn(II) complexes $[Zn_2(CH_3$**L2**$)(CH_3COO)_2](PF_6)$ and

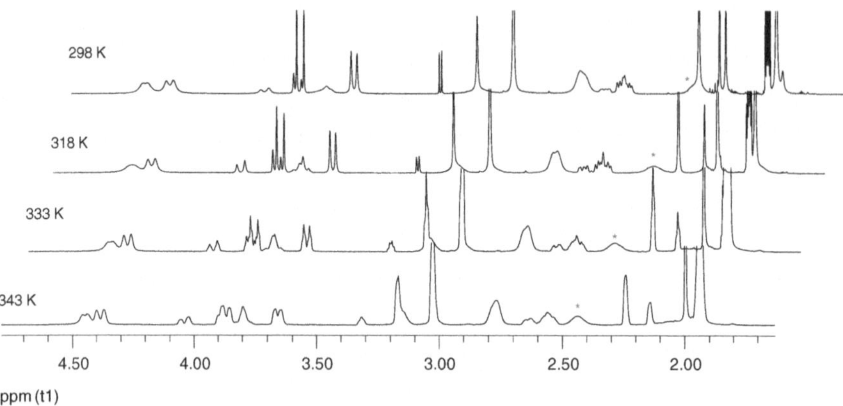

Fig. 4.16 Aliphatic region of $[Zn_2(CH_3\mathbf{L2})(CH_3COO)_2](PF_6)$ at different temperatures. The most significant difference was the shift of a water signal (*asterisk*)

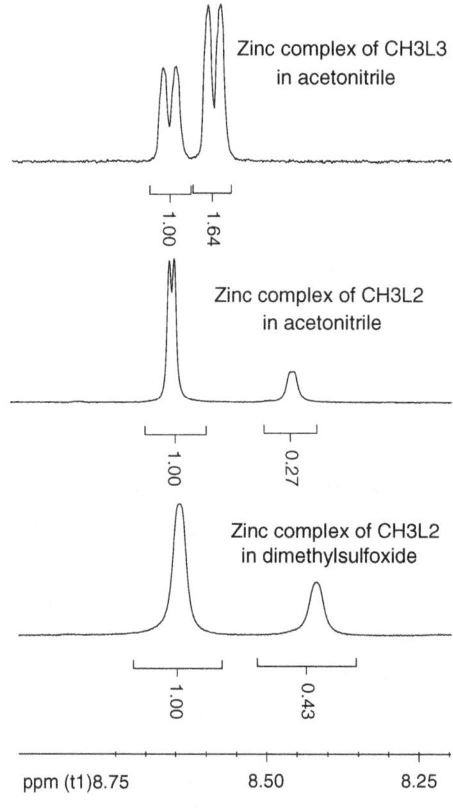

Fig. 4.17 Comparison of one aromatic signal in acetonitrile of $[Zn_2(CH_3\mathbf{L3})(CH_3COO)_2](PF_6)$ and $[Zn_2(CH_3\mathbf{L2})(CH_3COO)_2](PF_6)$, the ladder being also recorded in dimethylsulfoxide

$[Zn_2(NO_2\mathbf{L2})(CH_3COO)_2](PF_6)$ the previously reported Zn(II) complex $[Zn_2(CH_3H\mathbf{L1})(CH_3COO)(H_2O)](PF_6)$ with pendant alcohol ligand arms was investigated to study the effect of alkoxide nucleophiles in the reaction [30]. PNMR can be used to determine the extent of incorporation of ^{18}O-labeled water/hydroxide into the hydrolysis product of BDNPP as the ^{31}P-chemical shifts are very sensitive to structural changes arising from an ^{18}O bound to the phosphorus instead of an ^{16}O; the signal arising from ^{18}O is thus shifted to lower frequency [60–62]. In this experiment the substrate BDNPP was hydrolyzed in a 50:50 mixture of acetonitrile:buffer (50 mM, HEPES, pH 8) in the presence of one equivalent of ligand. In the isotopic labeling studies the buffer solution contained 50 % ^{18}O (97 % purity). For the unlabeled (^{16}O) experiment the solution was made up of 0.01 mmol of the respective complex in acetonitrile (0.3 mL), 100 mM HEPES buffer pH 8 (0.15 mL) and distilled water (0.15 mL). To this mixture, one equivalent BDNPP (5.1 mg) was added and the mixture was left for one week at room temperature prior to the spectrum being recorded. For the ^{18}O-labeled sample the solution was made up from a solution of 0.01 mmol of the respective complex in acetonitrile (0.3 mL), 100 mM HEPES buffer pH 8 (0.15 mL) and ^{18}O-water (97 %, 0.15 mL). To this, one equivalent BDNPP (5.1 mg) was added and the mixture was left for one week at room temperature prior to recording the spectrum. After one week the hydrolysis product of BDNPP, DNPP, was observed in the ^{31}P NMR spectrum. The complete disappearance of the BDNPP signal at −14.13 ppm was, however, not observed in this time frame. Hydrolysis of BDNPP catalyzed by $Zn_2(CH_3\mathbf{L2})(CH_3COO)_2](PF_6)$ in an unlabeled medium gives rise to the DNPP signal at −2.68 ppm (Fig. 4.18, bottom). For $Zn_2(CH_3\mathbf{L2})(CH_3COO)_2](PF_6)$, or any other complex that utilizes a nucleophile arising from a coordinated aqua ligand, in a 50:50 ^{18}O:^{16}O environment is it expected that DNPP will be found with 50 % ^{18}O incorporated, as the nucleophile has to be a Zn(II) bound water/hydroxide (from the solvent). This was in fact observed in the ^{31}P spectrum by a splitting of the DNPP-peak into a doublet with a peak separation of approximately 0.02 ppm (Fig. 4.18) [63]. The same observation was made for the reaction catalyzed by $[Zn_2(CH_3H\mathbf{L1})(CH_3COO)(H_2O)](PF_6)$ with one peak at −10.93 ppm in unlabeled water and the presence of two peaks (at −10.277 and −10.944 ppm) in 50 % ^{18}O labeled medium. Hydrolysis of BDNPP catalyzed by $[Zn_2(NO_2\mathbf{L2})(CH_3COO)_2](PF_6)$ in an unlabeled medium gives rise to the DNPP signal at −3.28 ppm. A splitting in the labeling experiment was observed similar to the other complexes in the ^{31}P-spectrum by a splitting of the DNPP-peak into a doublet with the additional second peak at −3.23 ppm. The difference in shifts is presumably due to the different coordination environments present in the samples. As only one equivalent BDNPP was used we propose that the hydrolyzed product, DNPP, is still associated with the complex after hydrolysis and is not exchanged for another substrate molecule as would be the case when using a super stoichiometric amount of substrate [64, 65].

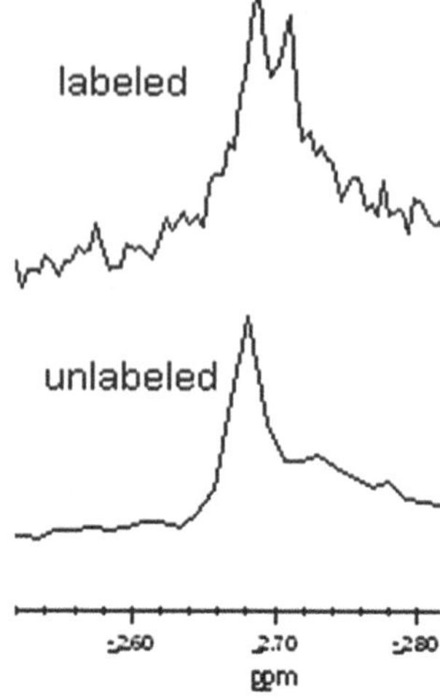

Fig. 4.18 ^{31}P NMR of the BDNPP hydrolysis product DNPP catalyzed by $[Zn_2(CH_3\mathbf{L2})(CH_3COO)_2](PF_6)$ in labeled and unlabeled media

4.2.8 Phosphatase-like Kinetics

Phosphatase-like activities of all complexes were measured using the activated substrate BDNPP over a pH range 5–10.5. The data were fitted to equations derived for monoprotic (Eq. 4.1, Fig. 4.19) or diprotic (Eq. 4.2 and Fig. 4.19) systems [67]. v_0 is the initial reaction rate, v_{max} refers to the maximum rate with 5 mM BDNPP as substrate, and K_{a1} and K_{a2} are the catalytically relevant protonation equilibrium constants.

$$v_0 = \frac{v_{max}}{1 + \frac{[H^+]}{K_a}} \tag{4.1}$$

$$v_0 = \frac{v_{max}}{\left(1 + \frac{[H^+]}{K_{a1}} + \frac{K_{a2}}{[H^+]}\right)} \tag{4.2}$$

For the hydrolysis of the activated substrate BDNPP the complexes $[Zn_2(CH_3\mathbf{L2})(CH_3COO)_2](PF_6)$, $[Zn_2(CH_3\mathbf{L3})(CH_3COO)_2](PF_6)$, $[Zn_2(NO_2\mathbf{L2})$-$(CH_3COO)_2](PF_6)$ and $[Zn_2(Br\mathbf{L2})$-$(CH_3COO)_2](PF_6)$ displayed very similar kinetically relevant pK_{a1} values (Table 4.6).

The rate (v_0) *versus* pH profiles for three of the Zn(II) complexes, $[Zn_2(CH_3\mathbf{L2})$-$(CH_3COO)_2](PF_6)$, $[Zn_2(CH_3\mathbf{L3})(CH_3COO)_2](PF_6)$, and $[Zn_2(Br\mathbf{L2})(CH_3COO)_2](PF_6)$, were consistent with the presence of two protonation equilibria and were fitted to bell-shaped profiles. The buffer range was insufficient to permit conclusive analysis of high pH data for $[Zn_2(NO_2\mathbf{L2})(CH_3COO)_2](PF_6)$. Therefore, the data were fitted to a sigmoidal curve, corresponding to only one protonation equilibrium.

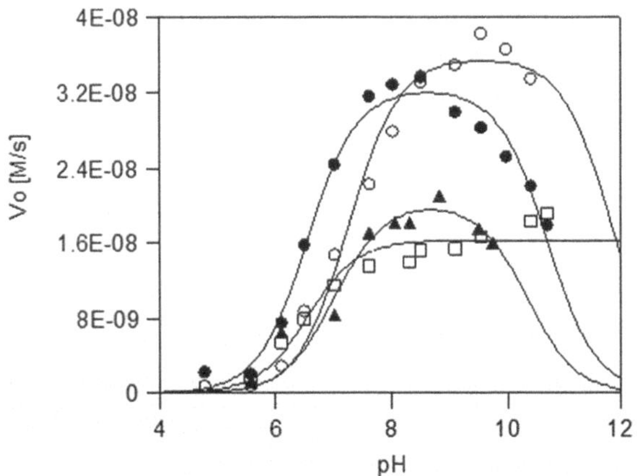

Fig. 4.19 pH dependence of the rate of BDNPP cleavage by $[Zn_2(CH_3\mathbf{L2})(CH_3COO)_2](PF_6)$ (*open circle*), $[Zn_2(NO_2\mathbf{L2})(CH_3COO)_2](PF_6)$ (*open square*), $[Zn_2(Br\mathbf{L2})(CH_3COO)_2](PF_6)$ (*filled triangle*) and $[Zn_2(CH_3\mathbf{L3})(CH_3COO)_2](PF_6)$ (*filled circle*)

Table 4.6 Kinetic data for $Zn_2(CH_3\mathbf{L2})(CH_3COO)_2](PF_6)$, $[Zn_2(CH_3\mathbf{L3})(CH_3COO)_2](PF_6)$, $[Zn_2(NO_2\mathbf{L2})(CH_3COO)_2](PF_6)$ and $[Zn_2(Br\mathbf{L2})(CH_3COO)_2](PF_6)$

	k_{cat} (s^{-1})	K_m (mM)	pK_{a1}, pK_{a2}
$[Zn_2(CH_3\mathbf{L2})(CH_3COO)_2](PF_6)$ [54]	$5.70 \pm 0.04 \times 10^{-3}$	20.8 ± 5.0	6.76, 11.00
$[Zn_2(NO_2\mathbf{L2})(CH_3COO)_2](PF_6)$	$0.76 \pm 0.04 \times 10^{-3}$	6.5 ± 1.4	6.55
$[Zn_2(Br\mathbf{L2})(CH_3COO)_2](PF_6)$	$1.90 \pm 0.04 \times 10^{-3}$	7.1 ± 4.00	6.56, 10.60
$[Zn_2(CH_3\mathbf{L3})(CH_3COO)_2](PF_6)$	$3.60 \pm 0.04 \times 10^{-3}$	18.9 ± 3.5	7.72, 10.83

Measurements of the dependence of the rate on substrate concentration for the di-Zn(II) complexes were carried out at the pH with the highest activity, being pH 8.5 for the $CH_3H\mathbf{L3}$, pH 9.5 for the $NO_2H\mathbf{L2}$ and pH 8.8 for the $BrH\mathbf{L2}$-complex, respectively. It was observed that the BDNPP hydrolysis catalyzed by the complexes followed Michaelis-Menten saturation behavior and the data were fit (Fig. 4.20) using non-linear least square analysis to give the Michaelis constant K_m, V_{max} and k_{cat} (V_{max}/[complex]) for each complex [67]. Here, [S] refers to the substrate concentration.

$$v_0 = \frac{v_{max}[S]}{K_m + [S]} \quad (4.3)$$

Two of the complexes exhibited similar catalytic behaviour ($[Zn_2(CH_3\mathbf{L2})(CH_3COO)_2](PF_6)$, $k_{cat} = 5.70 \pm 0.04 \times 10^{-3}\ s^{-1}$, $K_m = 20.8 \pm 5$ mM; [54] $[Zn_2(CH_3\mathbf{L3})(CH_3COO)_2](PF_6)$, $3.60 \pm 0.04 \times 10^{-3}\ s^{-1}$, $K_m = 18.9 \pm 3.5$ mM). However the catalytic efficiency (k_{cat}/K_m) of the former ($0.27 \times 10^{-3}\ s^{-1}\ mM^{-1}$) was comparable to the complex $[Zn_2(Br\mathbf{L2})(CH_3COO)_2](PF_6)$ ($0.27 \times 10^{-3}\ s^{-1}\ mM^{-1}$) with the same catalytic efficiency. The least efficient catalyst was $[Zn_2(NO_2\mathbf{L2})(CH_3COO)_2](PF_6)$ with a catalytic efficiency of $0.11 \times 10^{-3}\ s^{-1}\ mM^{-1}$. The dependence of the rate of hydrolysis of BDNPP (5 mM) on the complex concentration was linear from 20 to 160 μM for all complexes.

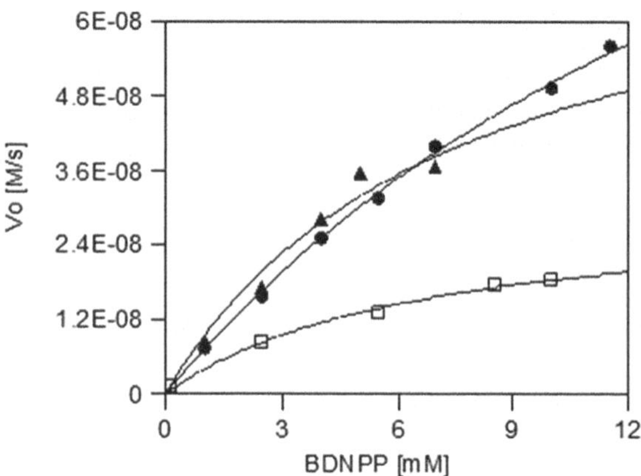

Fig. 4.20 Substrate concentration dependence of the catalytic rate for $[Zn_2(Br\mathbf{L2})(CH_3COO)_2](PF_6)$ (*filled triangle*), $[Zn_2(CH_3\mathbf{L3})(CH_3COO)_2](PF_6)$ (*filled circle*) and $[Zn_2(NO_2\mathbf{L2})(CH_3COO)_2](PF_6)$ (*open square*)

4.3 Discussion

4.3.1 Structures and Spectroscopy

All complexes display very similar coordination environments with two acetate moieties bridging the Zn(II) ions and completing the six-coordinate ligand sphere. The bidentate coordination of substrate/product analogs in $[Zn_4(CH_3\mathbf{L2})_2(NO_2C_6H_5OPO_3)_2(H_2O)_2](PF_6)_2$ and $[Zn_4(Br\mathbf{L2})_2(PO_3F)_2(H_2O)_2](PF_6)_2$ is in line with previous findings where BPNPP [48] and dibenzyl- [66] or diphenyl phosphate [46] had been used in crystallization studies with Zn(II) complexes. Analysis of the mass spectrum of $[Zn_2(CH_3\mathbf{L2})(CH_3COO)_2]PF_6$ measured in the presence of 25 equivalents of BPNPP showed $[Zn_2(CH_3\mathbf{L2})(PNPP)]^+$, which indicates not only that the complex is able to hydrolyze less activated substrates but also that the acetates are being fully replaced in the presence of substrate. The results of the NMR experiments suggest that in solution the ether arms of the ligands are loosely bound to the metal ion and can be easily replaced by other ligands, for example water and DMSO. These studies show that a catalytically competent dinuclear center with vacant coordination sites for substrate and nucleophiles is readily formed under kinetic conditions.

4.3.2 Mechanism of Hydrolysis

$[Zn_2(CH_3\mathbf{L2})(CH_3COO)_2](PF_6)$, $[Zn_2(CH_3\mathbf{L3})(CH_3COO)_2](PF_6)\cdot CH_3OH$, $[Zn_2(Br\mathbf{L2})(CH_3COO)_2](PF_6)$ and $[Zn_2(NO_2\mathbf{L2})(CH_3COO)_2](PF_6)$ contain an ether moiety, therefore the only nucleophile possible is a Zn-OH moiety. pK_{a1} is similar for all complexes and ranges from 6.55 to 7.72. The proposed mechanism is shown on the basis of $[Zn_2(CH_3\mathbf{L2})(CH_3COO)_2](PF_6)$ in Fig. 4.21c. The second

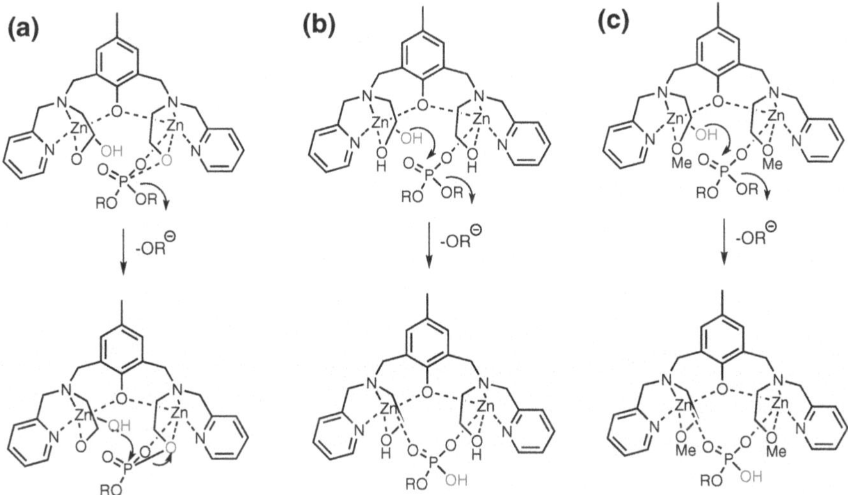

Fig. 4.21 The different nucleophiles involved in phosphoester cleavage in $[Zn_2(CH_3H$**L1**$)(CH_3COO)(H_2O)](PF_6)$ and $Zn_2(CH_3$**L2**$)(CH_3COO)_2](PF_6)$

pK_{a2} might be due to a second Zn(II) bound water molecule which, upon deprotonation, is competing for coordination with the substrate, and thus less effective ligand exchange and hydrolysis occurs. Another possibility is that the high pH supports the formation of $Zn(OH)_2$ as found in other Zn(II) complexes [48]. However, as no precipitate was observed during kinetic measurements the latter is unlikely. Only a slight change in activity was observed when using the ligand with bulky phenyl ether arms instead of methyl ethers. This suggests that in both cases the ether arms are not, or at best loosely, bound to the metals and are not involved in the hydrolytic process. The *para*-substituent of the bridging phenol affects the k_{cat}, lowering the activity in the case of the electron withdrawing substituents Br^- and NO_2^-. More aspects of inductive effects are discussed in the next section. Table 4.7 shows a summary of kinetic data for the complexes including $[Zn_2(CH_3H$**L1**$)(CH_3COO)(H_2O)](PF_6)$, a complex with alcohol ligand arms reported previously by Chen et al. [30]. For $[Zn_2(CH_3H$**L1**$)(CH_3COO)(H_2O)](PF_6)$ a mechanism whereby the alcohol arm acts as nucleophile in organophosphate hydrolysis was proposed by the authors (Fig. 4.21a) [30].

However when comparing the kinetically relevant pK_{a1} values of the ether ligand arm based complexes to $[Zn_2(CH_3H$**L1**$)(CH_3COO)(H_2O)](PF_6)$ one notices they are virtually the same for $[Zn_2(CH_3H$**L1**$)(CH_3COO)(H_2O)](PF_6)$ (6.60), $[Zn_2(Br$**L2**$)$-$(CH_3COO)_2](PF_6)$ (6.56) and $Zn_2(CH_3$**L2**$)(CH_3COO)_2](PF_6)$ (6.76). This suggests a similar nucleophile, which in case of the ether ligands can only be a terminal water molecule. Based on ^{18}O labeling studies conducted with $[Zn_2(CH_3H$**L1**$)(CH_3COO)$-$(H_2O)](PF_6)$ and $[Zn_2(CH_3$**L2**$)(CH_3COO)_2](PF_6)$ a plausible mechanism for these systems is proposed in Fig. 4.21b and c,

Table 4.7 Kinetic data for $Zn_2(CH_3\mathbf{L2})(CH_3COO)_2](PF_6)$, $[Zn_2(CH_3\mathbf{L3})(CH_3COO)_2](PF_6)$, $[Zn_2(NO_2\mathbf{L2})(CH_3COO)_2](PF_6)$, $[Zn_2(Br\mathbf{L2})(CH_3COO)_2](PF_6)$ and $[Zn_2(CH_3H\mathbf{L1})(CH_3COO)(H_2O)](PF_6)$

	k_{cat} (s^{-1})	K_m (mM)	pK_{a1}, pK_{a2}
$[Zn_2(CH_3\mathbf{L2})(CH_3COO)_2](PF_6)$ [54]	$5.70 \pm 0.04 \times 10^{-3}$	20.8 ± 5.0	6.76, 11.00
$[Zn_2(NO_2\mathbf{L2})(CH_3COO)_2](PF_6)$	$0.76 \pm 0.04 \times 10^{-3}$	6.5 ± 1.4	6.55
$[Zn_2(Br\mathbf{L2})(CH_3COO)_2](PF_6)$	$1.90 \pm 0.04 \times 10^{-3}$	7.1 ± 4.00	6.56, 10.60
$[Zn_2(CH_3\mathbf{L3})(CH_3COO)_2](PF_6)$	$3.60 \pm 0.04 \times 10^{-3}$	18.9 ± 3.5	7.72, 10.83
$[Zn_2(CH_3H\mathbf{L1})(CH_3COO)(H_2O)](PF_6)$ [30]	$1.07 \pm 0.04 \times 10^{-3}$	12.5 ± 7.0	6.60

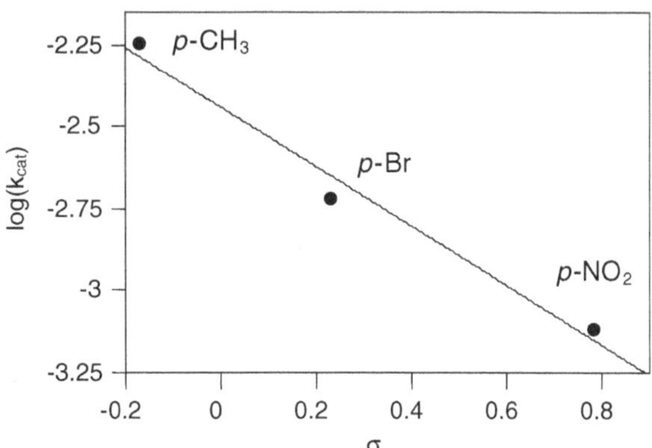

Fig. 4.22 Linear correlation for the Hammett parameter (σ) and $\log(k_{cat})$ for $Zn_2(CH_3\mathbf{L2})(CH_3COO)_2](PF_6)$, $Zn_2(NO_2\mathbf{L2})(CH_3COO)_2](PF_6)$ and $[Zn_2(Br\mathbf{L2})(CH_3COO)_2](PF_6)$

respectively. The previous mechanism proposed by Chen [30] would require the dual action of both the alkoxide, and subsequently, hydroxide nucleophiles (Fig. 4.21a).

4.3.3 Inductive Effects

As discussed above it was found that that the substituent in the *para*-position influences the rate of the BDNPP-hydrolysis. The Hammett parameter σ [68] for the three complexes $[Zn_2(CH_3\mathbf{L2})(CH_3COO)_2](PF_6).CH_3OH$, $[Zn_2(NO_2\mathbf{L2})(CH_3COO)_2](PF_6)$ and $[Zn_2(Br\mathbf{L2})(CH_3COO)_2](PF_6)$ was plotted against $\log(k_{cat})$ (Fig. 4.22).

The data shown in Fig. 4.22 demonstrates that introduction of an electron donating group in *para*-position enhances catalytic activity of the complex. This has been observed previously in a purple acid phosphatase model with a Fe(III)Zn(II) center [69]. However, neither for the PAP-model nor for any of the Zn(II) complexes a linear connection between σ and K_m could be determined. This indicates that the substituent in *para*-position is primarily affecting nucleophilic activation rather than binding of substrates to the metal ions.

4.4 Conclusion

The Zn(II) complexes presented in this chapter possess properties analogous to dinuclear metallohydrolases [19, 70, 71]. Complexes with $CH_3\mathbf{L2}^-$ and $CH_3\mathbf{L3}^-$ were more active as the previously reported $CH_3H\mathbf{L1}$ probably due to the ether arms being replaced by nucleophiles more easily than the coordinated alkoxide ligand arms. It was found that electron-donating *para*-substituents enhance the catalytic affinity of the complexes in BDNPP hydrolysis. Previous studies with the complex $[Zn_2(CH_3H\mathbf{L1})(CH_3COO)(H_2O)](PF_6)$ [30], and related complexes with a coordinated alcohol moiety, [17, 49, 72] suggested that the corresponding alkoxide was the nucleophile. The present study indicates, based on an analysis of the incorporation of ^{18}O into the hydrolysis product, that the three systems investigated here may employ a similar mechanistic strategy with a metal-bound hydroxide as nucleophile. The previously proposed model for the mechanism of $[Zn_2(CH_3H\mathbf{L1})]^{2+}$ may still be applicable but would likely require the dual action of both the alkoxide and hydroxide nucleophiles.

The next chapter will present an extensive investigation of the mechanistic pathways employed by two Cd(II) complexes in phosphoester and β-lactam hydrolysis.

References

1. Y. Gultneh, A.R. Khan, D. Blaise, S. Chaudhry, B. Ahvazi, B.B. Marvey, R.J. Butcher, J. Inorg. Biochem. **75**, 7–18 (1999)
2. E.-I. Ochiai, J. Chem. Educ. **65**, 943–946 (1988)
3. J. Riordan, B. Vallee (eds.), in *Methods in Enzymology* (Academic Press, United States, 1993). Print Book ISBN : 9780121821289
4. M.F. Summers, Coord. Chem. Rev. **86**, 43–134 (1988)
5. J. Weston, Chem. Rev. **105**, 2151–2174 (2005). *(Washington, DC, U.S.)*
6. E. Ghanem, F.M. Raushel, Toxicol. Appl. Pharmacol. **207**, 459–470 (2005)
7. P.J. O'Brien, D. Herschlag, Biochemistry **41**, 3207–3225 (2002)
8. E. Hough, L.K. Hansen, B. Birknes, K. Jynge, S. Hansen, A. Hordvik, C. Little, E. Dodson, Z. Derewenda, Nature **338**, 357–360 (1989)
9. C. Romier, R. Dominguez, A. Lahm, O. Dahl, D. Suck, Proteins: Struct., Funct., Bioinf. **32**, 414–424 (1998)
10. M.W. Crowder, J. Spencer, A.J. Vila, Acc. Chem. Res. **39**, 721–728 (2006)
11. N. Laraki, N. Franceschini, G.M. Rossolini, P. Santucci, C. Meunier, E. de Pauw, G. Amicosante, J.M. Frere, M. Galleni, Antimicrob. Agents Chemother. **43**, 902–906 (1999)
12. J.A. Cricco, E.G. Orellano, R.M. Rasia, E.A. Ceccarelli, A.J. Vila, Coord. Chem. Rev. **192**, 519–535 (1999)
13. C. Oefner, A. Douangamath, A. D'Arcy, S. Hafeli, D. Mareque, A. Mac Sweeney, J. Padilla, S. Pierau, H. Schulz, M. Thormann, S. Wadman, G.E. Dale, J. Mol. Biol. **332**, 13–21 (2003)
14. M. Klinkenberg, C. Ling, Y.H. Chang, Arch. Biochem. Biophys. **347**, 193–200 (1997)
15. R.A. Bradshaw, E. Yi, Essays Biochem. **38**, 65–78 (2002)
16. V.M. D'souza, R.C. Holz, Biochemistry **38**, 11079–11085 (1999)
17. E. Kimura, Y. Kodama, T. Koike, M. Shiro, J. Am. Chem. Soc. **117**, 8304–8311 (1995)

18. J.F. Fisher, S.O. Meroueh, S. Mobashery, J. Am. Chem. Soc. **105**, 395–424 (2005). (Washington, DC, U.S.)
19. N. Mitić, S.J. Smith, A. Neves, L.W. Guddat, L.R. Gahan, G. Schenk, Chem. Rev. **106**, 3338–3363 (2006). (Washington, DC, U.S.)
20. N.V. Kaminskaia, B. Spingler, S.J. Lippard, J. Am. Chem. Soc. **123**, 6555–6563 (2001)
21. Z. Wang, W. Fast, S.J. Benkovic, J. Am. Chem. Soc. **120**, 10788–10789 (1998)
22. C. Damblon, M. Jensen, A. Ababou, I. Barsukov, C. Papamicael, C.J. Schofield, L. Olsen, R. Bauer, G.C. Roberts, J. Biol. Chem. **278**, 29240–29251 (2003)
23. T. Koike, M. Takamura, E. Kimura, J. Am. Chem. Soc. **116**, 8443–8449 (1994)
24. A. Tamilselvi, G. Mugesh, J. Biol. Inorg. Chem. **13**, 1039–1053 (2008)
25. M.J. Hawk, R.M. Breece, C.E. Hajdin, K.M. Bender, Z. Hu, A.L. Costello, B. Bennett, D.L. Tierney, M.W. Crowder, J. Am. Chem. Soc. **131**, 10753–10762 (2009)
26. L. Hemmingsen, C. Damblon, J. Antony, M. Jensen, H.W. Adolph, S. Wommer, G.C. Roberts, R. Bauer, J. Am. Chem. Soc. **123**, 10329–10335 (2001)
27. H. Carlsson, M. Haukka, E. Nordlander, Inorg. Chem. **43**, 5681–5687 (2004)
28. M. Jarenmark, E. Csapo, J. Singh, S. Wockel, E. Farkas, F. Meyer, M. Haukka, E. Nordlander, J. Chem. Soc., Dalton Trans. **39**, 8183–8194 (2010)
29. M. Jarenmark, S. Kappen, M. Haukka, E. Nordlander, Dalton Trans. (8), 993–996 (2008)
30. J.W. Chen, X.Y. Wang, Y.G. Zhu, J. Lin, X.L. Yang, Y.Z. Li, Y. Lu, Z.J. Guo, Inorg. Chem. **44**, 3422–3430 (2005)
31. F. Meyer, H. Pritzkow, Eur. J. Inorg. Chem. **12**, 2346–2351 (2005)
32. G. Parkin, Chem. Rev. **104**, 699–767 (2004). (Washington, DC, U.S.)
33. R.R. Buchholz, M.E. Etienne, A. Dorgelo, R.E. Mirams, S.J. Smith, S.Y. Chow, L.R. Hanton, G.B. Jameson, G.Schenk, L. R. Gahan, Dalton Trans. **43**, 6045–6054 (2008)
34. M. Umayal, G. Mugesh, Inorg. Chim. Acta **372**, 353–361 (2011)
35. K. Selmeczi, C. Michel, A. Milet, I. Gautier-Luneau, C. Philouze, J.-L. Pierre, D. Schnieders, A. Rompel and C. Belle, Chem.–Eur. J. **13**, 9093–9106 (2007)
36. L.M. Berreau, Adv. Phys. Org. Chem. **41**, 79–181 (2006)
37. A. Tamilselvi, M. Nethaji and G. Mugesh, Chem.–Eur. J. **12**, 7797–7806 (2006)
38. B. Bauer-Siebenlist, F. Meyer, E. Farkas, D. Vidovic, S. Dechert, Chem.–Eur. J. **11**, 4349–4360 (2005)
39. B. Bauer-Siebenlist, S. Dechert, F. Meyer, Chem.–Eur. J. **11**, 5343–5352 (2005)
40. L.R. Gahan, S.J. Smith, A. Neves, G. Schenk, Eur. J. Inorg. Chem. **19**, 2745–2758 (2009)
41. G. Ambrosi, M. Formica, V. Fusi, L. Giorgi, M. Micheloni, Coord. Chem. Rev. **252**, 1121–1152 (2008)
42. H. Sakiyama, R. Mochizuki, A. Sugawara, M. Sakamoto, Y. Nishida, M. Yamasaki, J. Chem. Soc., Dalton Trans. **6**, 997–1000 (1999)
43. S.J. Lippard, C. He, J. Am. Chem. Soc. **122**, 184–185 (2000)
44. F. Meyer, P. Rutsch, Chem. Commun. **9**, 1037–1038 (1998)
45. C. Bazzicalupi, A. Bencini, E. Berni, A. Bianchi, P. Fornasari, C. Giorgi, B. Valtancoli, Inorg. Chem. **43**, 6255–6265 (2004)
46. C. Bazzicalupi, A. Bencini, A. Bianchi, V. Fusi, C. Giorgi, P. Paoletti, B. Valtancoli, D. Zanchi, Inorg. Chem. **36**, 2784–2790 (1997)
47. C. Vichard, T.A. Kaden, Inorg. Chim. Acta **337**, 173–180 (2002)
48. B. Bauer-Siebenlist, F. Meyer, E. Farkas, D. Vidovic, J.A. Cuesta-Seijo, R. Herbst-Irmer, H. Pritzkow, Inorg. Chem. **43**, 4189–4202 (2004)
49. C. Bazzicalupi, A. Bencini, E. Berni, A. Bianchi, V. Fedi, V. Fusi, C. Giorgi, P. Paolettti, B. Valtancoli, Inorg. Chem. **38**, 4115–4122 (1999)
50. M. Livieri, F. Mancin, U. Tonellato, J. Chin, J. Chem. Soc. Chem. Commun. **126**, 2862–2863 (2004)
51. J. Xia, Y.B. Shi, Y. Zhang, Q. Miao, W.X. Tang, Inorg. Chem. **42**, 70–77 (2003)
52. J. Burgess, *Metal ions in solution* (Halsted, Chichester, 1978)
53. N.V. Kaminskaia, C. He, S.J. Lippard, Inorg. Chem. **39**, 3365–3373 (2000)
54. K.E. Dalle, Honours Thesis, 2009

55. S. Petricek, A. Demsar, Polyhedron **29**, 3329–3334 (2010)
56. A. Crochet, K.M. Fromm, Z. Anorg, Allg. Chem. **636**, 1484–1496 (2010)
57. M.N. Burnett and C.K. Johnson, *ORTEP-III: Oak Ridge Thermal Ellipsoid Plot Program for Crystal Structure Illustrations*, 1996
58. K. Nakamoto, *Infrared and Raman Spectra of Inorganic and Coordination Compounds* (Wiley, New York, 1978)
59. M. Monroe, Molecular Weight Calculator (2004), http://omics.pnl.gov/software/MWCalculator.php. Accessed 20 Feb 2013
60. M.C. Mitchell, R.J. Taylor, T.P. Kee, Polyhedron **17**, 433–442 (1998)
61. R.K. Harris, B.E. Mann, *NMR and the Periodic Table*, (Academic, London, New York, 1978)
62. N.H. Williams, A.M. Lebuis, J. Chin, J. Am. Chem. Soc. **121**, 3341–3348 (1999)
63. M. Cohn, A. Hu, Proc. Natl. Acad. Sci. USA **75**, 200–203 (1978)
64. D.E.C. Corbridge, *Phosphorus: an Outline of its Chemistry, Biochemistry, and Technology*, (Elsevier, Amsterdam, New York, 1990)
65. J.B. Domingos, E. Longhinotti, T.A.S. Brandao, C.A. Bunton, L.S. Santos, M.N. Eberlin, F. Nome, J. Org. Chem. **69**, 6024–6033 (2004)
66. J.C. Mareque Rivas, R.T.M. de Rosales, S. Parsons, Dalton Trans. **23**, 4385–4386 (2003)
67. I.H. Segel, *Enzyme Kinetics: Behavior and Analysis of Rapid Equilibrium and Steady-State Enzyme Systems* (Wiley-Interscience, New York, 1975)
68. Y. Simon-Manso, J. Phys. Chem. A **109**, 2006–2011 (2005)
69. R.A. Peralta, A.J. Bortoluzzi, B. de Souza, R. Jovito, F.R. Xavier, R.A.A. Couto, A. Casellato, F. Nome, A. Dick, L.R. Gahan, G. Schenk, G.R. Hanson, F.C.S. de Paula, E.C. Pereira-Maia, SdP. Machado, P.C. Severino, C. Pich, T. Bortolotto, H. Terenzi, E.E. Castellano, A. Neves, M.J. Riley, Inorg. Chem. **49**, 11421–11438 (2010)
70. C.J. Jackson, P.D. Carr, J.W. Liu, S.J. Watt, J.L. Beck, D.L. Ollis, J. Mol. Biol. **367**, 1047–1062 (2007)
71. F. Ely, J.L. Foo, C.J. Jackson, L.R. Gahan, D. Ollis, G. Schenk, Curr. Top. Biomed. Res. **9**, 63–78 (2007)
72. M. Young, D. Wahnon, R. Hynes, J. Chin, J. Am. Chem. Soc. **117**, 9441–9447 (1995)

Chapter 5
Mechanistic Studies of Cd(II) Complexes as Phosphoesterase and Metallo-β-lactamase Models

5.1 Introduction

5.1.1 Cadmium as a Biological Relevant Metal

Cadmium gained its name from the Greek word for kadmeia which is an ancient name for Zn(II) oxide [1]. Cadmium is similar in some ways to Zn(II) and the most common oxidation state is Cd(II). In coordination compounds Cd(II) is often found hepta-coordinate [2, 3]. Cd(II) is often used as a sensor for structural and mechanistic studies of proteins in which Zn(II) is substituted by Cd(II) in their active centers. Often a higher hydrolytic activity is found than in the corresponding Zn(II) enzymes and complexes [2]. This might be due to the increased number of ligands (nucleophiles, substrate) that are possible to bind and exchange at the Cd(II) center. ^{113}Cd NMR spectroscopy is a method used for structure elucidation in biological, inorganic and organometallic cadmium-containing samples [5, 6]. The natural abundance of ^{113}Cd is 12.22 % [7] and it is very sensitive to environment changes due to the large surrounding electron cloud. It is also a spin half nucleus which generally yields sharp NMR signals [5]. Cd(II) has only limited biological relevance and is known to be extremely toxic to mammals. It is generally viewed as an element that is not used by nature. However, in 2005 Morel and co-workers reported an investigation of a metalloenzyme from *Thalassiosira weissflogii,* a marine phytoplankton, which specifically uses Cd(II) to achieve its biological function [8, 9]. With the aid of X-ray absorption near-edge spectroscopy and comparison with Cd(II) thiolate and imidazol complexes the authors suggested that the Cd(II) containing active site of this protein employed a geometry close to the tetrahedral coordination environment found in a Zn(II)-containing class of carbonic anhydrases in higher plants and, moreover, that Cd(II) was bound to a least one cysteine residue [9, 10]. A crystal structure of this enzyme reported by the same group in 2008 showed that indeed Cd(II) was coordinated by three amino acid residues—two cysteine and one histidine, respectively [9]. The tetrahedral coordination of Cd(II) was completed by a water molecule which is connected to an extensive and well-ordered water hydrogen bond network around the active site.

L. J. Daumann, *Spectroscopic and Mechanistic Studies of Dinuclear Metallohydrolases and Their Biomimetic Complexes*, Springer Theses, DOI: 10.1007/978-3-319-06629-5_5,

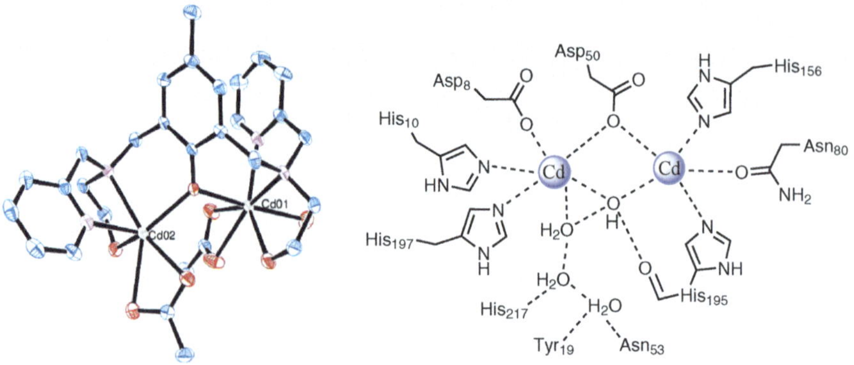

Fig. 5.1 The structure of the Cd(II) complex reported by Mirams et al. [2]. Counter ion and hydrogen atoms have been omitted for clarity. *Left* the dinuclear active site of GpdQ [20]

5.1.2 Cadmium Model Complexes

Biomimetic Cd(II) complexes which mimic either structural or functional aspects of their enzymatic counterpart are rare in the literature and mostly based on mimicking mononuclear enzymes like alcohol dehydrogenase and carbonic anhydrase [4, 10–16]. Di- and multinuclear Cd(II) complexes are mostly studied in terms of their NMR-shift, structure and coordination number [3, 4, 17–19]. Of peculiar interest are models that employ the same metals as their enzymatic analog. For GpdQ a comparative study of its Cd(II) derivative and a biomimetic has been reported [2]. The structure of the respective model is depicted in Fig. 5.1; the ligand mimics the N,O donor atoms of the active site of the enzyme GpdQ (Fig. 5.1, left).

The X-ray structure of the model complex shows that one Cd(II) ion is seven-coordinate with two nitrogen donors and an oxygen donor from one binding site of the ligand, an oxygen donor from the bridging oxygen of the ligand, two oxygen donors from bidentate acetate and an oxygen donor from a bound water molecule. This is an important feature, as a terminal water molecule is the likely nucleophile in GpdQ [20, 21]. Remarkably, potentiometric studies showed that the two Cd(II) ions show different binding affinities, with log $K_1 = 13.6$ and log $K_2 = 3.2$, mimicking the different affinities found for the native enzyme [2, 20, 21]. The phosphoesterase-like activity of the biomimetic was studied utilizing the substrate BDNPP and yielded a pK_a of 8.9, with $k_{cat} = 0.004\ s^{-1}$. In the same work, studies of Cd_2-GpdQ and the phosphodiester substrate BPNPP resulted in a k_{cat} of 15 s^{-1} and a catalytically relevant pK_a of 9.4 [2]. For both the biomimetic and the enzyme, a terminal hydroxide is therefore implicated as the catalytic nucleophile. Another mechanistic study of a Cd(II)-substituted phosphoesterase has been reported by Ely et al. [22]. The kinetically relevant pK_a ($\sim$4–5) for Cd_2-OpdA (organophosphate-degrading agent) in the hydrolysis of ethyl paraoxon suggested a bridging hydroxide as the nucleophile [22].

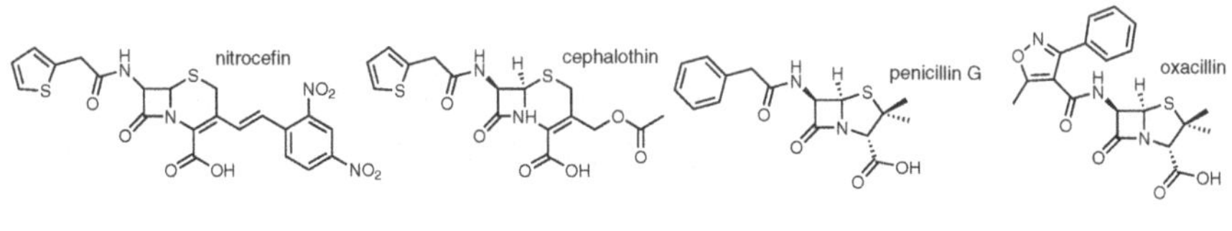

Fig. 5.2 Metallo-β-lactamase substrates

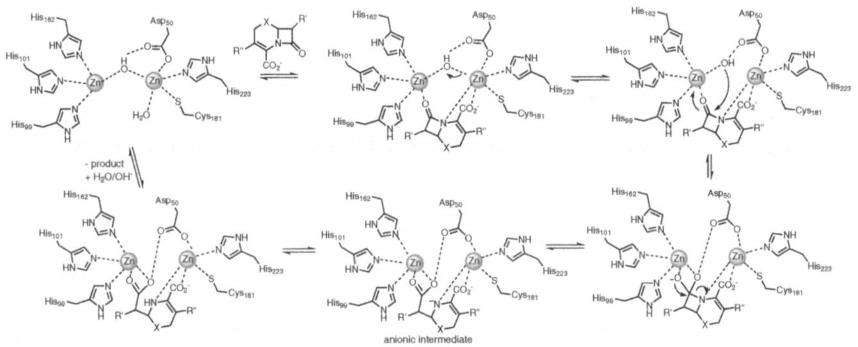

Fig. 5.3 The active site of the metallo-β-lactamase from *B. fragilis* and the proposed mechanism of cephalosporin hydrolysis [29]

5.1.3 Metallo-β-lactamase Enzymes and Related Biomimetics

An important class of hydrolytic metalloenzymes is the metallo-β-lactamase family. While class B β-lactamases (metallo-β-lactamases) are known to utilize Zn(II) ions in vivo to gain full catalytic activity some examples have been published where Cd(II) was used for crystallographic, spectroscopic and kinetic investigations [23–25]. Commonly used substrates for kinetic assays in MβL studies and some antibiotics are shown in Fig. 5.2. While nitrocefin is not used therapeutically it has been used extensively in MβL assays as it undergoes a distinctive color change upon hydrolysis which can be monitored spectrophotometrically [26]. Nitrocefin and cephalothin belong to the class of cephalosporin lactams with the latter being the first cephalosporin antibiotic to be commercially used [27, 28]. Another important class are the carbapenem antibiotics which differ in the ring system adjacent to the lactam ring and are not further discussed here. Oxacillin and penicillin G have been used therapeutically and in kinetic assays. Both have the penam ring, adjacent to the lactam cycle, in common [27].

The two metal ions in MβLs are usually bridged by a hydroxide and have a 4, 5 mixed coordination sphere. The mechanism proposed for dinuclear Zn(II) MβLs, based on UV-Vis stopped flow measurements, and model complex studies, is shown in Fig. 5.3 [29]. The first step in the catalytic cycle is the binding of the

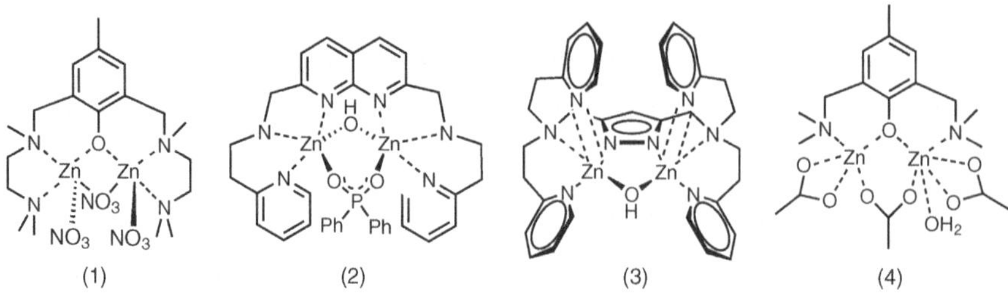

Fig. 5.4 Dinuclear MβL models with varying bridging and coordinating moieties [32, 37–39, 41]

cephalosporin substrate via its carboxylate and β-lactam carbonyl moiety. Subsequently, it is proposed that the bridging hydroxide shifts to a terminal position, oriented by a deprotonated aspartate, and attack of the β-lactam carbonyl follows [29, 30]. The tetrahedral intermediate collapses to produce an anionic species which has been observed for the nitrocefin hydrolysis by several MβLs and a Zn(II) model complex [26, 29, 31]. Protonation and release of the hydrolyzed substrate completes the catalytic cycle and an external water/hydroxide regenerates the active site.

Despite the valuable contributions model systems have made for the elucidation of the enzymes mechanism only few research groups have reported MβL models [32–38]. This is most likely due to the fact of the variety of different classes of MβL enzymes and a poor understanding of the action of catalysis. Some studies have been conducted to investigate both the MβL activity as well as phosphoesterase activity [36, 39, 40]. It was found that the nature of the nucleophile and the mechanism employed by these systems is greatly dependent on the ligand used [39, 41]. A selection of dinuclear MβL models is shown in Fig. 5.4. The complex (1) with a phenol based ligand from Kaminskaia et al. [41] hydrolyzed nitrocefin with $k_{cat} = 1.0 \times 10^{-2}\ min^{-1}$. The kinetically relevant pK_a was determined to be 7.5, typical for a bridging hydroxide. In complex (2) however, reported by the same group [39], the bridging hydroxide appeared to be acting as a base for an external water molecule, rather than being the nucleophile. The pK_a of 8.7 was thus assigned to a terminal water deprotonation. Interestingly the hydrolysis rate of complex (2) towards lactams was $3.4 \times 10^{-2}\ min^{-1}$, similar to complex (1). A range of complexes with pyrazolate based ligands such as (3) were reported by Meyer and co-workers [38, 32]. They investigated structure activity correlations in dinuclear zinc complexes. A comprehensive kinetic analysis of (3) was not conducted due to experimental limitations but a significant rate enhancement opposed to the hydrolysis of penicillin by free Zn(II) has been observed during in situ IR measurements [38]. Mugesh and co-workers used the substrate oxacillin and monitored hydrolysis by (4) with ^{1}H-NMR and reverse-phase HPLC [37]. A pK_a of around 8 for this system suggests the same nucleophile (terminal water) as in complex (2).

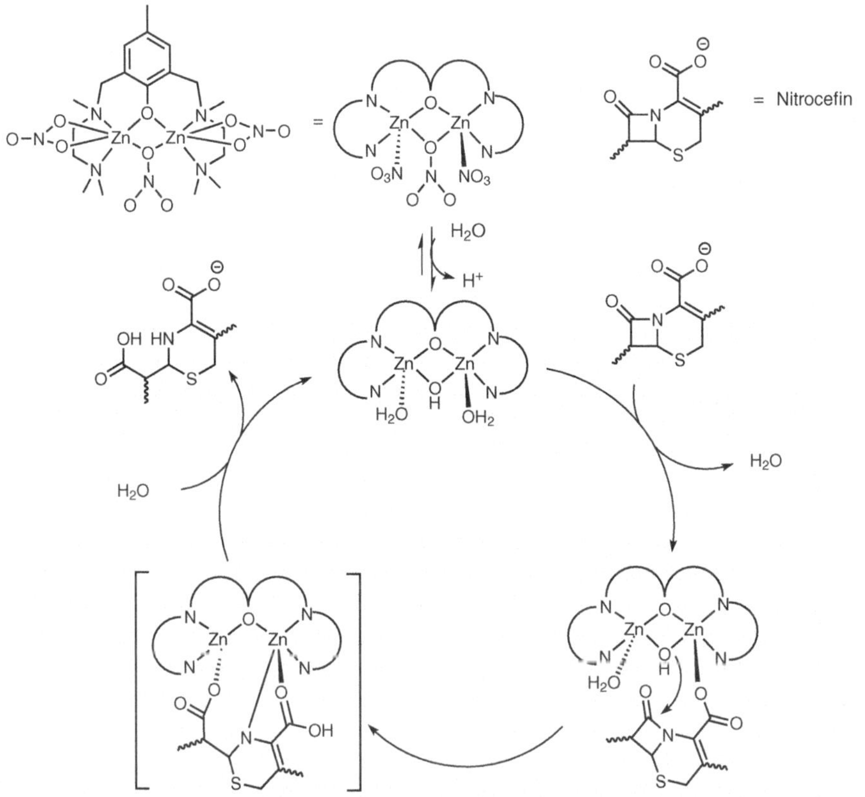

Fig. 5.5 Mechanism of nitrocefin cleavage by complex (1) where the bridging hydroxide acts as nucleophile [41]

A more detailed mechanistic description of the nitrocefin hydrolysis catalyzed by complex (1) is shown in Fig. 5.5 [41]. It has been found that an anchoring carboxylate group on the substrate is necessary for efficient hydrolysis. The binding of the carboxylate of nitrocefin comprises the first step in the mechanism and the optimal oriented substrate then undergoes (rate-limiting) nucleophilic attack of the bridging hydroxide [41]. Although coordination of the β-lactam carbonyl oxygen to Zn(II) was not detected, the authors state that this might be the case for the transition state [41]. After rate-limiting hydrolysis of the intermediate, cephalosporanoic acid is released and the catalyst is recovered.

A blue anionic intermediate has been observed by UV-Vis spectroscopy when the water levels of the assay were decreased to 20 % [26, 41]. The absorption maximum of the intermediate at 640 nm was similar as found previously in the MβL enzyme from *Bacillus fragilis* [26, 31, 42]. The blue color was proposed to arise from deprotonation of the amine nitrogen and resulting extended conjugation [26, 31, 42]. The species found in the enzyme catalyzed hydrolysis of nitrocefin

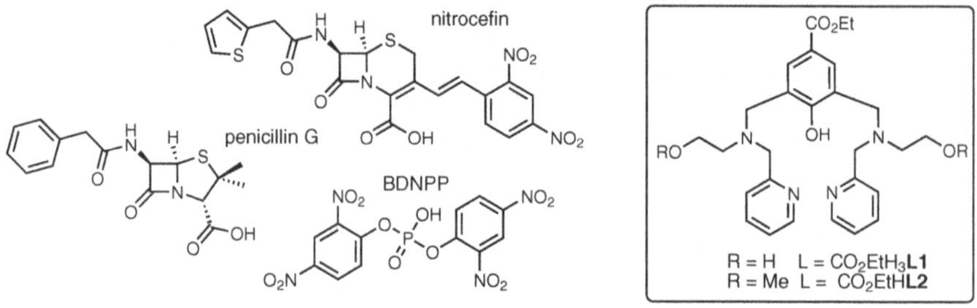

Fig. 5.6 Substrates used in this chapter for β-lactamase and phosphatase assays and the structures of the ligands CO_2EtH_3**L1** and CO_2EtH**L2** used to generate dinuclear Cd(II) complexes

absorbs at 665 nm and the difference in the absorption maximum may be due to influences of the surrounding enzyme pocket [41, 42]. Based on a recent publication by Dong et al. [43] (who had isolated a methyl parathion hydrolase (MPH) from *Pseudomonas* sp. which showed high similarity to the active site of a dinuclear metallo-β-lactamase), Mugesh et al. investigated the phosphotriesterase (PTE) activity of a metallo-β-lactamase and biomimetics [40]. The authors found that dinuclear Zn(II) complexes with good MβL activity were in fact poor functional mimics of PTE and vice versa [40]. Their observations also suggested that lactam substrates exhibit a different binding mode from phosphotriesters and furthermore that the hydrolysis product of the latter substrate (phosphodiesters) inhibit the PTE activity of lactamases and their mimics [40].

5.1.4 Target Ligands and Aims of This Chapter

To expand the study of Cd(II) model complexes this chapter reports the synthesis and structural characterization of a Cd(II) containing model system, based on the ligand CO_2EtH_3**L1** (ethyl 4-hydroxy-3,5-bis(((2-hydroxyethyl)(pyridin-2-yl-methyl)-amino)-methyl)-benzoate). The phosphoesterase-like activity of the complex and an investigation of its metallo-β-lactamase activity will be addressed. In addition a second complex was used in organophosphate and β-lactam hydrolysis to investigate the mechanism in the absence of alkoxide nucleophiles utilizing the ligand (CO_2EtH**L2**). The two ligands and the substrates employed are shown in Fig. 5.6.[1]

[1] Parts of this Chapter have been reprinted with permission from (L.J. Daumann et al., "*Cadmium(II) Complexes: Mimics of Organophosphate Pesticide Degrading Enzymes and Metallo-β-lactamases*" Inorg. Chem. **2012**, 51, 7669–7681). Copyright (2014) American Chemical Society.

5.2 Results

5.2.1 *Syntheses of the Complexes with the Ligands CO_2EtHL2 and CO_2EtH_3L1*

The complexes were synthesized by adding two equivalents of cadmium(II) acetate to a solution of the respective ligand in methanol and refluxing the mixture for 30 min (see Chap. 2). The complex of CO_2EtH_3**L1** was characterized as $[Cd_4(CO_2EtH_2\mathbf{L1})_2(CH_3COO)_{3.75}\ Cl_{0.25}(H_2O)_2]\text{-}(PF_6)_2$ indicating that a single deprotonation (of the phenolic oxygen) had occurred. The complex of the second ligand CO_2EtH**L2** could not be isolated as a solid and was thus generated in situ for kinetic and mass spectrometry experiments.

5.2.2 *Solid State Structure of $[Cd_4(CO_2EtH_2L1)_2(CH_3COO)_{3.75}Cl_{0.25}(H_2O)_2](PF_6)_2$*

X-ray quality crystals of $[Cd_4(CO_2EtH_2\mathbf{L1})_2(CH_3COO)_{3.75}Cl_{0.25}(H_2O)_2](PF_6)_2$ were isolated after combining of methanolic solutions of cadmium(II) acetate and CO_2EtH_3**L1** in the presence of sodium acetate and sodium hexafluorophosphate and subsequent slow evaporation of the solvent. The structure and atomic numbering scheme are illustrated in Fig. 5.7, crystal data are shown in Table 5.1 and selected bond lengths and angles are shown in Tables 5.2 and 5.3.

The complex crystallized as a dimer of dimer structure with disorder around the chloride and acetate ligands. The percentage of chloride in the complex was determined with microanalysis and refined in the crystal structure to be 0.25 equivalents (see Chap. 2). The dimers are connected through the carbonyl oxygen donor of the ethyl ester bonded to one of the two Cd(II) ions bridged by the $CO_2EtH_2\mathbf{L1}^-$ ligand. The dimers are therefore connected intermolecularly "head to tail" through the carbonyl oxygen donors. The dimeric unit has two Cd(II) ions coordinated by one ligand molecule. The seven coordinate Cd(1) has a distorted pentagonal bipyramidyl geometry whilst Cd(2) has a distorted six coordinate geometry. One formally bidentate acetate ligand is coordinated to Cd(1) with Cd(2) coordinated by disordered chloride and acetate ligands. The coordination sphere of Cd(1) is composed of two nitrogen donors (Cd(1)-N(1), 2.411(5) Å and Cd(1)-N(2), 2.319(5) Å), the protonated alcohol (Cd(1)-O(2), 2.475(4) Å), a water molecule (Cd(1)-O(8), 2.273(4) Å) a bidentate acetate ligand (Cd(1)-O(4), 2.384(4) Å and Cd(1)-2.428(5) Å), the phenoxide oxygen bridges both Cd(II) ions with (Cd(1)-O(1), 2.367(3) Å, Cd(2)-O(1), 2.252(3) Å), Cd(1)-O(1)-Cd(2) 128. 62(16)° and Cd(1)···Cd(2) 4.162 Å. For the second metal ion, the coordination sphere is composed of two nitrogen donors (Cd(2)-N(3), 2.367(4) Å and Cd(2)-N(4), 2.376(5) Å), the protonated alcohol (Cd(2)-O(3), 2.366(4) Å) and a

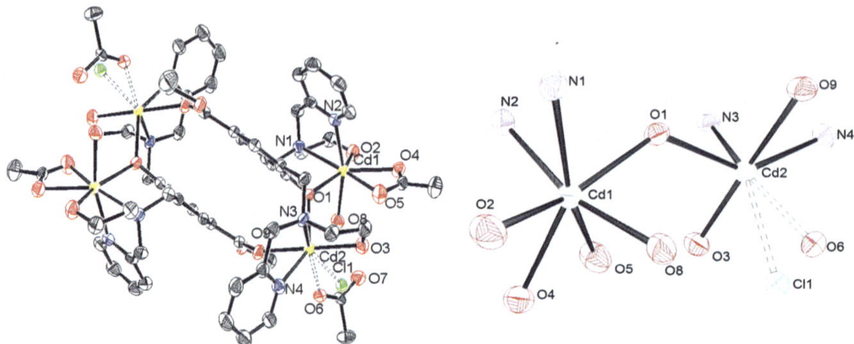

Fig. 5.7 Structure of the tetranuclear complex $[Cd_4(CO_2EtH_2\mathbf{L1})_2(CH_3COO)_{3.75}Cl_{0.25}(H_2O)_2](PF_6)_2$ and a view of the first coordination sphere. The disordered coordinating acetate oxygen and the chloride are shown with dashed bonds. Counter ions and hydrogen atoms have been omitted for clarity (25 % ellipsoid probability)

Table 5.1 Crystal data and refinement details for $[Cd_4(CO_2EtH_2\mathbf{L1})_2(CH_3COO)_{3.75}Cl_{0.25}(H_2O)_2](PF_6)_2$

Empirical formula	$C_{29}H_{37}Cd(2)F_6N_4O_7P$
Formula weight	923.40
Crystal system	Triclinic
Space group	P-1
Unit cell dimensions	a = 11.7360(7) Å α = 117.110(10)°
	b = 13.6870(10) Å β = 105.594(8)°
	c = 14.233(2) Å γ = 96.651(7)°
V (Å^3)	1882.9(3)
Z	2
Temperature (K)	293(2)
ρ_{calcd} (g cm^{-3})	1.629
Wavelength	1.5418 (Cu K_α)
Abs. coeff. μ (mm^{-1})	10.125
F(000)	920
No. reflns measured	14,646
No. Independent reflns.	5,615
No. obs	4,013
θ	3.73–60.56
Refinement method	Full-matrix least-squares on F^2
Goodness of fit on F^2	0.953
Final R indices ($I > 2\sigma(I)$)	R1 = 0.0364, wR2 = 0.0928
R indices (all data)	R1 = 0.0555, wR2 = 0.0995

disordered chloride (Cd(2)-Cl(1), 2.638(14) Å; 0.18) and monodentate acetate (Cd(2)-O(6) 2.180(5) Å; 0.82). The coordination sphere of Cd(2) is completed by an interaction with the carbonyl oxygen atom of the neighboring dimer (Cd(2)-O(9), 2.413(4) Å) resulting in the dimer of dimer configuration. For both

Table 5.2 Selected bond lengths (Å) for $[Cd_4(CO_2EtH_2\mathbf{L1})_2(CH_3COO)_{3.75}Cl_{0.25}(H_2O)_2](PF_6)_2$

N(1)-Cd(1)	2.411(5)	N(2)-Cd(1)	2.319(5)	N(3)-Cd(2)	2.367(4)
N(4)-Cd(2)	2.376(5)	O(1)-Cd(2)	2.252(3)	O(1)-Cd(1)	2.367(3)
O(2)-Cd(1)	2.475(4)	O(3)-Cd(2)	2.366(4)	O(4)-Cd(1)	2.384(4)
O(5)-Cd(1)	2.428(5)	O(8)-Cd(1)	2.273(4)	O(9)-Cd(2)	2.413(4)
O(6)-Cd(2)	2.180(5)	Cd(2)-Cl(1)	2.638(14)	Cd(1)-Cd(2)	4.162

Table 5.3 Selected angles (°) for $[Cd_4(CO_2EtH_2\mathbf{L1})_2(CH_3COO)_{3.75}Cl_{0.25}(H_2O)_2](PF_6)_2$

Cd(2)-O(1)-Cd(1)	128.62(15)	O(8)-Cd(1)-N(2)	164.24(15)
O(8)-Cd(1)-O(1)	82.88(13)	N(2)-Cd(1)-O(1)	84.67(14)
O(8)-Cd(1)-O(4)	97.22(15)	N(2)-Cd(1)-O(4)	98.22(16)
O(1)-Cd(1)-O(4)	130.62(14)	O(8)-Cd(1)-N(1)	95.60(15)
N(2)-Cd(1)-N(1)	73.42(16)	O(1)-Cd(1)-N(1)	83.79(13)
O(4)-Cd(1)-N(1)	144.48(15)	O(8)-Cd(1)-O(5)	93.21(17)
N(2)-Cd(1)-O(5)	93.66(18)	O(1)-Cd(1)-O(5)	78.17(14)
O(4)-Cd(1)-O(5)	52.47(15)	N(1)-Cd(1)-O(5)	158.75(16)
O(8)-Cd(1)-O(2)	82.87(15)	N(2)-Cd(1)-O(2)	103.20(16)
O(1)-Cd(1)-O(2)	148.57(14)	O(4)-Cd(1)-O(2)	78.98(15)
N(1)-Cd(1)-O(2)	69.93(15)	O(5)-Cd(1)-O(2)	130.51(15)
O(6)-Cd(2)-O(1)	125.01(16)	O(6)-Cd(2)-O(3)	92.28(17)
O(1)-Cd(2)-O(3)	95.70(14)	O(6)-Cd(2)-N(3)	148.49(18)
O(1)-Cd(2)-N(3)	85.29(14)	O(3)-Cd(2)-N(3)	74.43(14)
O(6)-Cd(2)-N(4)	86.56(19)	O(1)-Cd(2)-N(4)	140.37(15)
O(3)-Cd(2)-N(4)	107.36(16)	N(3)-Cd(2)-N(4)	71.06(16)
O(6)-Cd(2)-O(9)	86.51(16)	O(1)-Cd(2)-O(9)	83.12(14)
O(3)-Cd(2)-O(9)	177.39(14)	N(3)-Cd(2)-O(9)	107.74(14)
N(4)-Cd(2)-O(9)	74.90(16)	O(6)-Cd(2)-Cl(1)	26.1(3)
O(1)-Cd(2)-Cl(1)	104.5(3)	O(3)-Cd(2)-Cl(1)	77.8(3)
N(3)-Cd(2)-Cl(1)	151.3(3)	N(4)-Cd(2)-Cl(1)	111.5(3)

Cd(1) and Cd(2), the Cd-O bond lengths from the pendant alcohol appear typical as do the Cd-N bond lengths. The alcohol arm of the ligand is protonated and the charge is balanced by two PF_6^- anions.

5.2.3 *Infrared Spectrum of $[Cd_4(CO_2EtH_2L1)_2(CH_3COO)_{3.75}Cl_{0.25}(H_2O)_2](PF_6)_2$*

The infrared spectrum of $[Cd_4(CO_2EtH_2L1)_2(CH_3COO)_{3.75}Cl_{0.25}(H_2O)_2](PF_6)_2$ supports the findings in the crystal structure and features, apart from typical ligand and hexafluorophosphate bands, distinct carbonyl stretches from the ethyl ester of the ligand and the acetate groups. The ester carbonyl stretch at 1,670 cm^{-1} is at

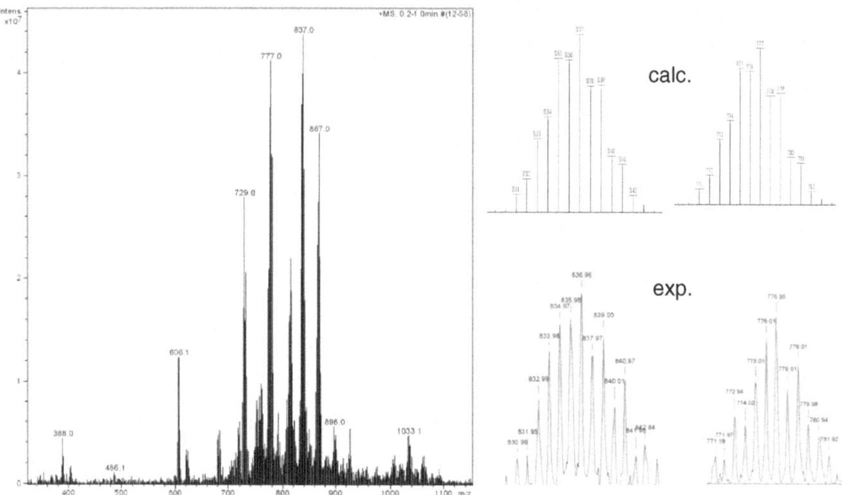

Fig. 5.8 Mass spectrum of $[Cd_2(CO_2EtH_2\mathbf{L1})(CH_3COO)_2]^+$ in methanol (*left*) and calculated splitting pattern (*top right*) of two $Cd(II)_2$ species and recorded splitting pattern (*below right*)

lower frequency than observed for the free ligand (1,704 cm^{-1}) attributed to the weakening of the C=O bond upon coordination to Cd(II) [44]. The asymmetric and symmetric bands from unidentate acetato ligand were observed at 1,604 and 1,373 cm^{-1}, respectively, with the $\Delta\nu_{asym\text{-}sym} = 231\ cm^{-1}$ [44]. Infrared bands from the bidentate non-bridging acetate ligands were assigned to 1,547 and 1,413 cm^{-1}, respectively, with $\Delta\nu_{asym\text{-}sym} = 134\ cm^{-1}$ [44].

5.2.4 Mass Spectrometry of $[Cd_4(CO_2EtH_2L1)_2(CH_3COO)_{3.75}Cl_{0.25}(H_2O)_2](PF_6)_2$ and $[Cd_2(CO_2EtL2)(CH_3COO)_2]^+$

The mass spectrum of the complex with $CO_2EtH_3\mathbf{L1}$ in methanol displays an isotope pattern typical of a $Cd(II)_2$ species, the most prominent pattern assigned to the species $[C_{31}H_{39}Cd_2N_4O_9]^+$ (*m/z* found 837.0, calc. *m/z* 837.08 (100 %)) assigned as $[Cd_2(CO_2EtH_2\mathbf{L1})(CH_3COO)_2]^+$ (Fig. 5.8). Loss of one acetate results in $[Cd_2(CO_2EtH_2\mathbf{L1})(CH_3COO)\text{-}H]^+$ (found *m/z* 777.0, calc. *m/z* 777.1 (100 %)) whilst the coordination of methanol results in $[Cd_2(CO_2EtH_2\mathbf{L1})(CH_3COO)_2(CH_3OH)]^+$ (found *m/z* 867.0, calc. *m/z* 868.11 (100 %), 867.1 (66 %)). A half mass peak at 605.1 (calc. *m/z* 606.1 (100 %), 605.2 (83 %)) is attributed to a complex formed by two ligands and two Cd(II) ions. Interestingly the base peak of the spectrum measured in acetonitrile/water 1:1 corresponds to doubly positive charged complex formed by two ligands and two Cd(II) ions ($[Cd_2(CO_2EtH_2\mathbf{L1})_2]^{2+}$; found *m/z* 605.1, calc. *m/z* 606.1 (100 %), 605.2 (82.8 %)).

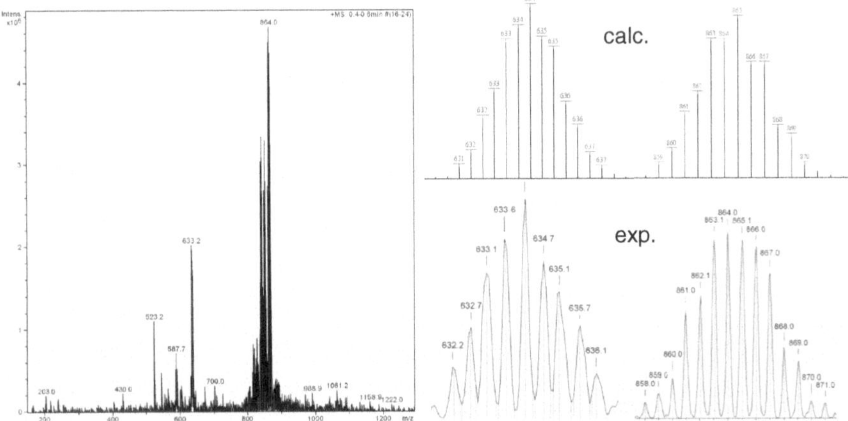

Fig. 5.9 Mass spectra of $[Cd_2(CO_2Et\mathbf{L2})(CH_3COO)_2]^+$ in methanol (*left*) (*right*, *inset* recorded splitting pattern (*black*) of the Cd_2 species at m/z = 634.2 and calculated (*blue*))

The mass spectrum of the complex with $CO_2EtH\mathbf{L2}$ in methanol showed the typical $Cd(II)_2$ pattern and the peak at *m/z* 864.0 (100 %), 863.0 (90 %), calc. *m/z* 865.1 (100 %), 863.1 (88 %), 864.11 (87 %) was attributed to a $[Cd_2(CO_2Et\mathbf{L2})(CH_3COO)_2]^+$ species (Fig. 5.9, left). The base peak of the spectrum in acetonitrile/water 1:1 corresponds to a doubly positive charged complex formed by two ligands and two Cd(II) ions ($[Cd_2(CO_2Et\mathbf{L2})_2]^{2+}$; found m/z 634.2, calc. *m/z* 634.2 (100 %), 633.68 (92 %)).

5.2.5 Binding of Organophosphates to $[Cd_2(CO_2EtH_2L1)(CH_3COO)_2]^+$ in Solution

To monitor the substrate binding and species present during catalysis, mass spectrometry and ^{113}Cd NMR spectroscopy were conducted with mixtures of the complex with phosphoester substrates (Fig. 5.10).

After addition of $[Cd_2(CO_2EtH_2\mathbf{L1})(CH_3COO)_2]^+$ to a solution of PNPP (a product mimic) in acetonitrile, the mass spectrum of the mixture shows free ligand and a peak at *m/z* 936.0 arising from $[Cd_2(CO_2EtH_2\mathbf{L1})(PNPP)]^+$ (calc. *m/z* (100 %) 936.0), in addition to the *m/z* 605.0 peak attributed to $[C_{54}H_{66}Cd_2N_8O_{10}]^{2+}$ suggesting that this latter peak is not representative of the catalytically active species but rather a species produced under the conditions of the mass spectrometer (Fig. 5.10).

Using diphenyl phosphate (DPP), a substrate mimic lacking the activating nitro groups and is therefore not hydrolyzed by the model complex, under the same experimental conditions resulted in a peak at *m/z* 1215.9 assigned to

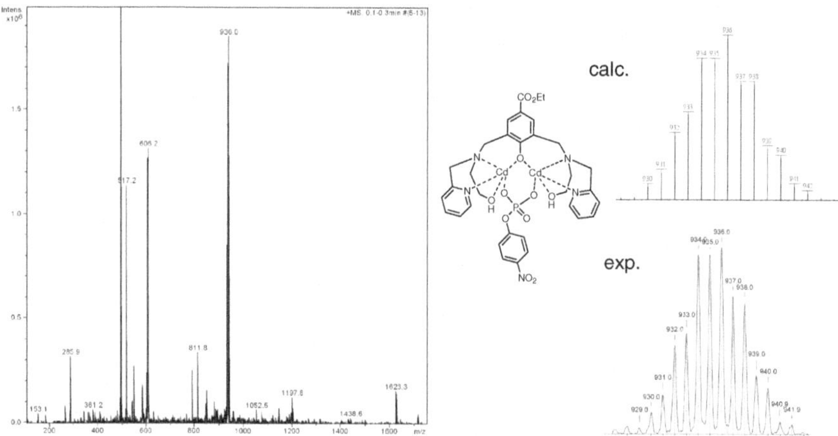

Fig. 5.10 Mass spectrum of $[Cd_2(CO_2EtH_2\mathbf{L1})(CH_3COO)_2]^+$ in acetonitrile/water 1:1 in the presence of excess PNPP, structure of the major species $[Cd_2(CO_2EtH_2\mathbf{L1})(PNPP)]^{+}$ recorded and calculated splitting pattern on the *right*

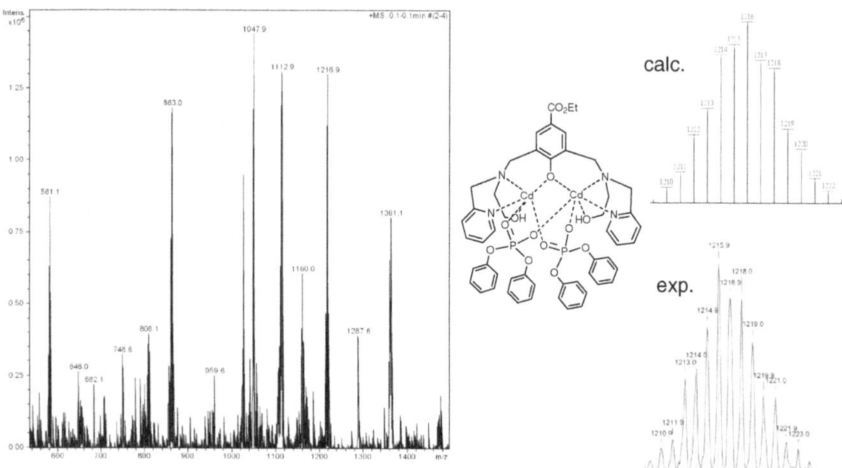

Fig. 5.11 Mass spectrum of $[Cd_2(CO_2EtH_2\mathbf{L1})(CH_3COO)_2]^+$ in acetonitrile/water 1:1 in the presence of excess DPP, major species $[Cd_2(CO_2EtH_2\mathbf{L1})(DPP)_2]^+$, recorded and calculated splitting pattern on the *right*

$[Cd_2(CO_2EtH_2\mathbf{L1})(DPP)_2]^+$ (Fig. 5.11, calc. *m/z* 1216.1 (100 %), 1215.1 (88.6 %)). The fact that the $[Cd_2(CO_2EtH_2\mathbf{L1})]^{2+}$ species was not observed suggests that the catalytically active species is comprised of one ligand and two Cd(II) ions, with the acetates readily replaced by substrate and solvent molecules/nucleophiles.

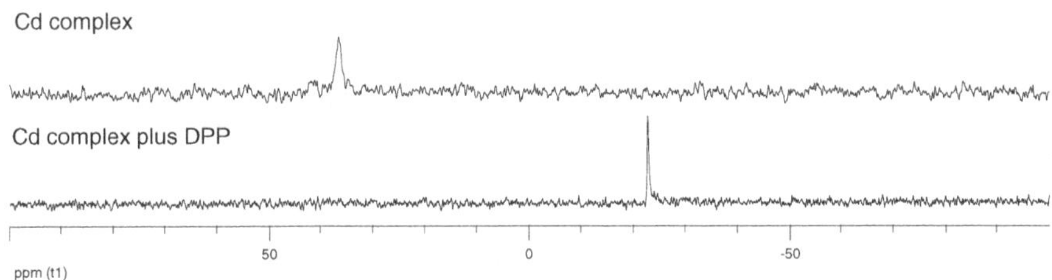

Fig. 5.12 ^{113}Cd NMR of $[Cd_2(CO_2EtH_2\mathbf{L1})(CH_3COO)_2]^+$ in CD_3CN/D_2O 1:1 in the absence (*top*) and presence (*bottom*) of a substrate mimic

^{113}Cd NMR was employed to investigate the coordination environment of the Cd(II) ions in the complex. Only one resonance at −36.4 ppm was detected, suggesting that the two Cd(II) ions have the same coordination environment in solution. When ten equivalents of the substrate analog DPP were added to the complex, structural changes influencing the chemical shift were apparent in the NMR spectrum. The ^{113}Cd resonance of the complex was shifted to −22.6 ppm (Fig. 5.12). It is proposed that the substrate is replacing acetonitrile solvent molecules thus reducing the number of nitrogen donors bound to Cd(II). The chemical shifts of Cd(II) are known to be very sensitive to the chemical environment [5]. The fact that only one signal is observed suggests a bidentately, symmetrically bound phosphodiester.

5.2.6 Binding of β-Lactams to $[Cd_2(CO_2EtH_2L1)(CH_3COO)_2]^+$ in Solution

UV-Vis spectroscopy, in solution IR measurements, ^{13}C NMR and mass spectrometry have been employed to investigate the course of lactam binding/hydrolysis to/by $[Cd_2(CO_2EtH_2\mathbf{L1})(CH_3COO)_2]^+$. β-lactams can bind to the metal complexes through the carboxylate residue, β-lactam amide, sulfur atom or terminal amides [41, 45]. In order to investigate which group is involved in metal binding of nitrocefin in solution IR measurements were conducted. No change of carboxylate carbonyl absorption was observed in this solvent system upon addition of $[Cd_2(CO_2EtH_2\mathbf{L1})(CH_3COO)_2]^+$. Usually a shift of the carbonyl absorption of 20–50 cm^{-1} to lower wavenumbers would be expected upon coordination of the lactam to the metal ion [41]. Meyer et al. reported an infrared experiment where only in the presence of TRIS buffer this shift was observed in penicillin hydrolysis by a dinuclear Zn(II) complex, however not in a DMSO/water mixture [38]. The investigation of hydrolysis of cephalothin, a more 'nitrocefin-like' substrate, in pure DMSO by Kaminskaia et al. showed that no significant change of the β-lactam carbonyl absorption was observed in IR measurements suggesting that

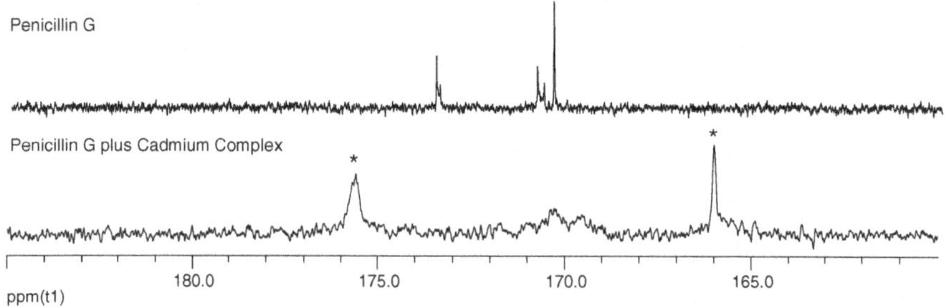

Fig. 5.13 Carbonyl region of the ^{13}C NMR spectrum of penicillin G in d^6-DMSO/d^6-acetone (1:1) with and without one equivalent of $[Cd_2(CO_2EtH_2\mathbf{L1})(CH_3COO)_2]^+$, signals due to the ester and acetate from the complex are marked with an *asterisk*

this substrate was binding solely via its carboxylate group to transition metals [39]. In a more recent study the authors suggest that nitrocefin, penicillin G and cephalothin, despite their different structures, bind similarly to metal complexes [26]. It should be noted that in the experiment with $[Cd_2(CO_2EtH_2\mathbf{L1})(CH_3COO)_2]^+$ a characteristic color change from yellow to orange was observed, indicating hydrolysis of the nitrocefin substrate. Changes in the IR spectrum, however, could only be visualized in the difference spectra as the amide and carboxylate bands overlap. During the course of nitrocefin hydrolysis by $[Cd_2(CO_2EtH_2\mathbf{L1})(CH_3COO)_2]^+$ the band at 1,673 cm^{-1} which was attributed to the lactam carbonyl group disappeared.

Another way to investigate the binding of lactam substrates to metal complexes is ^{13}C NMR. In an attempt to monitor the binding of the carboxylate group of penicillin (this substrate was used due to the availability of reference values in the literature) [41] to $[Cd_2(CO_2EtH_2\mathbf{L1})(CH_3COO)_2]^+$ equimolar amounts of penicillin and complex were mixed in DMSO/acetone and after 1 day the ^{13}C NMR recorded. Intriguingly, the lactam carbonyl resonance of penicillin (173.5 ppm) had disappeared suggesting that the complex did cleave the lactam bond of this substrate under these (non-buffered, water-free) conditions. In addition an additional resonance had appeared between 169.1 and 171.7 ppm. This signal did not arise from an acetate or ethyl ester of the complex (by comparison to the free complex spectrum). The chemical shift is however indicative for a newly formed ester moiety and not for a carboxylic acid (Fig. 5.13). The spectrum exhibited a large signal to noise ratio and thus the assignment of this signal is not unambiguous. A statement about the carboxylate shift of the substrate upon metal binding is therefore not made here.

A blue intermediate was observed in the UV-Vis spectrum when $[Cd_2(CO_2EtH_2\mathbf{L1})\text{-}(CH_3COO)_2]^+$ was added to a solution of nitrocefin in dry acetonitrile (Fig. 5.14). The absorption maximum at 640 nm was similar as found in a previously reported model system and the MβL enzyme from *Bacillus fragilis* [26, 31, 42]. The blue color is suggested to arise from deprotonation of the amine nitrogen and resulting extended conjugation [26, 31]. Protonation of the

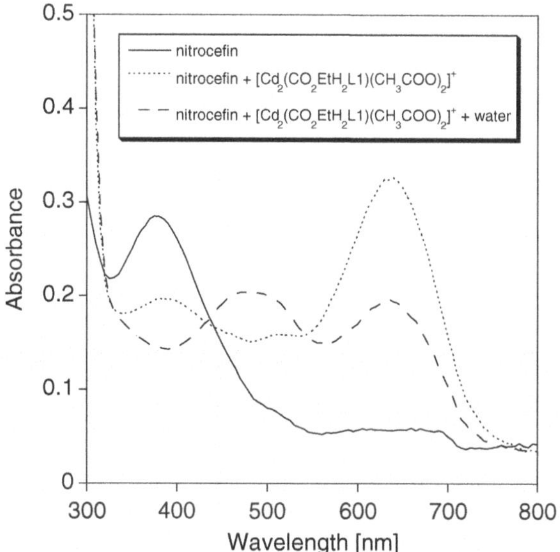

Fig. 5.14 UV-Vis spectra that show the interconversion of nitrocefin into the blue species and hydrolyzed nitrocefin

carboxylate groups does not affect the UV-Vis spectrum [26]. Upon addition of water however, the blue intermediate was converted by protonation into hydrolyzed nitrocefin as shown in the spectra in Fig. 5.14.

This intermediate was further analyzed by mass spectrometry. A peak at *m/z* 1313.8 is attributed to a species comprised of one ligand, two Cd(II) ions, two acetonitrile solvent molecules and the blue intermediate covalently bound to the complex via the alcohol arm of the ligand (calc. *m/z* 1315.1 (100 %), 1314.1 (84 %), proposed structure and spectrum shown in Fig. 5.15). A peak at *m/z* 1748.7 was assigned to a similar species as above but with an additional not hydrolyzed nitrocefin bound and no solvent molecules coordinated (calc. *m/z* 1749.1 (100 %), 1748.1 (80 %)). Further addition of nitrocefin to the mass spectrum sample caused the latter peak to become the major peak in the spectrum. The finding of a covalently bound species and an ester rather than a carboxylic acid as hydrolysis product supports the results of the ^{13}C NMR experiments conducted with the substrate penicillin. Interestingly a blue species was not observed when $[Cd_2(CO_2Et\mathbf{L2})(CH_3COO)_2]^+$ was added to a solution of nitrocefin in dry acetonitrile. However, complex with one or two intact nitrocefin molecules bound could be detected in the mass spectrum (found *m/z* 1320.9 and 1776.7).

5.2.7 Metallo-β-lactamase-like Activity

Nitrocefin (Fig. 5.2) was employed as substrate for the metallo-β-lactamase assays as the hydrolysis can be conveniently monitored by UV-Vis at 390 nm. The pH-dependence data of the hydrolytic activity of $[Cd_2(CO_2EtH_2\mathbf{L1})(CH_3COO)_2]^+$

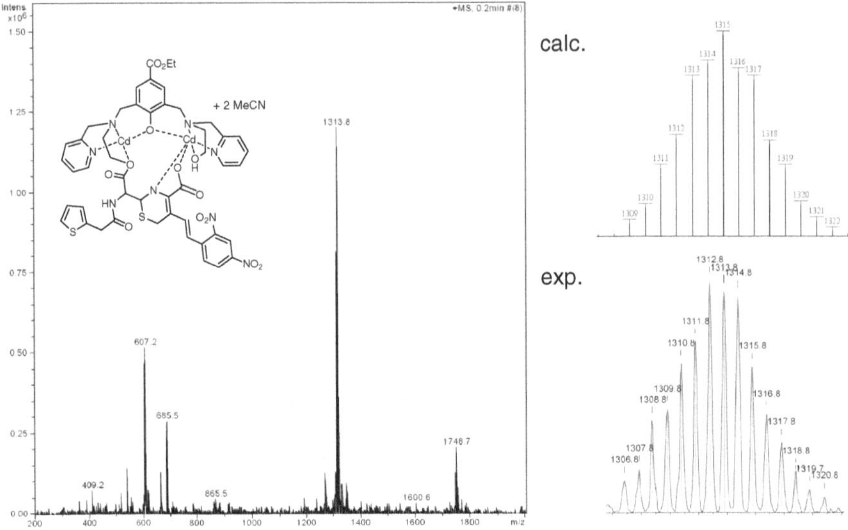

Fig. 5.15 Mass spectrum of the blue intermediate measured in dry acetonitrile, *right* calculated splitting pattern (*top*) and recorded (*bottom*)

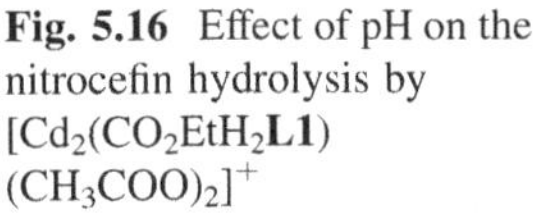

Fig. 5.16 Effect of pH on the nitrocefin hydrolysis by $[Cd_2(CO_2EtH_2\mathbf{L1})(CH_3COO)_2]^+$

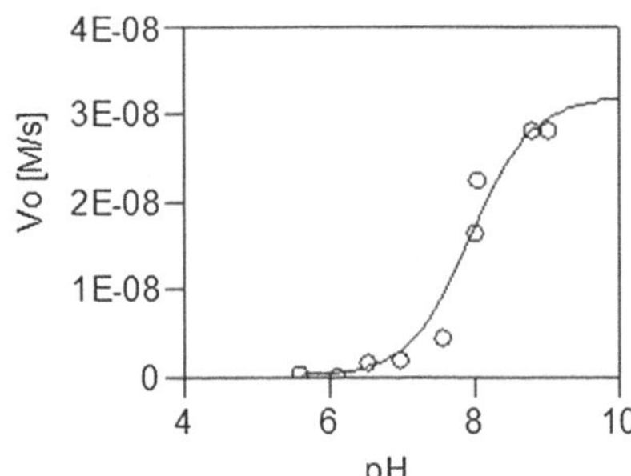

were fitted to Eq. 2.3 (see Chap. 2) and displayed one protonation event important for hydrolysis (Fig. 5.16). The pK_a of 7.95 ± 0.12 is significantly lower than for the OP-hydrolysis.

The substrate dependence was conducted at pH 8 and at physiological temperature (37 °C) and resulted in $V_{max} = 6.97 \times 10^{-8}$, with $k_{cat} = 1.39 \times 10^{-2}\ s^{-1}$ and $K_M = 0.11$ mM (Fig. 5.17).

Substrate concentrations above 0.06 mM resulted in significantly decreased activity (substrate inhibition). It was in fact shown with UV-Vis titration experiments that after two equivalents of nitrocefin, the third equivalent is only hydrolyzed very slowly and not to completion. In Fig. 5.19 the UV-Vis spectra are shown for the first 10 min of the reaction with a total of one or two equivalents nitrocefin, respectively. Interestingly the first equivalent is hydrolyzed quickly and whereas the fourth equivalent is barely hydrolyzed, with rates similar to auto hydrolysis.

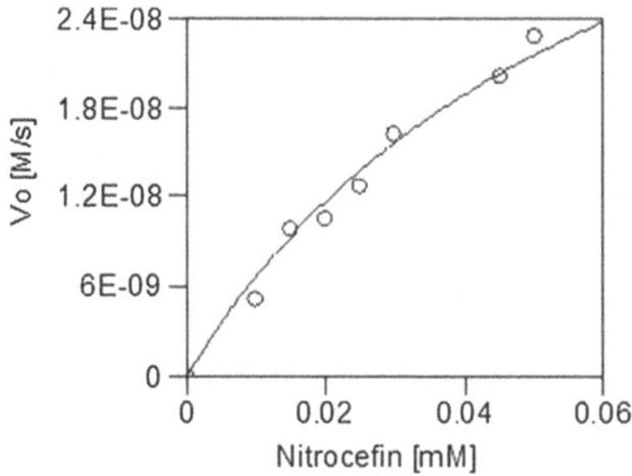

Fig. 5.17 Effect of concentration of nitrocefin on its hydrolysis by $[Cd_2(CO_2EtH_2\mathbf{L1})(CH_3COO)_2]^+$

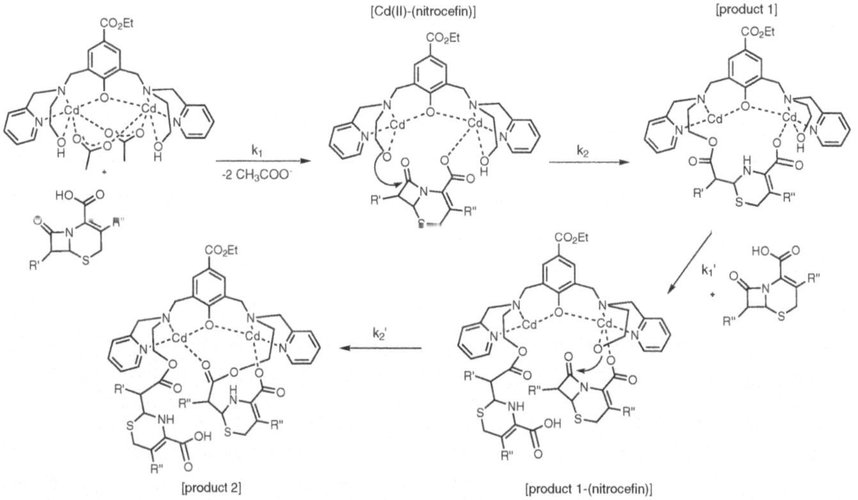

Fig. 5.18 Proposed course of hydrolysis of two nitrocefin molecules by $[Cd_2(CO_2EtH_2L1)(CH_3COO)_2]^+$

The spectral changes for the first and second equivalent of substrate were fitted separately to a model involving two consecutive reactions (Fig. 5.18), where the [Cd(II)-(nitrocefin)] species represent a bound substrate complex and [product 1] represents a complex with the substrate hydrolyzed but still coordinated to the metal ion (product inhibition, Fig. 5.18). The software ReactLab was used to fit the data [47]. Analysis of the data (300–700 nm; [complex] 0.05 mM, [substrate] 0.05 mM; pH 8) based on this model resulted in $k_1 = 4.8 \times 10^3 \pm 8.7\ M^{-1}\ min^{-1}$, $k_2 = 4.3 \times 10^{-2} \pm 1.4 \times 10^{-4}\ min^{-1}$ and $k_1' = 1.5 \times 10^3 \pm 9.5\ M^{-1}\ min^{-1}$, $k_2' = 6.1 \times 10^{-2} \pm 3 \times 10^{-4}\ min^{-1}$ [47]. The data predict that the addition of the second substrate molecule is less facile and k_2 and k_2' are of the same order as expected if the reaction represents the hydrolysis of the lactam ring by the

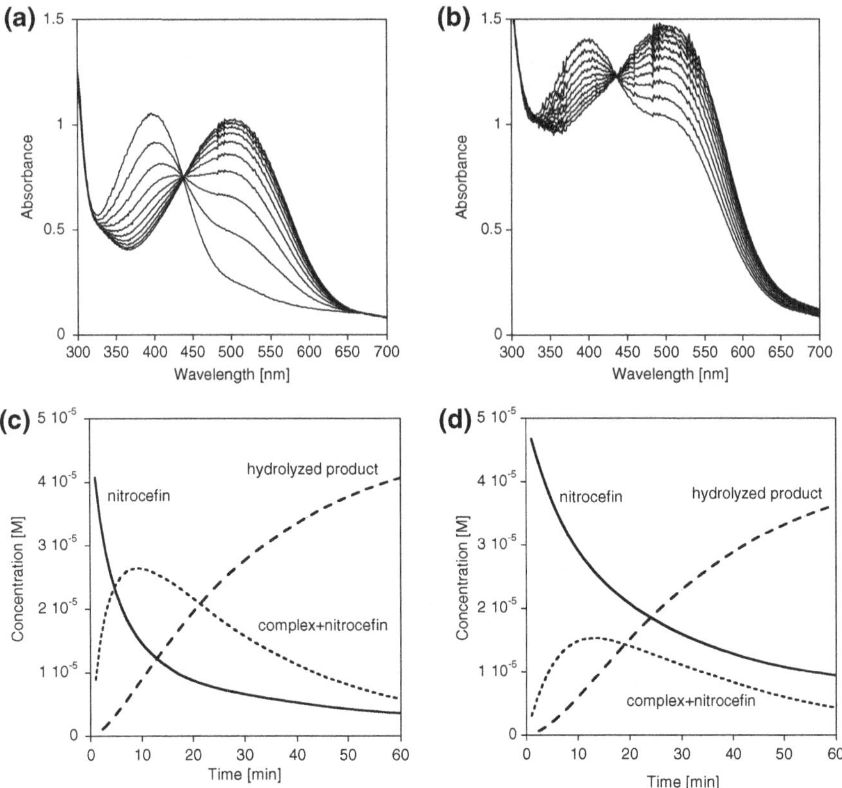

Fig. 5.19 UV-vis spectra of the hydrolysis of successively one (**a**) and two equivalents (**b**) of nitrocefin by $[Cd_2(CO_2EtH_2L1)(CH_3COO)_2]^+$; pH 8, buffer:MeCN 1:1 (first 10 min of the reaction are shown, initial concentration of complex and nitrocefin were 0.05 mM; spectra recorded at 1 min intervals); 37 °C (**c**) species distribution for first equivalent (**d**) species distribution for second equivalent

coordinated nucleophile. Addition of, successively, a third and fourth equivalent of nitrocefin resulted in very slow spectral changes, indistinguishable from those observed for the auto hydrolysis reaction. The complex with the ether arm ligand $[Cd_2(CO_2Et\mathbf{L2})\text{-}(CH_3COO)_2]^+$ showed no activity in the hydrolysis of nitrocefin over the pH range studied (pH 4.5–10).

5.2.8 Phosphatase-like Activity

The phosphatase-like activity promoted by the complex $[Cd_2(CO_2EtH_2\mathbf{L1})(CH_3COO)_2]^+$ in 50:50 acetonitrile/water was measured at pH 10.4 with a BDNPP concentration of 5 mM. The [complex] dependence was linear in the region 0.005–0.05 mM. Dependence of pH on activity was measured from pH 7–10.7

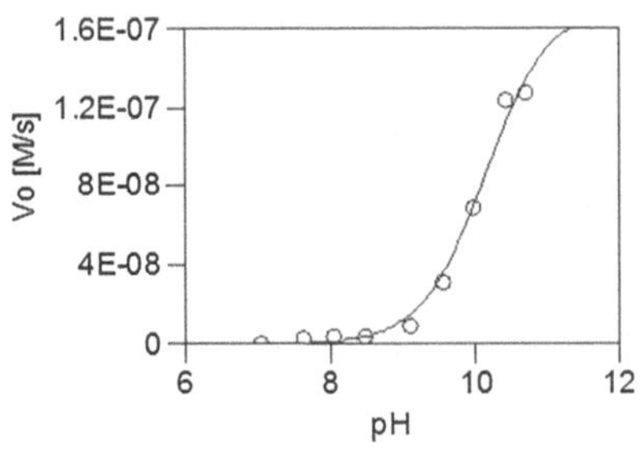

Fig. 5.20 Effect of the solution pH on BDNPP hydrolysis rate by $[Cd_2(CO_2EtH_2\mathbf{L1})(CH_3COO)_2]^+$

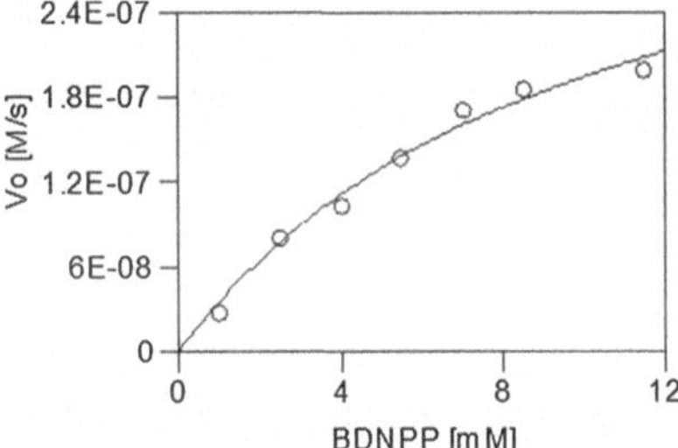

Fig. 5.21 Effect of BDNPP concentration on the rate of its hydrolysis by $[Cd_2(CO_2EtH_2\mathbf{L1})(CH_3COO)_2]^+$

resulting in a typical monoprotic curve exhibiting a limiting rate at high pH (Fig. 5.20). The data were fit to Eq. 2.3 (Chap. 2). The fit resulted in $pK_{a1} = 10.11$ and $v_{max} = 1.71 \times 10^{-7}\ M\ s^{-1}$.

The initial rate of BDNPP cleavage, determined at pH 10.4 as a function of substrate concentration, reveals typical saturation behavior (Fig. 5.21). The data were fitted by non-linear regression, using the Michaelis-Menten equation (Eq. 2.5, Chap. 2). Here $K_m = 9.4 \pm 2.1$ mM, $v_{max} = 3.76 \times 10^{-7}\ M\ s^{-1}$, with $k_{cat} = 9.4 \pm 0.2 \times 10^{-3}\ s^{-1}$.

For comparative purposes the pH dependence was also measured with the in situ generated complex $[Cd_2(CO_2Et\mathbf{L2})(CH_3COO)_2]^+$ and resulted in a pK_a of ~8.7, different from the complex with the alcohol arm ligand $EtCO_2H_3\mathbf{L1}$. Additionally an ^{18}O-labeling experiment was conducted to investigate the nature of the nucleophile in phosphoester hydrolysis. ^{31}P NMR can be used to determine the extent of incorporation of ^{18}O-labeled water/hydroxide into the hydrolysis product of BDNPP as phosphorus signals display a high sensitivity to isotopic shifts from an ^{18}O bound to the phosphorus instead of an ^{16}O (see also Chap. 4). After 3 days the hydrolysis product of BDNPP, DNPP, was observed in the ^{31}P-NMR spectrum. The complete disappearance of the BDNPP signal at −14.56 ppm was, however, not observed in this time frame. For the complex $[Cd_2(CO_2EtH_2\mathbf{L1})(CH_3COO)_2]^+$ the DNPP signal was observed at −3.15 ppm. If the nucleophile is a Cd(II) bound water/hydroxide (from the solvent) we expect to find in a 50:50 ^{18}O:^{16}O environment DNPP with 50 % ^{18}O incorporated. This was in fact observed in the ^{31}P-spectrum by a splitting of the DNPP-peak into a doublet with a peak separation of approximately 0.07 ppm.

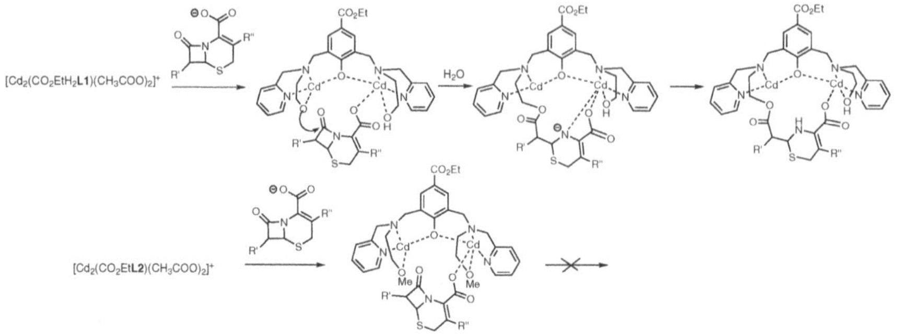

Fig. 5.22 Mechanism proposed for nitrocefin hydrolysis via the formation of the *blue* intermediate by $Cd_2(CO_2EtH_2\mathbf{L1})(CH_3COO)_2]^+$ and binding of nitrocefin to $Cd_2(CO_2Et\mathbf{L2})(CH_3COO)_2]^+$

5.3 Discussion

5.3.1 Mechanism of β-Lactam Hydrolysis

The nucleophile in β-lactam hydrolysis by biomimetic Zn(II) complexes has been reported previously to be either a bridging hydroxide or a terminal water molecule. A bridging hydroxide was ruled out as nucleophile [26]. It is unlikely that this is the nucleophile in the reported complex as the two Cd(II) ions are separated by 4.162 Å and thus too far away to establish a single hydroxide bridge between the two metal centers. Crystal structures with pyrazolate based ligands reported by Meyer et al. support this proposal [47, 48]. There are a number of findings that led to the proposal that the metal bound alcohol is the nucleophile (Fig. 5.22):

(i) In mass spectrometry experiments, a peak at *m/z* 1313.8 was attributed to a species comprised of one ligand, two Cd(II) ions, two acetonitrile solvent molecules and hydrolyzed nitrocefin.

(ii) Bazzicalupi et al. reported that a Zn(II)-bound alcohol (pK_a 6.9) is deprotonated at slightly lower pH than a Zn(II)-bound water and moreover being the more potent nucleophile [49]. Transferring this assignment of the catalytically relevant pK_a in the Cd(II) complex, is not unambiguous as Cd(II) is known to be less acidic due to its larger size and greater softness. The pK_a of 7.95, however, is likely to be attributed to one of the alcohol arms of the ligand bound to the Cd(II). Meyer et al. found using infrared spectroscopy that, in the reaction of a dinuclear Zn(II) complex with penicillin G in methanol, ester formation due to alcoholysis of Zn(II)-bound methoxide had occurred [38]. The esterified hydrolyzed penicillin was not detected in mass spectrometry experiments, however a methoxide bound to the Zn(II) complex showed that an alkoxide nucleophile was indeed present under kinetic/mass spectrometry conditions. Methanolysis of the substrate nitrocefin by two Zn(II) ions has also been reported by Brown et al. [50]. The authors found that in anhydrous methanol, two Zn(II) ions act cooperatively to hydrolyze the substrate.

(iii) Furthermore, the $[Cd_2(CO_2Et\mathbf{L2})(CH_3COO)_2]^+$ complex with the ether arms was inactive towards lactam substrates. This complex should however have sufficient vacant coordination sites to accommodate a terminal water/hydroxide as nucleophile in addition to the substrate. That up to two nitrocefin molecules can bind to $[Cd_2(CO_2Et\mathbf{L2})(CH_3COO)_2]^+$ was shown in mass spectrometry experiments. Hence, the lack of activity in the ether arm complex is not attributed to steric hindrance or poor substrate affinity but rather due to the absence of alkoxide nucleophiles.

(iv) The addition of two equivalents of nitrocefin to $[Cd_2(CO_2EtH_2\mathbf{L1})(CH_3COO)_2]^+$ and subsequent incubation at 37 °C for 1 h, resulted in significantly decreased ability of the complex to degrade the organophosphate substrate BDNPP. This suggests that the bound, hydrolyzed nitrocefin is blocking coordination of phosphoester substrate.

(v) Finally, the rapid decrease in activity after the complete hydrolysis of one equivalent nitrocefin suggests the hydrolyzed substrate is bound irreversibly to the complex.

5.3.2 Mechanism of Phosphodiester Hydrolysis

The absence of a metal bound alcohol in $[Cd_2(CO_2Et\mathbf{L2})(CH_3COO)_2]^+$ allows the assignment of the kinetically relevant pK_a to a metal-bound terminal water molecule as found in related Zn(II) complexes [49, 51]. The higher pK_a opposed to the Zn(II) systems is attributed to the less acidic character of the Cd(II) ion [52]. Interestingly, previous studies of similar Zn(II) complexes have shown that the pK_as of the ether- and alcohol-ligand catalyzed reaction are often found to be similar within error suggesting that the same nucleophile is present [51]. It is known that the ether arms are only partially coordinated in solution whereas this has not been reported for the alcohol arms which should be bound more tightly [51, 53]. Thus in the ether arm mechanism, a nucleophile (terminal water/hydroxide) or substrate molecule could replace an ether arm, whereas in the mechanism with $EtCO_2H_3\mathbf{L1}$ the alcohol arms are bound more tightly restricting the geometry of the complex somewhat so that the pK_a of terminal water molecule is shifted to a more alkaline region. Another scenario is possible—the hydrolysis by the complex with the alcohol arm ligand could follow a mechanism where the alkoxide acts as general base. Kaminskaia et al. reported a Zn(II) complex with a bridging hydroxide where the nucleophile was an external water ($pK_a = 7.06$) activated by the bridging hydroxide as base [39]. It is possible that the alcohol arm of $EtCO_2H_3\mathbf{L1}$ is acting here as general base. The pK_a of 10.1, however, would be unusually high for a Cd(II)-bound alcohol. The ^{18}O-labeling experiment suggests

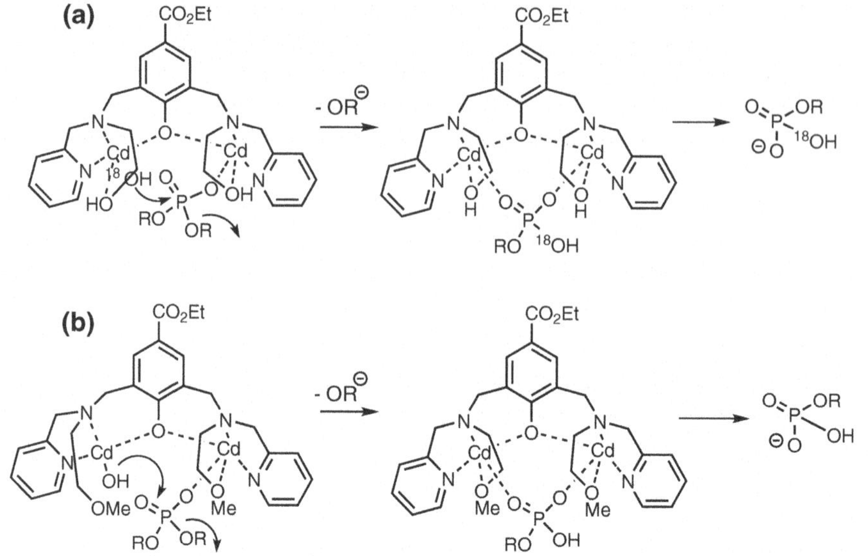

Fig. 5.23 Proposed mechanism for the BDNPP hydrolysis of the two complexes

that the nucleophile in OP hydrolysis by $[Cd_2(CO_2EtH_2\mathbf{L1})\text{-}(CH_3COO)_2]^+$ is not a ligand centered alkoxide but rather a water molecule which is terminally bound to one Cd(II) ion. It is thus proposed that in both complexes the acting nucleophile is terminally bound water (Fig. 5.23).

The difference in pH is attributed to subtle structural and electronic differences in the complexes. In $[Cd_2(CO_2Et\mathbf{L2})(CH_3COO)_2]^+$ a terminally Cd(II)-bound water ($pK_a = 8.7$) is replacing one of the ether arms, in $[Cd_2(CO_2EtH_2\mathbf{L1})(CH_3COO)_2]^+$ however the alcohol arm is bound tightly to the Cd(II) ions and thus the terminal water has to compete with the ligand donor atoms for coordination.

Other mechanistic scenarios are possible. One would be where the alcohol arm of the ligand is actively involved in the mechanism. In this case it could be envisaged that the alcohol arm is deprotonated at pH 10.1 and acts as the primary nucleophile to hydrolyze the substrate BDNPP. The ligand is then recovered by an external water molecule and the substrate is released. A second scenario is possible where, upon deprotonation, the alkoxide acts as a general base to activate an external water molecule which then hydrolyzes the substrate. Both scenarios would be consistent with the finding of a 50 % ^{18}O labeled DNPP (Fig. 5.24).

The higher activity and the difference in pH dependence of the complex $[Cd_2(CO_2EtH_2\mathbf{L1})(CH_3COO)_2]^+$ opposed to the previously reported complex of Mirams et al. may be attributed to the ethyl ester group in *para*-position of the bridging phenolic oxygen [2]. The complex reported by Mirams et al. with the same donor atom set but a methyl group in *para*-position of the bridging phenolic oxygen, exhibited a pK_a around 9 [2], suggesting that the ethyl ester has an effect on the polarization of the nucleophile, shifting the catalytically relevant pK_a.

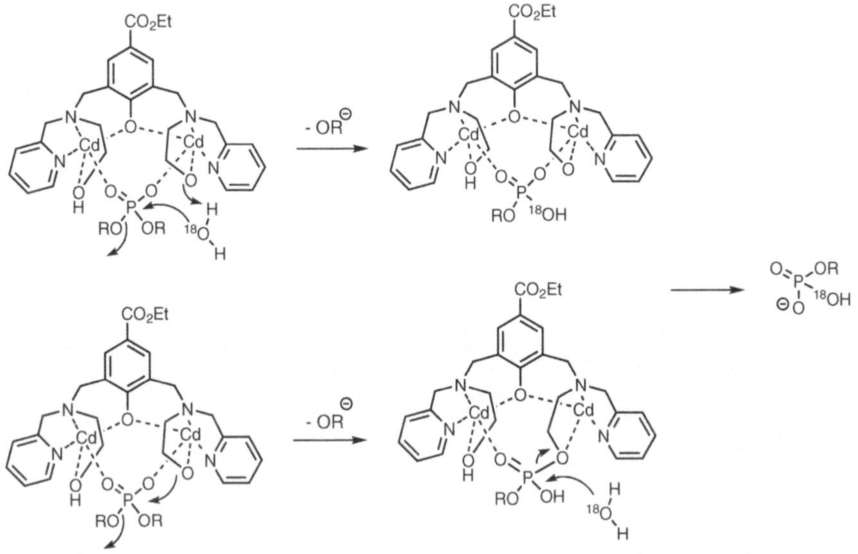

Fig. 5.24 Possible alternative mechanisms involving the alkoxide as a nucleophile that could lead ^{18}O being incorporated in the hydrolysis product

In a paper by Peralta et al. a linear correlation between the pK_a values attributed to deprotonation of the Fe(III)-bound water for complexes with different substituents, methyl, H, Br and NO_2 respectively, was found [54]. This study shows that subtle structural changes in the ligand can influence catalytic activity and the acidity of the Cd(II) ions in nucleophilic activation.

5.4 Conclusion

The preparation and characterization of Cd(II) model complexes is essential for the understanding of Cd(II) substituted enzymes and their mechanisms. Both Cd(II) complexes discussed in this chapter are excellent biomimetics of phosphoesterases. The kinetically relevant pK_a values that were determined for the BDNPP hydrolysis by $[Cd_2(CO_2EtH_2\mathbf{L1})(CH_3COO)_2]^+$ and $[Cd_2(CO_2Et\mathbf{L2})(CH_3COO)_2]^+$ correspond to a terminal water molecule as the hydrolysis-initiating nucleophile and are similar to the pK_a found in the Cd_2-GpdQ enzyme suggesting that enzyme and model complex employ a similar mechanistic strategy. Additionally, the complex $[Cd_2(CO_2EtH_2\mathbf{L1})(CH_3COO)_2]^+$ serves as a model for metallo-β-lactamases. The mechanism, however, seems to differ from enzymatic catalysis. Nitrocefin was hydrolyzed by the complex via a ligand centered alkoxide nucleophile. A blue intermediate, which has been observed in enzymatic nitrocefin hydrolysis was observed and characterized.

The next chapter will introduce five Co(II) complexes that serve as phosphoesterase and metallo-β-lactamase mimics. A spectroscopic and kinetic characterization will be presented.

References

1. M. Eagleson, *Concise Encyclopedia Chemistry* (Walter de Gruyter, New York, 1994)
2. R.E. Mirams, S.J. Smith, K.S. Hadler, D.L. Ollis, G. Schenk, L.R. Gahan, J. Biol. Inorg. Chem. **13**, 1065–1072 (2008)
3. V. Aletras, N. Hadjiliadis, D. Stabaki, A. Karaliota, M. Kamariotaki, I. Butler, J.C. Plakatouras, S. Perlepes, Polyhedron **16**, 1399–1402 (1997)
4. K.Y. Choi, Y.M. Jeon, K.C. Lee, H. Ryu, M. Suh, H.S. Park, M.J. Kim, Y.H. Song, J. Chem. Crystallogr. **34**, 591–596 (2004)
5. M.F. Summers, Coord. Chem. Rev. **86**, 43–134 (1988)
6. E. Kolehmainen, in *Encyclopedia of Spectroscopy and Spectrometry*, ed. by L. John (Academic Press, Oxford, 1999), pp. 834–843
7. J.K. Bohlke, J. Phys. Chem. Ref. Data **34**, 57 (2005)
8. T.W. Lane, M.A. Saito, G.N. George, I.J. Pickering, R.C. Prince, F.M. Morel, Nature **435**, 42 (2005)
9. Y. Xu, L. Feng, P.D. Jeffrey, Y. Shi, F.M.M. Morel, Nature **452**, 56–61 (2008)
10. A. Dolega, K. Baranowska, J. Gajda, S. Kazmierski, M.J. Potrzebowski, Inorg. Chim. Acta **360**, 2973–2982 (2007)
11. L.M. Berreau, Adv. Phys. Org. Chem. **41**, 79–181 (2006)
12. F.E. Jacobsen, S.M. Cohen, in *Using cobalt(II) and cadmium(II) Substituted Model Complexes to Improve zinc(II)-metalloprotein Inhibitor Design*. 229th ACS National Meeting, in San Diego, CA, 2005
13. J. Hsieh, M.A. Viktora, D. Rabinovich, in *Mononuclear Cadmium Complexes with Sulfur-rich Coordination Environments*. 56th Southeast Regional Meeting, 2004
14. A. Dolega, K. Baranowska, D. Gudat, A. Herman, J. Stangret, A. Konitz, M. Smiechowski, S. Godlewska, Eur. J. Inorg. Chem. **2009**(24), 3644–3660 (2009)
15. A. Dolega, Wiad. Chem. **64**, 389–411 (2010)
16. K. Pladzyk, D. Baranowska, S. Gudat, M. Godlewska, J. Wieczerzak, M. Chojnacki, K. Bulman, Januszewicz, A. Dolega, Polyhedron **30**, 1191–1200 (2011)
17. K. Byriel, L. Gahan, C. Kennard, J. Latten, P. Healy, Aust. J. Chem. **46**, 713–719 (1993)
18. E. Tomat, L. Cuesta, V.M. Lynch, J.L. Sessler, Inorg. Chem. **46**, 6224–6226 (2007)
19. M.A. Harvey, S. Baggio, M.T. Garland, R. Baggio, J. Coord. Chem. **58**, 243–253 (2005)
20. K.S. Hadler, E.A. Tanifum, S.H. Yip, N. Mitić, L.W. Guddat, C.J. Jackson, L.R. Gahan, K. Nguyen, P.D. Carr, D.L. Ollis, A.C. Hengge, J.A. Larrabee, G. Schenk, J. Am. Chem. Soc. **130**, 14129–14138 (2008)
21. K.S. Hadler, L.R. Gahan, D.L. Ollis, G. Schenk, J. Inorg. Biochem. **104**, 211–213 (2010)
22. F. Ely, K.S. Hadler, L.R. Gahan, L.W. Guddat, D.L. Ollis, G. Schenk, Biochem. J. **432**, 565–573 (2010)
23. M. Damblon, A. Jensen, I. Ababou, C. Barsukov, C.J. Papamicael, L. Schofield, R. Olsen, Bauer, G.C. Roberts, J. Biol. Chem. **278**, 29240–29251 (2003)
24. L. Hemmingsen, C. Damblon, J. Antony, M. Jensen, H.W. Adolph, S. Wommer, G.C. Roberts, R. Bauer, J. Am. Chem. Soc. **123**, 10329–10335 (2001)
25. N.O. Concha, B.A. Rasmussen, K. Bush, O. Herzberg, Protein Sci. **6**, 2671–2676 (1997)
26. N.V. Kaminskaia, B. Spingler, S.J. Lippard, J. Am. Chem. Soc. **123**, 6555–6563 (2001)
27. C. Cojocel, *Beta-Lactam Antibiotics* (Springer, Boston, 2008)
28. L.E. Asbel, M.E. Levison, Infect. Dis. Clin. North Am. **14**, 435–447 (2000)

29. M.W. Crowder, J. Spencer, A.J. Vila, Acc. Chem. Res. **39**, 721–728 (2006)
30. N.P. Sharma, C. Hajdin, S. Chandrasekar, B. Bennet, K.-W. Yang, M.W. Crowder, Biochemistry **45**, 10729–10738 (2006)
31. Z. Wang, W. Fast, S.J. Benkovic, J. Am. Chem. Soc. **120**, 10788–10789 (1998)
32. F. Meyer, H. Pritzkow, Eur. J. Inorg. Chem. **2005**(12), 2346–2351 (2005)
33. J. Weston, Chem. Rev. **105**, 2151–2174 (2005). (Washington, DC)
34. A. Tamilselvi, G. Mugesh, J. Biol. Inorg. Chem. **13**, 1039–1053 (2008)
35. G. Parkin, Chem. Rev. **104**, 699–767 (2004). (Washington, DC)
36. M. Umayal, G. Mugesh, Inorg. Chim. Acta **372**, 353–361 (2011)
37. A. Tamilselvi, M. Nethaji, G. Mugesh, Chem. Eur. J. **12**, 7797–7806 (2006)
38. B. Bauer-Siebenlist, S. Dechert, F. Meyer, Chem. Eur. J. **11**, 5343–5352 (2005)
39. N.V. Kaminskaia, C. He, S.J. Lippard, Inorg. Chem. **39**, 3365–3373 (2000)
40. A. Tamilselvi, G. Mugesh, Chem. Eur. J. **16**, 8878–8886 (2010)
41. N.V. Kaminskaia, B. Spingler, S.J. Lippard, J. Am. Chem. Soc. **122**, 6411–6422 (2000)
42. Z. Wang, W. Fast, S.J. Benkovic, Biochemistry **38**, 10013–10023 (1999)
43. Y.J. Dong, M. Bartlam, L. Sun, Y.F. Zhou, Z.P. Zhang, C.G. Zhang, Z.H. Rao, X.E. Zhang, J. Mol. Biol. **353**, 655–663 (2005)
44. K. Nakamoto, *Infrared and Raman Spectra of Inorganic and Coordination Compounds* (Wiley, NY, 1978)
45. M. Asso, R. Panossian, M. Guiliano, Spectrosc. Lett. **17**, 271–278 (1984)
46. M. Maeder, ReactLab KINETICS
47. B. Bauer-Siebenlist, F. Meyer, E. Farkas, D. Vidovic, S. Dechert, Chem. Eur. J. **11**, 4349–4360 (2005)
48. B. Bauer-Siebenlist, F. Meyer, E. Farkas, D. Vidovic, J. A. Cuesta-Seijo, R. Herbst-Irmer, H. Pritzkow, Inorg. Chem. **43**, 4189–4202 (2004)
49. C. Bazzicalupi, A. Bencini, E. Berni, A. Bianchi, V. Fedi, V. Fusi, C. Giorgi, P. Paoletti, B. Valtancoli, Inorg. Chem. **38**, 4115–4122 (1999)
50. P.J. Montoya-Pelaez, R.S. Brown, Inorg. Chem. **41**, 309–316 (2002)
51. L.J. Daumann, K.E. Dalle, G. Schenk, R.P. McGeary, P.V. Bernhardt, D.L. Ollis, L.R. Gahan, Dalton Trans. **41**, 1695–1708 (2012)
52. E. Kimura, T. Koike, T. Shiota, Y. Iitaka, Inorg. Chem. **29**, 4621–4629 (1990)
53. J.W. Chen, X.Y. Wang, Y.G. Zhu, J. Lin, X.L. Yang, Y.Z. Li, Y. Lu, Z.J. Guo, Inorg. Chem. **44**, 3422–3430 (2005)
54. R.A. Peralta, A.J. Bortoluzzi, B. de Souza, R. Jovito, F.R. Xavier, R.A.A. Couto, A. Casellato, F. Nome, A. Dick, L.R. Gahan, G. Schenk, G.R. Hanson, F.C.S. de Paula, E.C. Pereira-Maia, S.d.P. Machado, P.C. Severino, C. Pich, T. T. Bortolotto, H. Terenzi, E.E. Castellano, A. Neves, M.J. Riley, Inorg. Chem. **49**, 11421–11438 (2010)

Chapter 6
Spectroscopic and Mechanistic Studies of Co(II) Phosphoesterase and Metallo-β-lactamase Biomimetics

6.1 Introduction

6.1.1 *Cobalt: An Excellent Spectroscopic Probe for Protein Active Sites*

Cobalt is a transition metal that is less abundant in nature than most other first row transition metals [1]. However, as a cofactor for Vitamin B_{12} it is an essential trace element for humans [2]. Cobalt is usually found to be coordinated by a corrin ligand backbone in natural systems and one of the few true organometallic compounds in nature is the Co(III)-alkyl group in Vitamin B_{12} (the only known vitamin to contain a metal ion) [3]. A few Co(II) containing enzymes such as prolidase, nitrile hydratase, bromoperoxidase or glucose isomerase have been isolated from bacteria [2]. However, Co(II) is extensively used as spectroscopic probe for Zn(II)-containing enzyme active sites due to its unique spectroscopic and magnetic properties. Often the catalytic activity in the cobalt substituted protein is retained [4]. The active site geometry employed by Co(II) is often found to be virtually identical to the Zn(II) enzyme in crystallographic studies [4]. The most common oxidation states are Co(II) and Co(III), with the Co(II) always having unpaired electrons rendering the ion paramagnetic [1]. Octahedral or tetrahedral geometries are easily distinguished in the electronic absorption spectra, with the absorption band of the octahedral Co(II) displaying a characteristic splitting from spin-orbit coupling [5]. Structures and inhibitor/substrate interactions in Zn(II) enzymes such as carboanhydrase, carboxypeptidase A, alcohol dehydrogenase or alkaline phosphatase have been analyzed with cobalt substitution but also Cu(I)/(II) and Fe(II)/(III) enzymes have been analyzed with this method [4].

L. J. Daumann, *Spectroscopic and Mechanistic Studies of Dinuclear Metallohydrolases and Their Biomimetic Complexes*, Springer Theses, DOI: 10.1007/978-3-319-06629-5_6,

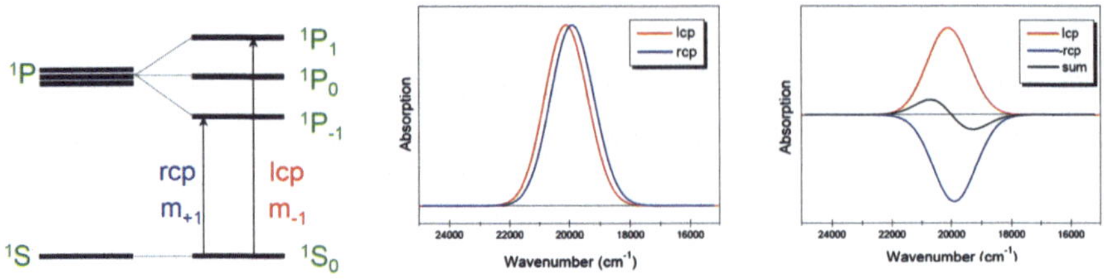

Fig. 6.1 The allowed transition from the [1]S state to the [1]P state (which is split in the presence of a magnetic field, Zeeman splitting). Pictures taken from Ref. [11]

6.1.2 Magnetic Circular Dichroism

MCD has been shown to be a useful method to probe the geometric and electronic structure of biological metal systems [6–9]. MCD measures the differential absorption of left and right circular polarized light, similar to circular dichroism (CD), but in the presence of a magnetic field [10]. Application of a magnetic field parallel to the direction of light causes all matter to exhibit MCD activity so it is not restricted to paramagnetic materials [10]. The electronic levels split due to the magnetic field (Zeeman effect) and both excited and electronic ground state can be probed, while other methods such as EPR only probe the electronic ground state [10]. The formalism of MCD intensity is illustrated in Eq. 6.1

$$\frac{\Delta A}{E} = \gamma \beta H \left[A_1 \left(\frac{-\delta f(E)}{\delta E} \right) + \left(B_0 + \frac{C_0}{kT} \right) f(E) \right] \quad (6.1)$$

where ΔA denotes the differential absorption of left and right circularly polarized light, E is hν, β is the Bohr magneton, k the Boltzman constant, H refers to the applied magnetic field, $f(E)$ is a bandshape function, $\frac{-\delta f(E)}{\delta E}$ the first derivative, T is the temperature, γ represents for a number of spectroscopic constants and finally A_1, B_0 and C_0 are the three intensity defining components which give rise to the A-, B- and C-terms, respectively [10]. The selection rules for A- and C-terms require $\Delta M_L = \pm 1$ and $\Delta M_S = 0$. A-terms give rise to derivative-shaped bands and are not dependent on temperature [10, 11]. They require orbital degeneracy either in the ground or excited state [10]. An example is given in Fig. 6.1, where a transition from the [1]S state to the degenerate [1]P state is shown and the resulting MCD spectrum is depicted in black on the right.

C-terms are related to the A-terms and are inversely temperature dependent and thus dominate the MCD spectra of bioinorganic compounds which are often acquired at low temperatures [10]. The temperature dependency is due to the higher population of the Zeeman-split ground state, as the population of the lowest state rises with lower temperatures, the C-term intensity increases. For a C-term to be observed, orbital degeneracy is required in the ground state (opposed to the

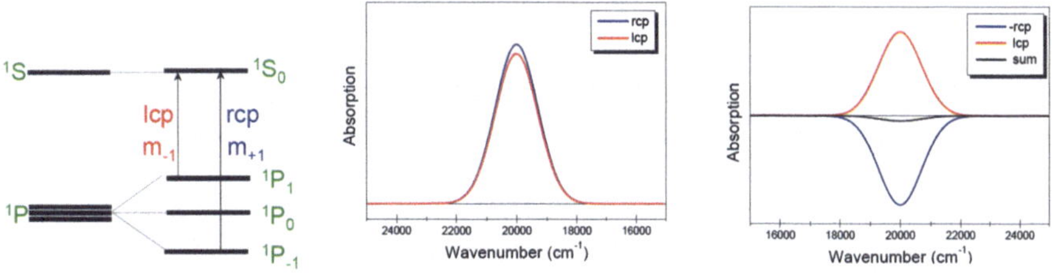

Fig. 6.2 The simulation of a C-term transition for a system with a degenerate ground state. Pictures taken from Ref. [11]

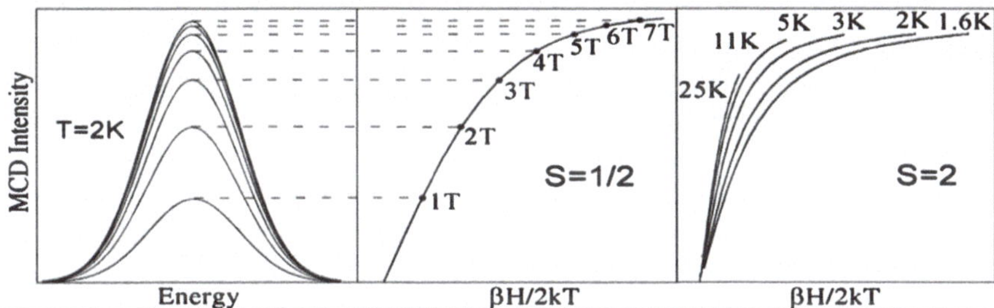

Fig. 6.3 VTVH MCD, picture taken from Ref. [13]. Copyright (2003) National Academy of Sciences, U.S.A

required degeneracy in the excited state for A-terms), which occurs via spin-orbit coupling and an absorption band shape is observed (Fig. 6.2) [10].

B-terms arise from field induced mixing of excited and ground states to other states with Zeeman interaction [10]. States with similar energy are prone to mix more. These transitions have absorption-band shape and are temperature independent. However, if the two mixed states are degenerate in energy, an A-term is observed. When the states are similar in energy then a derivative shaped signal is observed (pseudo A-term) [10].

Ground state electronic structure information such as zero field splitting (ZFS), g-tensors or spin state can be obtained, in addition to magnetic coupling interaction information for dimetallic sites, by analyzing variable field, variable temperature (VTVH) data [5, 10, 12]. For a Kramer system (half integer spin, e.g. $S = 1/2$) the VTVH saturation curves show a linear increase at low magnetic fields. At some H/T, however, the intensity of a C-term saturates. Furthermore, all VTVH-isotherms of a Kramer system superimpose [13]. Kramer systems with $S > 1/2$, however, have sublevels which can split under the applied magnetic field. Once they become thermally available the ground state gets depopulated resulting in a nesting behavior of the isotherms [13]. This is also observed in dimeric systems where the ZFS, super exchange coupling or transition polarizations affect the saturation magnetization curves. Examples are shown in Fig. 6.3. The left shows how the MCD signal increases with increasing magnetic field.

The middle of Fig. 6.3 displays the overlaying isotherms for S = 1/2 Kramers system and on the right the nesting behaviour for a S = 2 non-Kramers system at different temperatures is shown [13]. The steepness of the magnetisation curves is influenced by g [14]. MCD spectroscopy can be used to probe different metal binding sites. This was shown for the enzyme GpdQ, which has two distinct binding sites, a 6-coordinate site with high affinity and a 5-coordinate site with low affinity [6, 15]. Hadler et al. were able to show that only in the presence of substrate a dinuclear Co(II)Co(II) center formed in GpdQ; the authors based this on the observation that a 5-coordinate transition arose at 570 nm when phosphate was present; distinctively different from a 6 coordinate transition (500 nm) [15]. Also the geometric and electronic structure of Co(II)-substituted GpdQ and the related phosphoesterase enzyme OpdA has been extensively explored using MCD spectroscopy [16].

MCD of model complexes is extremely useful for the interpretation of features found in enzyme spectra. In small models the coordination geometry can be easily controlled and distinct transitions for 4- (negative band a 600 ± 50 nm and less intense positive band at 525 ± 50 nm), 5- (negative bands between 650 and 450 nm) and 6-coordinate Co(II) (sharp negative band around 500 ± 50 nm) can be assigned [5]. Band intensities can also be an indicator for the coordination number of a Co(II) ion [5]. While 4- and 5-coordinate Co(II) displays bands of substantial intensity at room temperature the transitions from 6-coordinate Co(II) are up to 10 times less intense. At very low temperature, however, the latter are often more intense than 5-coordinate transitions. A number of dinuclear Co(II) complexes have been studied by MCD [12, 14, 17], and with the aid of VTVH MCD magnetic coupling parameters have been obtained [14].

6.1.3 Magnetic Properties of Co(II)

High spin (perfect) octahedral Co(II) displays $^4T_{1g}$ symmetry in the ground state. The magnetic moments of complexes with T ground terms are often found to show considerable temperature dependence and their interpretation is difficult due to a partial quenching of the angular orbital momentum [18, 19]. Fitting of experimental susceptibility data and explaining the magnetic properties is a challenging endeavor and only a few research groups have described the magnetism of dinuclear Co(II) compounds [14, 19–23]. Due to large spin orbit coupling these systems exhibit substantial zero field splitting causing high magnetic anisotropy. D is the axial zero field splitting (ZFS) parameter E the rhombic ZFS parameter. Zero field splitting parameters (D) can be measured by magnetic susceptibility methods, EPR or VTVH MCD and their magnitude provides valuable information about the coordination number and geometry of paramagnetic metal ions. As determined by VTVH MCD for Co(II) in a 4-coordinate environment D is usually below 13 cm^{-1}, for 5-coordinate between 2.6 and 50 cm^{-1} and for 6-coordinate above 50 cm^{-1} [5]. Another important parameter in dinuclear Co(II) systems is the magnetic exchange coupling (J). Larrabee et al. reported a range of dinuclear Co(II)Co(II), Co(II)Co(III) and Co(III)Co(III) complexes to provide insight into the electronic structure and reactivity of cobalt substituted *Ec*MetAp (methionine aminopeptidase from *Escherichia coli*) [14, 17]. By studying these models the authors were able to develop a method to distinguish between μ-aqua and μ-hydroxo species in related enzymes. In the dinuclear Co(II) (d^7, high spin) complex weak magnetic exchange coupling via the aqua/hydroxyl bridge is predicted. By comparing the J values obtained by VTVH–MCD measurements the authors found that μ-aqua complexes display, in general (given other structural properties are similar), a greater J value and can thus be distinguished from μ-hydroxo complexes [12]. A second mixed valence complex was studied which mimicked a non-coupling event as Co(III) is diamagnetic, which makes it an excellent model for Co_2-EcMetAP fumagillin complex where EXAFS studies have found that one Co(II) is displaced upon inhibitor binding [17]. A Co(II) complex employing a mixed 5- and 6-coordination sphere which proved to be an excellent spectroscopic model for not only Co(II) substituted methionine aminopeptidase but as well for the enzyme GpdQ and other dinuclear metalloenzymes was reported by Larrabee [12]. The authors gained insight about the potential nucleophile in hydrolysis of phosphate esters by comparing the J values [12]. The trend that μ-hydroxo structures have positive J values and are ferromagnetically coupled and μ-aqua bridged derivatives exhibit weak antiferromagnetic (negative J values) or no coupling, was observed [12].

The magnetic exchange coupling can also be calculated using Density Functional Theory (DFT) based on the broken symmetry approach, developed by Noodleman [24]. The formalism for a model with two interacting metal centers A and B is as follows:

$$E(S_{\max}) - E(BS) = -S^2_{\max} J \tag{6.2}$$

$E(S_{\max})$ refers to the energy of the state with maximum number of unpaired spins (high spin), $E(BS)$ to the energy of the broken symmetry state (antiferromagnetic state, spin in magnetic orbitals of metal A is antiparallel to B) [24, 25]. Hence, J can be obtained by calculating the energy difference between the highest spin state and the broken symmetry spin state. Extensive use has been made of the broken symmetry approach to calculate the exchange coupling parameter J in di- and trinuclear metal systems [25, 26], it should be noted however, that the accuracy of the calculations when compared to experimentally found values of J is greatly influence by the functional used [25, 26].

6.1.4 Phosphoesterase Models

One of the first synthetic compounds capable of phosphate ester cleavage via a dinuclear metal center was reported in 1984 by Sargeson (Fig. 6.4a) [27].

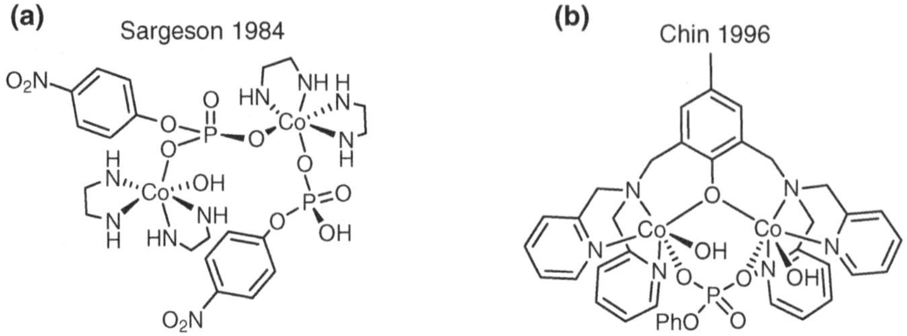

Fig. 6.4 Phosphate ester cleavage by dinuclear Co(III) complexes

The presence of a second Co(III) center was suggested to be responsible for a 26-fold increase in the rate of hydrolysis compared to that observed for the corresponding mononuclear complex [27]. It was proposed that the coordinated PNPP was cleaved by an intramolecular attack of a terminal bound hydroxide. Since Sargeson's early work other examples of functional or spectroscopic model complexes for several classes of phosphoesterases enzymes have been reported [28–36], however, only a few use cobalt. The Co(III)Co(III) complex with a phenolate based ligand (Fig. 6.4b) showed activity in phosphomonoester cleavage and a bridging coordination of the substrate was confirmed with ^{31}P NMR [37].

6.1.5 Target Ligands and Aims of this Chapter

Five dinucleating ligands will be used to synthesize Co(II) complexes.[1] The ligands contain either an alcohol- (**L1** ligand) or methyl- (**L2** ligands) moiety as pendant arms, in addition to the pyridine residues (Fig. 6.5).

CO_2EtH_3**L1** and CO_2EtH**L2** only differ in R″ and allow comparison of the phosphoesterase-like activity in the presence and absence of alcohol nucleophiles. The model substrate BDNPP will be used. Also the ability of the complexes derived from these two ligands to hydrolyze substrates other than organophosphates will be investigated using the β-lactam substrates nitrocefin and penicillin G.

The ligands CO_2EtH**L2**, CH_3H**L2**, NO_2H**L2** and BrH**L2** bear the same donor atom set and only differ in the R′ substituent in *para*-position of the phenolic oxygen that forms a bridge between the two Co(II) ions in the complex. Magnetic coupling

[1] Parts of this Chapter have been reprinted with permission from (L.J. Daumann et al., *Synthesis, Magnetic Properties, and Phosphoesterase Activity of Dinuclear Cobalt(II) Complexes. Inorganic Chemistry* **2013** 52(4), 2029–2043). Copyright (2013) American Chemical Society and in L.J. Daumann et al., "*Dinuclear Cobalt(II) Complexes as Metallo-β-lactamase Mimics.* Eur. J. Inorg. Chem. **2013**, 17, 3082–3089.

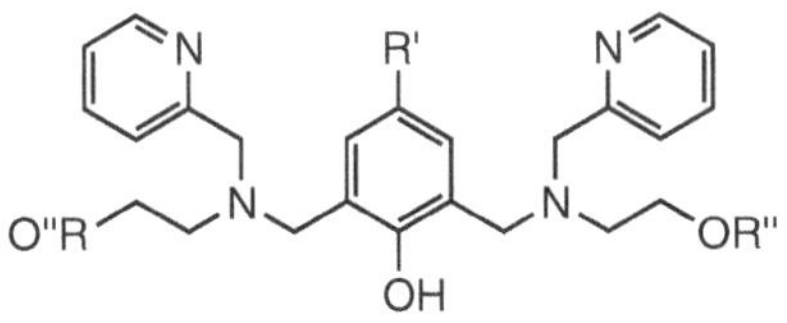

R' = CO_2Et	R" = H	CO_2EtH_3**L1**
R' = CO_2Et	R" = Me	CO_2EtH**L2**
R' = CH_3	R" = Me	CH_3H**L2**
R' = NO_2	R" = Me	NO_2H**L2**
R' = Br	R" = Me	BrH**L2**

Fig. 6.5 Ligands employed to generate dinuclear Co(II) complexes

strongly depends on the nature of the bridging atom, the M(II)-O-M(II) angle and the separation of the two metal centers [38]. Changes in the electron density at the phenolic oxygen via different electron withdrawing residues in *para*-position are proposed to affect the magnetic coupling and/or kinetic parameters. This chapter reports an investigation of the magnetic properties of dinuclear Co(II) complexes using (VTVH) MCD, solid state magnetic susceptibility measurements and Evans NMR. In addition DFT calculations (broken symmetry approach) will be conducted to help with the interpretation of obtained magnetic coupling constants.

6.2 Results

6.2.1 Syntheses of the Co(II) Complexes

The Co(II) complexes were in general synthesized by adding two equivalents of Co(II) acetate to a methanolic solution of the respective ligand (CO_2EtH_3**L1**, CO_2EtH**L2**, CH_3H**L2**, NO_2H**L2** and BrH**L2**), followed by addition of solid sodium hexafluorophosphate (see Chap. 2). No inert atmosphere conditions were necessary. All complexes were air stable and contain two Co(II) ions. The Co(III)Co(II) and Co(III)Co(III) derivatives of the complexes could not be obtained by direct synthesis using $Na_3Co(NO_2)_6$ nor were they observed in CV-experiments suggesting that the ligands do not support Co(III) due to a lack of nitrogen donors [1].

6.2.2 Crystal Structures of the Co(II) Complexes

Crystals of the Co(II) complexes suitable for X-ray crystallography were obtained by slow evaporation of methanolic solutions of the complex solutions. Bond lengths and angles are comparable to Co(II) compounds reported in the literature (Tables 6.1 and 6.2) [14, 21, 22]. In all cases the complexes crystallized with one negatively charged ligand (phenolic oxygen deprotonated) and two acetates bridging both metal ion centers, completing the hexacoordinate coordination sphere, with the charge balanced by a hexafluorophosphate ion (Fig. 6.6). For clarity the first coordination sphere around the triply bridged μ-phenoxo-(bis)μ-acetato Co(II) core

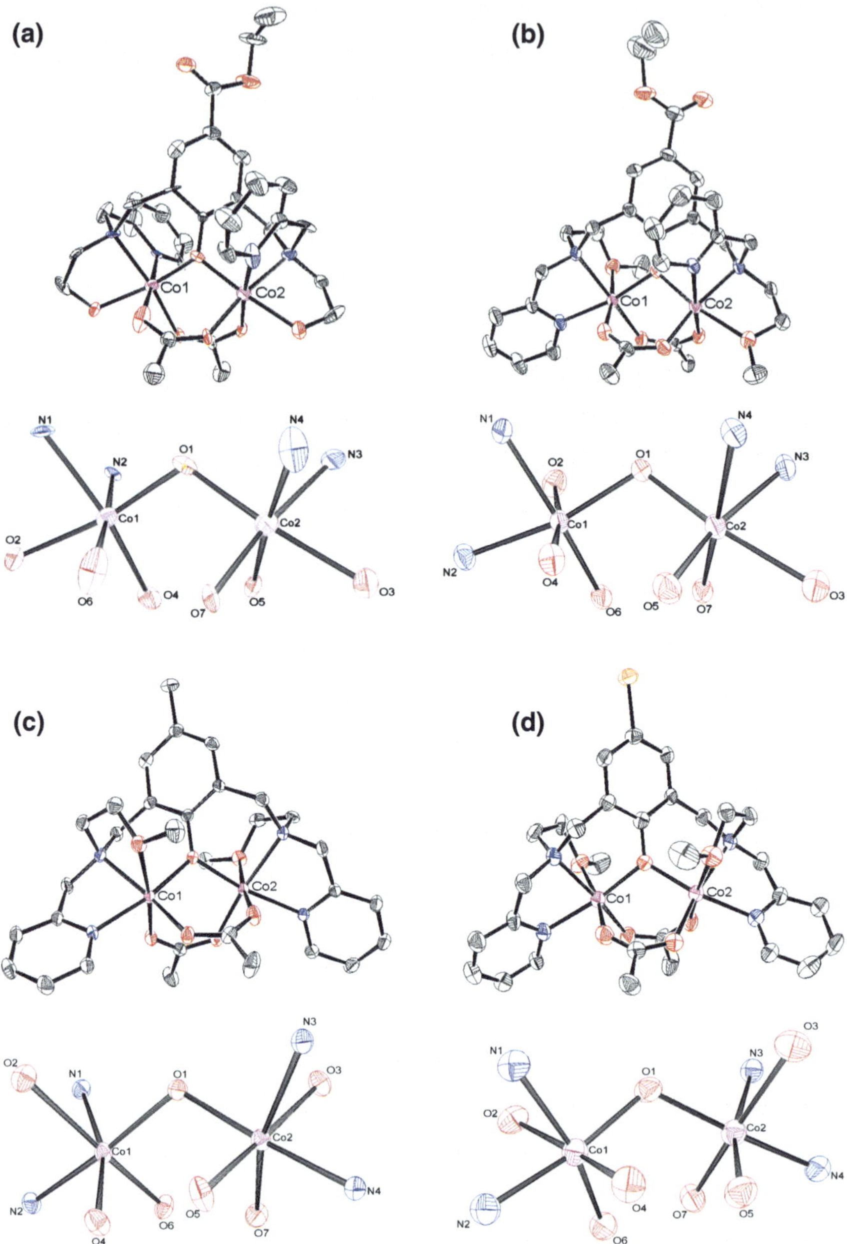

Fig. 6.6 Structures of **a** $[Co_2(CO_2EtH_2\mathbf{L1})(CH_3COO)_2](PF_6)$, **b** $[Co_2(CO_2Et\mathbf{L2})(CH_3COO)_2](PF_6)$, **c** $[Co_2(CH_3\mathbf{L2})(CH_3COO)_2](PF_6)$ and **d** $[Co_2(Br\mathbf{L2})(CH_3COO)_2](PF_6)$ and views of the respective first coordination spheres. Counter ions and hydrogen atoms have been omitted for clarity (25 % ellipsoid probability in all ORTEP plots)

is also displayed in Fig. 6.6. Intriguingly the complexes crystallized in different isomeric forms.

To simplify the structural discussion the phenol back bone and the two Co(II) ions are considered to be 'in-plane' on the page from the readers point of view and the ligand arms can be either 'in-plane' or 'out of plane' (towards the viewer or in the opposite direction). For the complex with CO_2EtH_3**L1** the pyridines are coordinating the Co(II) out of plane and are *trans* to each other whilst for CO_2EtH**L2** where the coordinating alcohol has been substituted with a methyl ether, an asymmetric arrangement of the ligand arms is observed. The latter complex exhibited significant disorder around the ethyl ester so the ethyl group was refined in two separate positions. One pyridine arm has changed position with the methyl ether arm in the complex with CO_2EtH**L2** deviating from the pseudo C_2 symmetry found in $[Co_2(CO_2EtH_2\mathbf{L1})(CH_3COO)_2]PF_6$.

The complexes $[Co_2(CH_3\mathbf{L2})\text{-}(CH_3COO)_2]PF_6$ and $[Co_2(Br\mathbf{L2})(CH_3COO)_2]PF_6$ employ the same coordination geometry as found in the respective Zn(II) complexes (Chap. 4) with the pyridine arms *trans* to each other and coordinating in-plane and an overall pseudo C_2 symmetry. A crystal structure of the complex of the ligand NO_2H**L2** could not be obtained since the crystals did not diffract. The structure of this complex was thus calculated using DFT. The starting parameters for the calculation were from the complex with CH_3H**L2**. Geometry optimizations were also conducted for all remaining structures using the Amsterdam Density Functional program and the Becke-Perdew functional [39].

6.2.3 *Infrared-Spectroscopy*

The infrared spectra of all complexes showed a shift of the asymmetric and symmetric acetate stretch to higher frequencies (respective to free acetate $\nu_{asym} = 1{,}578\ cm^{-1}$, $\nu_{sym} = 1{,}414\ cm^{-1}$) which indicates bridging bidentate binding for acetate ($[Co_2(CO_2EtH_2\mathbf{L1})(CH_3COO)_2]PF_6$, $[Co_2(CO_2Et\mathbf{L2})(CH_3COO)_2]PF_6$, $[Co_2(CH_3\mathbf{L2})(CH_3COO)_2]PF_6$, $[Co_2(NO_2\mathbf{L2})(CH_3COO)_2]PF_6$ and $[Co_2(Br\mathbf{L2})(CH_3COO)_2]PF_6$, $\nu_{asym} = 1605, 1601, 1596, 1595$ and $1599\ cm^{-1}$ and $\nu_{sym} = 1442, 1421, 1424, 1429$ and $1421\ cm^{-1}$) [40]. The separation $\Delta\nu_{asym\text{-}sym}$ of both stretches confirms a symmetric bidentate coordination of the acetates for all Co(II) complexes. This is in accord with the crystal structures [40]. The hexafluorophosphate stretch and deformation bands were found in all spectra around 830 and 555 cm^{-1}.

6.2.4 *Mass Spectrometry of the Complexes*

Mass spectral data recorded in methanol revealed the presence of one negatively charged ligand, two divalent cobalt ions and two acetates for all complexes. Partial

Table 6.1 Crystallographic data of the Co(II) complexes $[Co_2(CO_2EtH_2\mathbf{L1})(CH_3COO)_2](PF_6)$, $[Co_2(CO_2Et\mathbf{L2})(CH_3COO)_2](PF_6)$, $[Co_2(CH_3\mathbf{L2})(CH_3COO)_2](PF_6)$ and $[Co_2(Br\mathbf{L2})(CH_3COO)_2](PF_6)$

	$[Co_2(CO_2EtH_2\mathbf{L1})(CH_3COO)_2](PF_6)$	$[Co_2(CO_2Et\mathbf{L2})(CH_3COO)_2](PF_6)$	$[Co_2(CH_3\mathbf{L2})(CH_3COO)_2](PF_6)$	$[Co_2(Br\mathbf{L2})(CH_3COO)_2](PF_6)$
Empirical formula	$C_{31}H_{39}Co_2F_6N_4O_9P$	$C_{33}H_{43}Co_2F_6N_4O_9P$	$C_{31}H_{41}Co_2F_6N_4O_7P$	$C_{30}H_{38}BrCo_2F_6N_4O_7P$
Formula weight	874.49	902.54	844.51	909.38
Wavelength (Å)	0.71073 (Mo K_α)	0.71073 (Mo K_α)	0.71073 (Mo K_α)	1.5418 (Cu K_α)
Crystal system	Orthorhombic	Monoclinic	Monoclinic	Triclinic
Space group	P2 1nb	P1 21/c1	P1 21/c1	P-1
a (Å)	11.815(3)	14.1820(7)	10.7830(3)	8.9780(4)
b (Å)	12.327(2)	14.3220(5)	14.8640(6)	13.0285(7)
c (Å)	25.575(6)	19.4160(8)	22.9880(6)	16.5846(9)
α (°)	90	90	90	98.929(5)
β (°)	90	94.152(4)	102.293(3)	89.826(4)
γ (°)	90	90	90	104.481(4)
Vol (Å^3)	3724.83(14)	3933.3(3)	3600.0(2)	1854.21(16)
Z	4	4	4	2
μ (mm^{-1})	1.019	0.968	1.047	9.385
F(000)	1,792	1,856	1,736	920
ρ (Mg/m^3)	1.559	1.524	1.558	1.629
Reflns col.	9,152	16,703	15,340	14,717
Ind. Reflns (R_{int})	4,692 (0.1194)	6,900 (0.0485)	6,332 (0.0437)	5,559 (0.0566)
θ range (°)	2.84–225	2.88–25	2.89–25	3.55–60.52
GOOF on F^2	0.637	0.838	0.801	1.125
Final R indices [$I > 2\sigma(I)$]	R1 = 0.0486, wR2 = 0.0641	R1 = 0.0512, wR2 = 0.1118	R1 = 0.0483, wR2 = 0.0713	R1 = 0.0839, wR2 = 0.2694
R indices (all data)	R1 = 0.1839, wR2 = 0.0792	R1 = 0.1071, wR2 = 0.1221	R1 = 0.0911, wR2 = 0.0741	R1 = 0.1252, wR2 = 0.2888

Table 6.2 Selected bond lengths (Å) and angles (°) for [$Co_2(CO_2EtH_2$**L1**)$(CH_3COO)_2](PF_6)$, [$Co_2(CO_2Et$**L2**)$(CH_3COO)_2](PF_6)$, [$Co_2(CH_3$**L2**)$(CH_3COO)_2](PF_6)$, [$Co_2(Br$**L2**)$(CH_3COO)_2](PF_6)$ and [$Co_2(NO_2$**L2**)$(CH_3COO)_2](PF_6)$ as obtained from X-ray crystallography and DFT-calculations

	[$Co_2(CO_2Et$**L1**) $(CH_3COO)_2](PF_6)$		[$Co_2(CH_3$**L2**) $(CH_3COO)_2](PF_6)$		[$Co_2(Br$**L2**)$(CH_3COO)_2](PF_6)$		[$Co_2(CO_2Et$**L2**) $(CH_3COO)_2](PF_6)$		[$Co_2(NO_2$**L2**) $(CH_3COO)_2](PF_6)$
	Found	Calc.	Found	Calc.	Found	Calc.	Found	Calc.	Calc.
N(1)-Co(1)	2.168(9)	2.209	2.161(2)	2.190	2.172(9)	2.176	2.175(4)	2.179	2.184
N(2)-Co(1)	2.145(10)	2.109	2.142(3)	2.134	2.111(9)	2.114	2.114(4)	2.117	2.119
N(3)-Co(2)	2.120(10)	2.201	2.149(3)	2.184	2.142(8)	2.174	2.154(3)	2.176	2.198
N(4)-Co(2)	2.070(11)	2.115	2.118(2)	2.130	2.113(9)	2.114	2.182(4)	2.122	2.129
O(1)-Co(1)	1.992(8)	2.032	2.014(2)	2.021	2.014(7)	2.035	2.016(3)	2.081	2.070
O(1)-Co(2)	2.041(8)	2.037	2.0026(19)	2.040	2.009(7)	2.035	2.044(3)	2.074	2.037
O(2)-Co(1)	2.119(8)	2.281	2.294(3)	2.230	2.223(8)	2.318	2.235(3)	2.311	2.315
O(3)-Co(2)	2.173(8)	2.355	2.261(2)	2.340	2.266(9)	2.308	2.221(3)	2.316	2.394
O(4)-Co(1)	2.046(8)	2.047	2.017(2)	2.039	2.088(9)	2.028	2.105(4)	2.028	2.111
O(5)-Co(2)	2.064(8)	2.109	2.073(2)	2.115	2.024(8)	2.136	2.008(3)	2.102	2.043
O(6)-Co(1)	2.190(9)	2.218	2.114(3)	2.137	2.010(8)	2.132	2.013(3)	2.102	2.036
O(7)-Co(2)	2.013(7)	2.059	2.011(3)	2.043	2.087(8)	2.027	2.068(3)	2.023	2.093
Co(1)-O(1)-Co(2)	114.6(3)	113.3	112.81(10)	112.71	113.4(3)	113.07	112.25(14)	112.51	111.22
Co(1)•••Co(2)	3.394	3.393	3.346	3.380	3.363	3.395	3.372	3.455	3.390

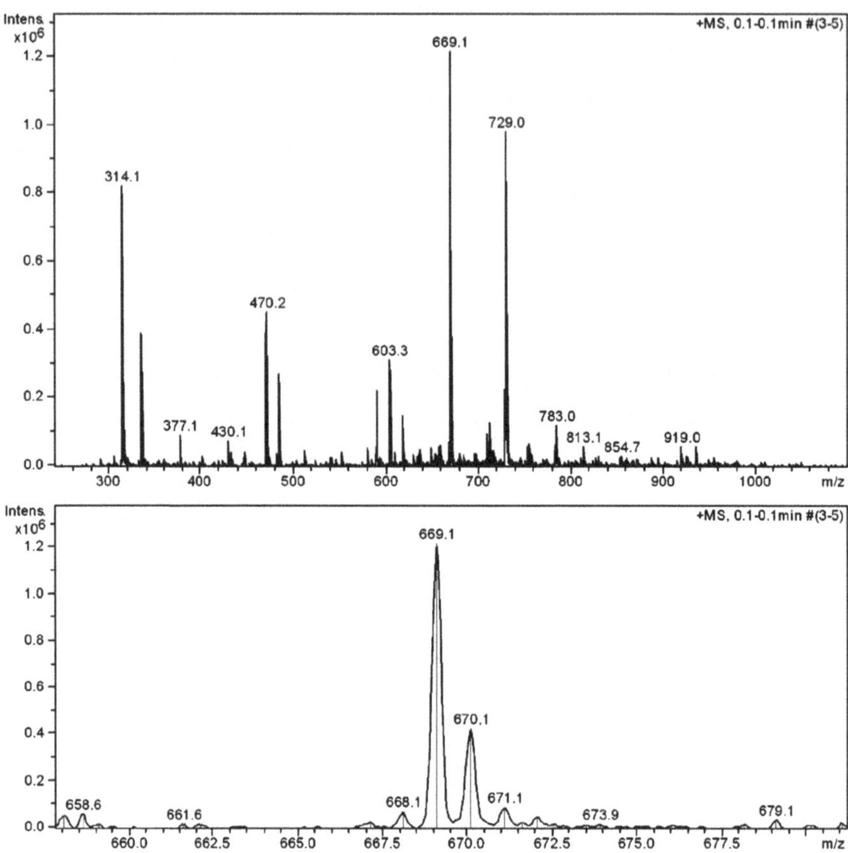

Fig. 6.7 ESI mass spectrum of $[Co_2(CO_2EtH_2\mathbf{L1})(CH_3COO)_2](PF_6)$ in acetonitrile (*top*) and enlarged main peak (*below*)

loss of the acetates and coordination of methanol solvent molecules was observed. The mass spectrum of $[Co_2(CO_2EtH_2\mathbf{L1})(CH_3COO)_2]PF_6$ in methanol showed a number of distinct peaks at *m/z* 729.16 ($[Co_2(CO_2EtH_2\mathbf{L1})(CH_3COO)_2]^+$ calc. *m/z* 729.1), 669.1 ($[Co_2(CO_2EtH\mathbf{L1})(CH_3COO)]^+$ calc. *m/z* 669.1), *m/z* 609.1 ($[Co_2(CO_2Et\mathbf{L1})]^+$ calc. *m/z* 609.1) and a doubly positive charged complex at m/z 305.1 ($[Co_2(CO_2EtH\mathbf{L1})]^{2+}$ calc. *m/z* 305.1). The complex $[Co_2(Br\mathbf{L2})(CH_3COO)_2]PF_6$ showed a species at *m/z* 734.9 which was attributed to $[Co_2(Br\mathbf{L2})(CH_3COO)(MeO)]^+$ (calc. *m/z* 735.1). The complex formed by the ligand $CO_2EtH\mathbf{L2}$ in methanol showed two peaks at m/z 757.4 and 729.4 which were assigned to $[Co_2(CO_2Et\mathbf{L2})(CH_3COO)_2]^+$ (calc. *m/z* 757.2) and $[Co_2(CO_2Et\mathbf{L2})(CH_3COO)(MeO)]^+$ (calc. *m/z* 729.2). The third ligand ($CH_3H\mathbf{L2}$) also showed the formation of a complex comprised of one ligand, two Co(II) ions and two acetates in methanolic solution ($[Co_2(CH_3\mathbf{L2})(CH_3COO)_2]^+$ found *m/z* 699.2; calc. *m/z* 699.2). Interestingly the complex with ($NO_2H\mathbf{L2}$) showed also water present: *m/z* 784.1

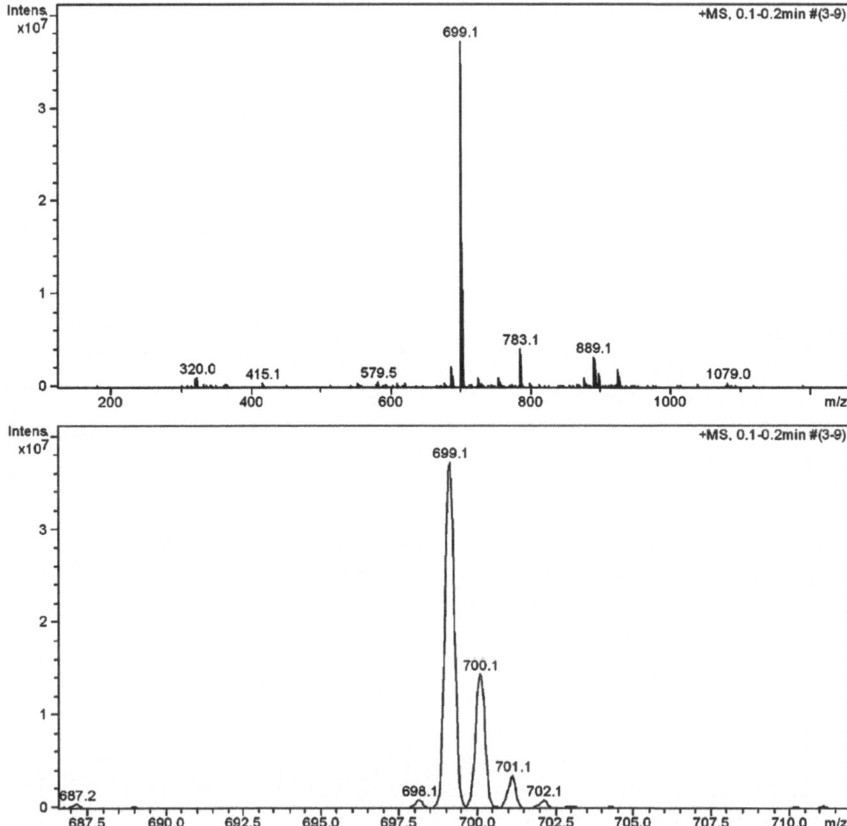

Fig. 6.8 ESI mass spectrum of $[Co_2(CO_2Et\mathbf{L2})(CH_3COO)_2](PF_6)$ in acetonitrile (*top*) and enlarged main peak (*below*)

$[Co_2(NO_2\mathbf{L2})(CH_3COO)_2(H_2O)_3]^+$; calc. *m/z* 784.2. The spectra were also recorded in acetonitrile as this solvent is the more relevant for kinetic measurements. The mass spectrum of $[Co_2(CO_2EtH_2\mathbf{L1})(CH_3COO)_2](PF_6)$ showed similar peaks (*m/z* = 729.0, 669.1) as observed in the methanol spectrum in addition to a new doubly charged ion (*m/z* 335.1, calc. *m/z* 335.1 $[Co_2(CO_2EtH_2\mathbf{L1})(CH_3COO)]^{2+}$, Fig. 6.7). For $[Co_2(CO_2Et\mathbf{L2})(CH_3COO)_2](PF_6)$ peaks were found at *m/z* 757.1 and 349.0 (calc. *m/z* 757.2 and 349.1 for $[Co_2(CO_2Et\mathbf{L2})(CH_3COO)_2]^+$ and $[Co_2(CO_2Et\mathbf{L2})(CH_3COO)]^{2+}$, respectively, Fig. 6.8). Similar species were also found for $[Co_2(CH_3\mathbf{L2})(CH_3COO)_2](PF_6)$ (*m/z* 699.1 and 320.0, calc. *m/z* 699.16 and 320.07 for $[Co_2(CH_3\mathbf{L2})(CH_3COO)_2]^+$ and $[Co_2(CH_3\mathbf{L2})(CH_3COO)]^{2+}$, respectively, Fig. 6.9) and $[Co_2(Br\mathbf{L2})(CH_3COO)_2]$-$(PF_6)$ (*m/z* 765.0 and 352.9, calc. *m/z* 763.1 (100 %), 765.06 (99 %) and 352.02 (100 %), 353.02 (97 %) for $[Co_2(Br\mathbf{L2})(CH_3COO)_2]^+$ and $[Co_2(Br\mathbf{L2})(CH_3COO)]^{2+}$). The latter complex also showed a species at m/z 847.1 which is proposed to arise from additional coordination of two

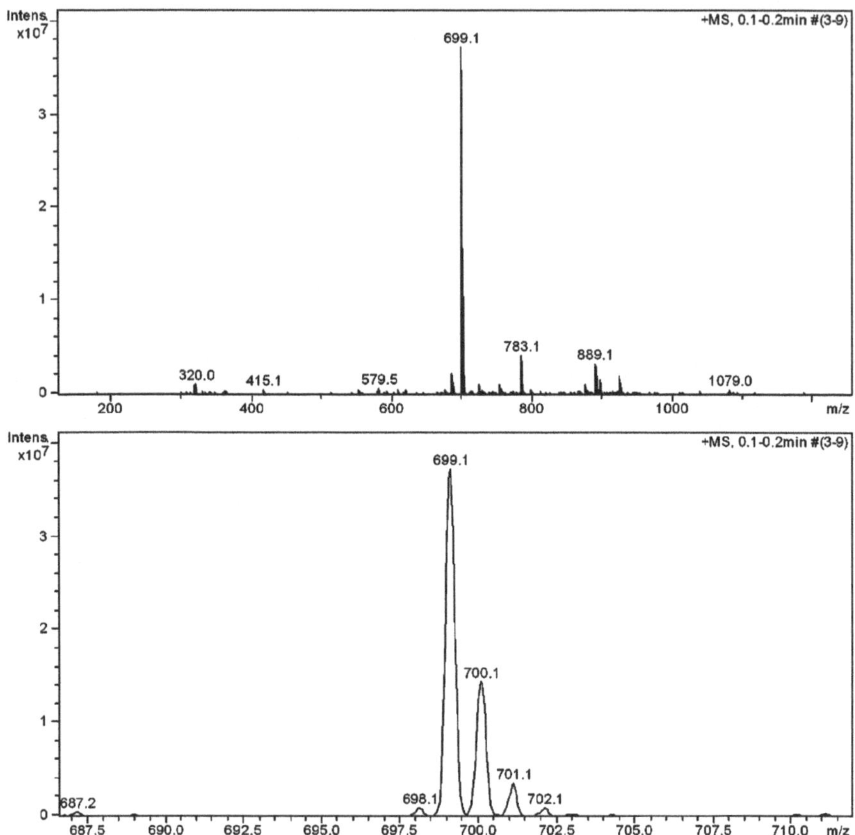

Fig. 6.9 ESI mass spectrum of $[Co_2(CH_3\mathbf{L2})(CH_3COO)_2](PF_6)$ in acetonitrile (*top*) and enlarged main peak (*below*)

acetonitrile molecules (calc. m/z 845.1 (100 %), 847.1 (98 %) $[Co_2(Br\mathbf{L2})(CH_3COO)_2(\text{acetonitrile})_2]^+$, Fig. 6.10) The complex $[Co_2(NO_2\mathbf{L2})(CH_3COO)_2](PF_6)$ showed one distinct peak at m/z 730.1 (calc. m/z 730.1, $[Co_2(NO_2\mathbf{L2})(CH_3COO)_2]^+$, Fig. 6.11).

6.2.5 UV-Vis Spectroscopy

6.2.5.1 In Solution

UV-Vis spectra were recorded for all complexes in acetonitrile (used for kinetic measurements) and ethanol (used in MCD spectroscopy). Bands typical for 6-coordinate Co(II) were present in solution. Figure 6.12 shows the UV-Vis

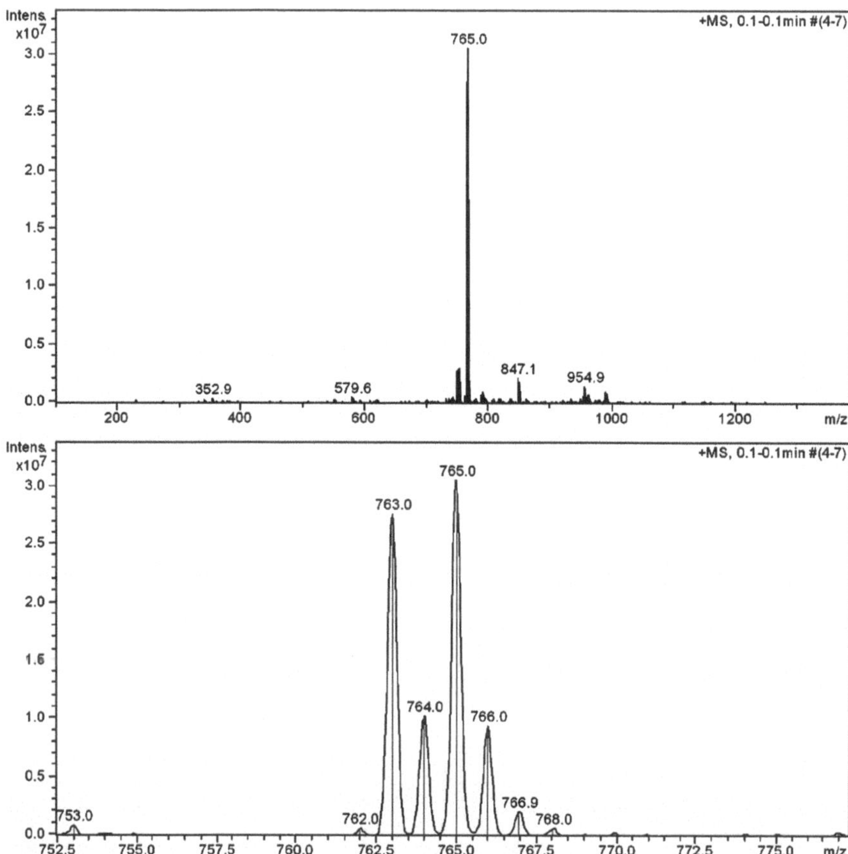

Fig. 6.10 ESI mass spectrum of $[Co_2(Br\mathbf{L2})(CH_3COO)_2](PF_6)$ in acetonitrile (*top*) and enlarged main peak (*below*)

spectra of $[Co_2(CO_2Et\mathbf{L2})(CH_3COO)_2]^+$ in acetonitrile (blue) and ethanol (black). A broad transition in the NIR region around 1,100 nm is apparent in addition to 2–3 bands between 400 and 600 nm which are typical for octahedral high-spin Co(II) complexes [21]. Table 6.3 lists the observed transitions and extinction coefficients for all complexes.

While the extinction coefficients for the first transition around 470 nm range from 38 to 56 M^{-1} cm^{-1} for the **L2** ligands the complex $[Co_2(CO_2EtH_2L1)(CH_3COO)_2]PF_6$ displays an almost two fold higher $\varepsilon = 78–99\ M^{-1}\ cm^{-1}$ for this transition. It is possible that in the complex with $CO_2EtH_2\mathbf{L1}$ a higher distortion from octahedral symmetry exists in solution, partially cancelling the Laporte rule [1, 21]. For an assignment to specific transitions refer to the next (Sect. 6.2.5.2).

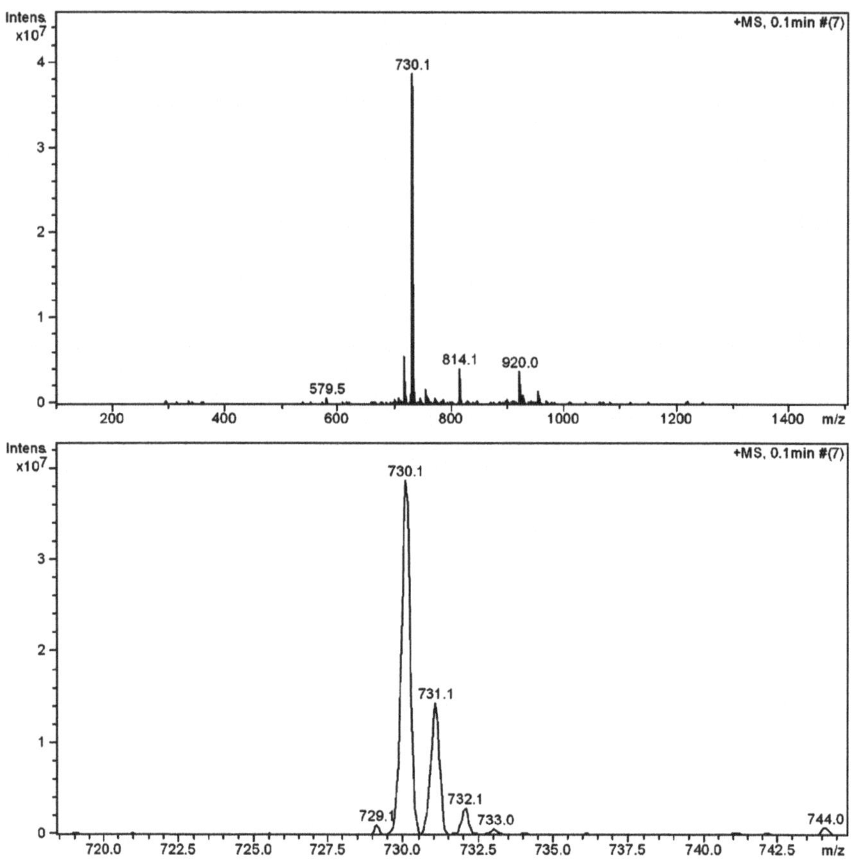

Fig. 6.11 ESI mass spectrum of $[Co_2(NO_2\mathbf{L2})(CH_3COO)_2](PF_6)$ in acetonitrile (*top*) and enlarged main peak (*below*)

6.2.5.2 Diffuse Reflectance Spectroscopy

The UV-Vis-NIR diffuse reflectance spectra (Fig. 6.13) of the Co(II) complexes showed a characteristic broad NIR-band around 1,100 nm corresponding to the $^4T_{1g} \rightarrow {}^4T_{2g}$ d–d transition of hexacoordinate Co(II) that was observed in some of the spectra measured in solution [14].

The visible absorption bands between 400 and 600 nm arise from the $^4T_{1g} \rightarrow {}^4T_{1g}(P)$ transition, which is split into three bands due to distortion from O_h symmetry [41]. These bands are better resolved than in the respective solution spectra. Two distinct phenolate O → Co(II) bands present below 400 nm were assigned to $\pi \rightarrow t_{2g}$ and $\pi \rightarrow e_g$ LMCT transitions [14]. Interestingly, the diffuse reflectance spectra exhibit close similarities to the spectra collected in solution.

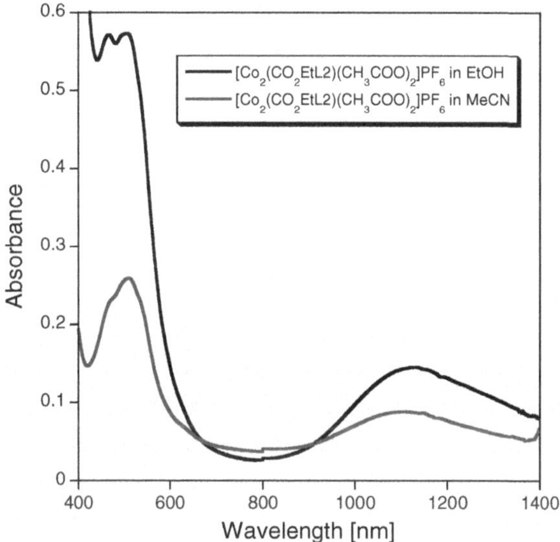

Fig. 6.12 UV-Vis-NIR spectra of $[Co_2(CO_2Et\mathbf{L2})(CH_3COO)_2]^+$ in ethanol (0.010 M) and acetonitrile (0.001 M)

Table 6.3 Observed absorption bands in different solvents for the Co(II) complexes

Complex	Solvent	λ_1 (nm)	ε (M^{-1} cm^{-1})	λ_2 (nm)	ε (M^{-1} cm^{-1})	λ_3 (nm)	ε (M^{-1} cm^{-1})
$[Co_2(CO_2EtH_2\mathbf{L1})(CH_3COO)_2]PF_6$	Acetonitrile	481	99	508	98	1,138	21
$[Co_2(CO_2EtH_2\mathbf{L1})(CH_3COO)_2]PF_6$	Ethanol	473	78	509	79	1,142	18
$[Co_2(CO_2Et\mathbf{L2})(CH_3COO)_2]PF_6$	Acetonitrile	473	56	509	63	1,104	22
$[Co_2(CO_2Et\mathbf{L2})(CH_3COO)_2]PF_6$	Ethanol	464	54	508	50	1,122	14
$[Co_2(CH_3\mathbf{L2})(CH_3COO)_2]PF_6$	Acetonitrile	467	38	511	42	1,090	17
$[Co_2(CH_3\mathbf{L2})(CH_3COO)_2]PF_6$	Ethanol	472	43	516	49	1,124	17
$[Co_2(NO_2\mathbf{L2})(CH_3COO)_2]PF_6$	Acetonitrile	–	–	–	–	1,118	9
$[Co_2(NO_2\mathbf{L2})(CH_3COO)_2]PF_6$	Ethanol	–	–	–	–	1,146	12
$[Co_2(Br\mathbf{L2})(CH_3COO)_2]PF_6$	Acetonitrile	464	52	507	54	–	–
$[Co_2(Br\mathbf{L2})(CH_3COO)_2]PF_6$	Ethanol	465	49	510	54	–	–

Hence the structures found in the solid state (crystal structure, IR) are close to those in solution, which is in agreement with the findings from mass spectrometry supporting the formation of $[Co_2(H_{-1}\mathbf{LX})(CH_3COO)_2]^+$ species in solution.

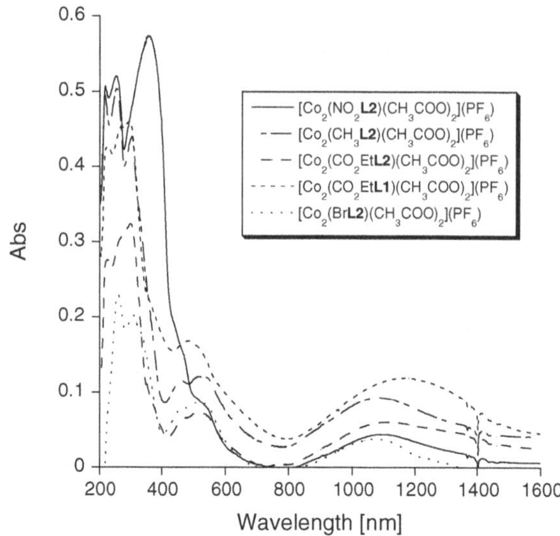

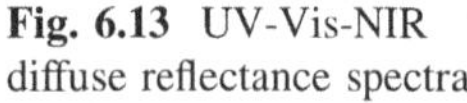
Fig. 6.13 UV-Vis-NIR diffuse reflectance spectra

6.2.6 *Magnetism*

6.2.6.1 Solid State Susceptibility Measurements

All complexes showed significant temperature dependence of the magnetic moment. Fits of the data were accomplished using the MagSaki software [20, 23]. This software calculates the magnetic susceptibility considering axial distortion, spin-orbit coupling, and anisotropic exchange interaction of two octahedral, axially distorted Co(II) centers [20, 23]. The software utilizes four variables: the axial-splitting parameter Δ, an orbital reduction factor κ, the spin-orbit coupling parameter λ, and the magnetic exchange coupling parameter J. From those variables the axial ZFS parameter D and anisotropic g-factors can be calculated. In Table 6.4 the parameters obtained from the fitting of the data at 500 and 1,000 Oe, respectively are displayed.

Figure 6.14 shows the experimental data and the obtained fits at 500 Oe. The μ_{eff} values per dimer at 300 K are between 6.5 and 7.2 BM. They are constant between 100 and 300 K for the complexes $[Co_2(CO_2EtH_2\mathbf{L1})(CH_3COO)_2]PF_6$, $[Co_2(CO_2Et\mathbf{L2})(CH_3COO)_2]PF_6$ and $[Co_2(CH_3\mathbf{L2})(CH_3COO)_2]PF_6$ indicating antiferromagnetic coupling [42]. The $\mu_{s(spin\text{-}only)}$ (3.87 BM per Co(II), 5.47 per dimer) or $\mu_{S+L(spin+orbital\ momentum)}$ (5.19 BM per Co(II)) values are not appropriate approximations for the μ_{eff} values of the complexes at room temperature as high spin Co(II) compounds have $^4T_1$ ground terms which cause an orbital contribution to the magnetic moment.

Table 6.4 Parameters obtained after fitting the data with the MagSaki software [23]

	Field (Oe)	J^a (cm^{-1})	λ	κ	ν	Δ	g_z	g_x	D (cm^{-1})
$[Co_2(CO_2EtH_2$**L1**$)(CH_3COO)_2]PF_6$	500	−0.66	−121	0.72	−4.4	383	2.3	4.7	97.9
	1,000	−0.69	−111	0.68	−3.3	249	2.5	4.6	102.0
$[Co_2(CO_2Et$**L2**$)(CH_3COO)_2]PF_6$	500	−0.10	−122	0.93	−5.6	635	2.1	4.9	105.6
	1,000	−0.08	−125	0.93	−5.6	651	2.1	4.9	108.2
$[Co_2(CH_3$**L2**$)(CH_3COO)_2]PF_6$	500	−0.67	−112	0.86	−5	481	2.2	4.8	98.3
	1,000	−0.69	−106	0.83	−4.6	404	2.2	4.8	95.7
$[Co_2(NO_2$**L2**$)(CH_3COO)_2]PF_6$	500	0.45	−175	0.74	−4.3	550	2.3	4.7	146.3
	1,000	0.38	−173	0.75	−5	648	2.2	4.7	132.5
$[Co_2(Br$**L2**$)(CH_3COO)_2]PF_6$	500	3.09	−103	0.85	−2.4	210	2.8	4.7	137.8
	1,000	3.13	−102	0.84	−2.4	205	2.8	4.7	134.9

[a] Hamiltonian $H = J \cdot S_1 \cdot S_2$

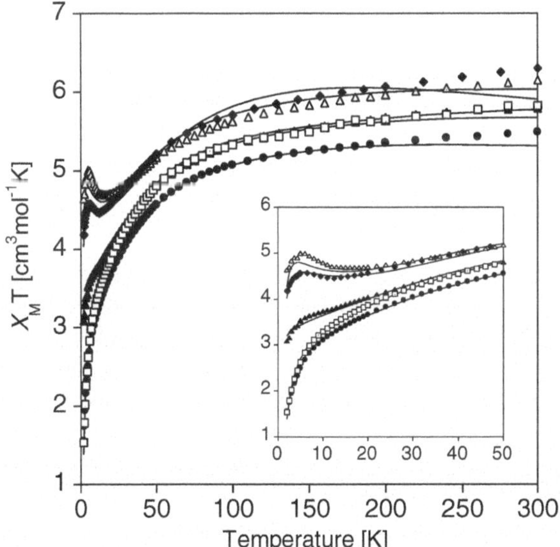

Fig. 6.14 Plot of $\chi_M T$ versus T for powdered samples of the complexes $[Co_2(CO_2EtH_2$**L1**$)(CH_3COO)_2](PF_6)$ (*filled circle*), $[Co_2(CO_2Et$**L2**$)(CH_3COO)_2](PF_6)$ (*filled triangle*), $[Co_2(CH_3$**L2**$)(CH_3COO)_2](PF_6)$ (*open square*), $[Co_2(Br$**L2**$)(CH_3COO)_2](PF_6)$ (*filled diamond*), $[Co_2(NO_2$**L2**$)(CH_3COO)_2](PF_6)$ (*open triangle*). The *solid lines* represent the best fit obtained (for the range of 2–300 K). *Inset* low temperature data for the complexes

6.2.6.2 Solution Susceptibility Measurements

The magnetic moments at room temperature in solution were determined with the Evans method (corrections were applied for the use of a superconducting magnet) [43]. The values (Table 6.5) are typical for two high-spin Co(II) in an octahedral environment [14].

Table 6.5 Magnetic moments in methanolic solution

Magnetic moment	($[Co_2(CO_2EtH_2\mathbf{L1})(CH_3COO)_2]^+$	($[Co_2(CO_2Et\mathbf{L2})(CH_3COO)_2]^+$	($[Co_2(CH_3\mathbf{L2})(CH_3COO)_2]^+$	($[Co_2(NO_2\mathbf{L2})(CH_3COO)_2]^+$	($[Co_2(Br\mathbf{L2})(CH_3COO)_2]^+$
μ (BM)	6.14	6.18	6.09	7.31	6.09

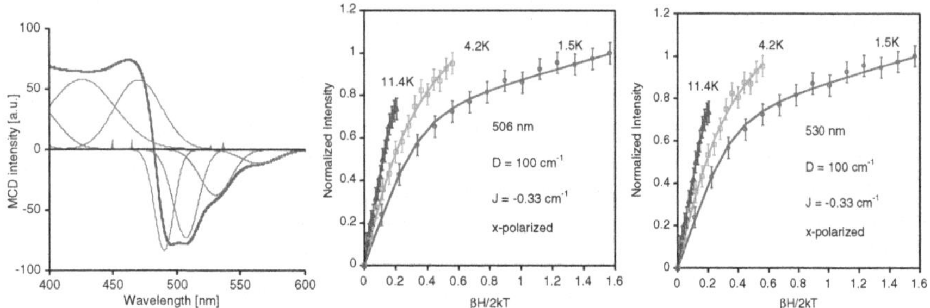

Fig. 6.15 MCD spectrum of $[Co_2(CO_2EtH_2\mathbf{L1})(CH_3COO)_2](PF_6)$ as solid mull at 1.5 K and 7.0 T. Gaussian deconvolved spectrum is identical with the recorded spectrum except for noise. *On the right* magnetization plots from VTVH analysis of the bands at 506 and 530 nm

6.2.6.3 Magnetic Circular Dichroism

MCD spectroscopy has been used to determine the nature of the magnetic exchange coupling and the coordination geometries of all complexes in ethanol solutions and in the solid state. Ethanol was used for the measurements for its ability to form an excellent glass at low temperatures. The spectra were dominated by temperature dependent C-terms. Lower temperature leads to more intense transitions due to the higher (Boltzmann) population of the magnetic ground state at low temperatures. The MCD intensity of the spectra at 50 K was decreased dramatically and thus no VTVH data above this temperature were collected for all systems.

The Gaussian deconvoluted spectra can be found in Figs. 6.15, 6.16, 6.17, 6.18, 6.19, 6.20, 6.21 and 6.22 (the $[Co_2(NO_2\mathbf{L2})(CH_3COO)_2]PF_6$ and $[Co_2(Br\mathbf{L2})(CH_3COO)_2]PF_6$ solid mull spectra were too noisy and are thus not presented). The Gaussians were fitted with the software Grams AI 8.0 in the spectra where the x-axis units are converted to wave numbers [44]. In the fitting process the bandwidths of the d–d transitions were restrained to lie between 600 and 1,000 cm^{-1} and a minimum of Gaussians were fitted to achieve a satisfactory composite spectrum. The analysis of VTVH-MCD intensity behavior of the spectra has also been conducted. The fitting of the VTVH data was achieved with the Fortran software VTVH 2.1.1 [45]. The g values obtained by the SQUID measurements were added and fixed in the input file for the software (for details about the fitting process please see Chap. 2).

Table 6.6 shows the parameters obtained for a selection of transitions. The fits are in remarkably good agreement with the data obtained from magnetic susceptibility measurements. Here the magnetic exchange coupling parameter J also

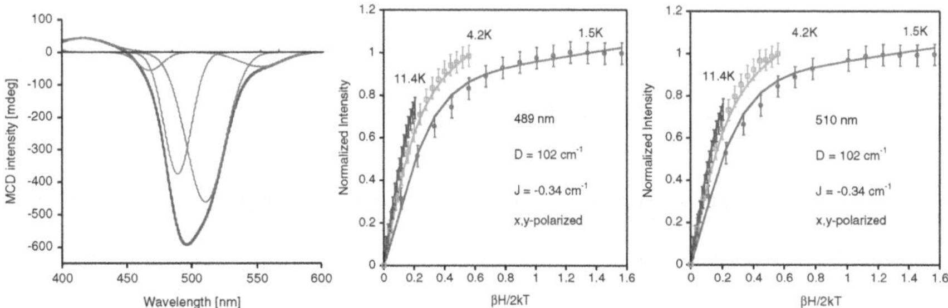

Fig. 6.16 MCD spectrum of $[Co_2(CO_2EtH_2\mathbf{L1})(CH_3COO)_2](PF_6)$ in ethanol at 1.5 K and 7.0 T (pathlength 0.62 cm, 16 mM). Gaussian deconvolved spectrum is identical with the recorded spectrum except for noise. Only the Gaussians for the d–d transitions are displayed. *On the right* magnetization plots from VTVH analysis of the bands at 489 and 510 nm

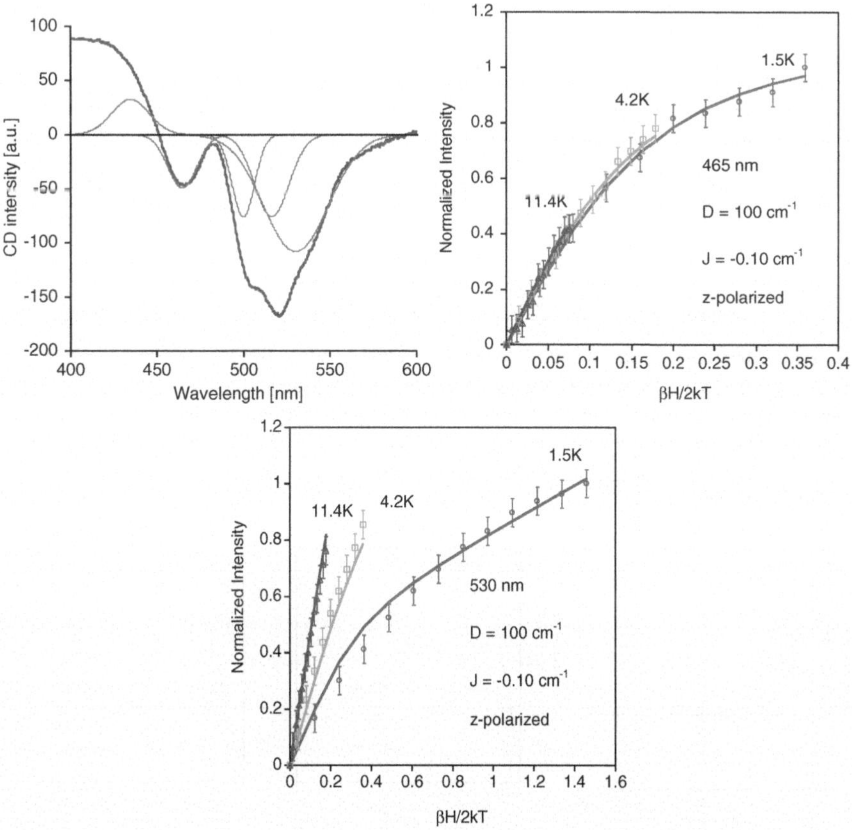

Fig. 6.17 MCD spectrum of $[Co_2(CO_2Et\mathbf{L2})(CH_3COO)_2](PF_6)$ as solid mull at 1.5 K and 7.0 T. Gaussian deconvolved spectrum is identical with the recorded spectrum except for noise. Only the Gaussians for the d–d transitions are displayed. *On the right* magnetization plots from VTVH analysis of the bands at 465 and 530 nm

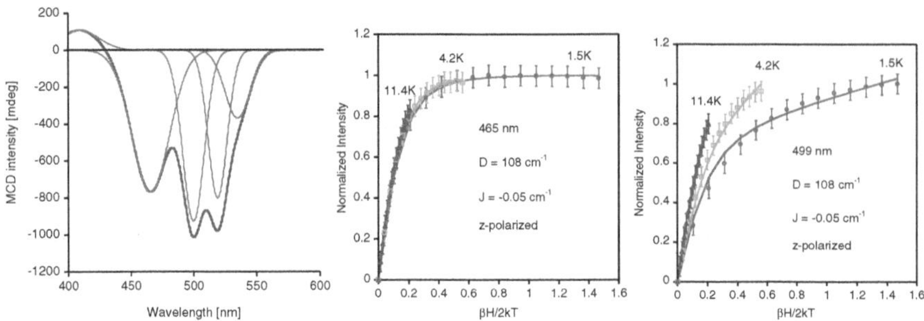

Fig. 6.18 MCD spectrum of $[Co_2(CO_2Et\mathbf{L2})(CH_3COO)_2](PF_6)$ in ethanol at 1.5 K and 7.0 T (pathlength 0.62 cm, 7 mM). Gaussian deconvolved spectrum is identical with the recorded spectrum except for noise. Only the Gaussians for the d–d transitions are displayed. *On the right* magnetization plots from VTVH analysis of the bands at 465 and 499 nm

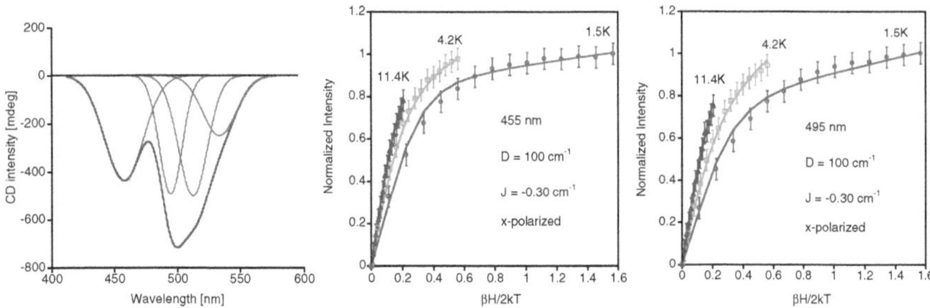

Fig. 6.19 MCD spectrum of $[Co_2(CH_3\mathbf{L2})(CH_3COO)_2](PF_6)$ ethanol at 1.5 K and 7.0 T (pathlength 0.62 cm, 34 mM). Gaussian deconvolved spectrum is identical with the recorded spectrum except for noise. Only the Gaussians for the d–d transitions are displayed. *On the right* magnetization plots from VTVH analysis of the bands at 455 and 495 nm

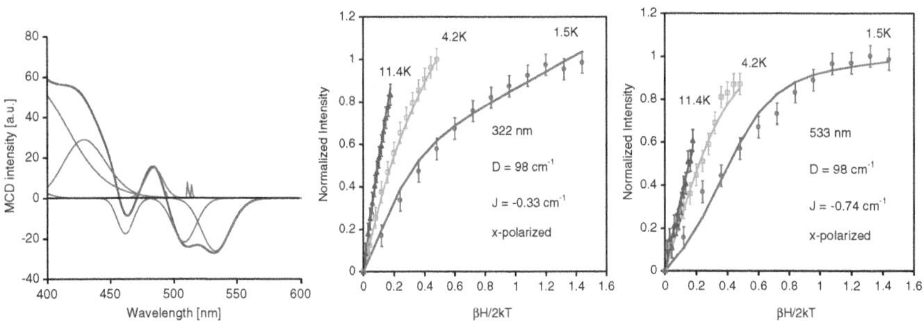

Fig. 6.20 MCD spectrum of $[Co_2(CH_3\mathbf{L2})(CH_3COO)_2](PF_6)$ as solid mull at 1.5 K and 7.0 T. Gaussian deconvolved spectrum is identical with the recorded spectrum except for noise. *On the right* magnetization plots from VTVH analysis of the bands at 533 and 322 nm

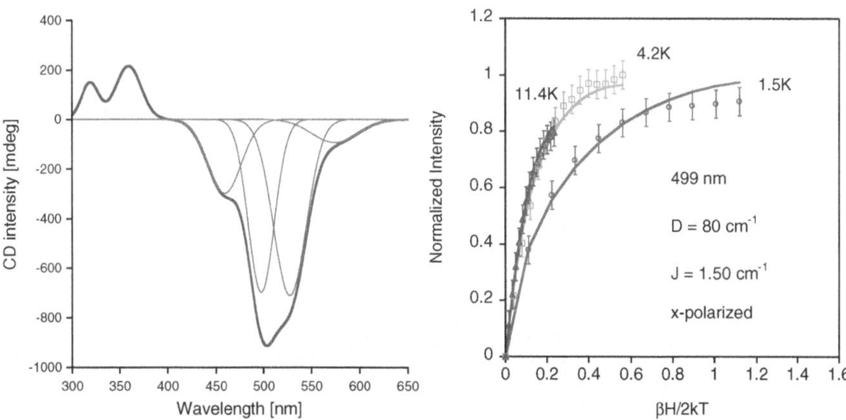

Fig. 6.21 MCD spectrum of $[Co_2(Br\mathbf{L2})(CH_3COO)_2]^+$ ethanol at 1.5 K and 7.0 T (pathlength 0.62 cm, 10 mM). Gaussian deconvolved spectrum is identical with the recorded spectrum except for noise. *On the right* magnetization plot from VTVH analysis of the 499 nm band

suggests antiferromagnetic coupling for the $[Co_2(CO_2EtH_2\mathbf{L1})(CH_3COO)_2]PF_6$, $[Co_2(CO_2Et\mathbf{L2})(CH_3COO)_2]PF_6$ and $[Co_2(CH_3\mathbf{L2})(CH_3COO)_2]PF_6$ complexes and ferromagnetic coupling for $[Co_2(NO_2\mathbf{L2})(CH_3COO)_2]PF_6$ and $[Co_2(Br\mathbf{L2})(CH_3COO)_2]PF_6$.

6.2.6.4 DFT Calculations

DFT calculations were conducted with the program ADF [39, 46] to confirm the magnetic coupling constants obtained by magnetic susceptibility and VTVH-MCD measurements. This was done in collaboration with G. Cavigliasso and R. Stranger at the Research School of Chemistry, Australian National University. Initially, single point calculations were conducted, however, after establishing the methods, only geometry optimized calculations were conducted to confirm coupling parameters. The approach was based on the broken symmetry theory [25, 26]. Table 6.7 shows the J values obtained for different functionals (BP = Becke Perdew or B3-LYP = Becke, three-parameter, Lee-Yang-Parr). The calculations were conducted for both functionals with and without relativistic corrections. In two cases the energies of the calculations did not reach convergence and a coupling constant could not be estimated ($[Co_2(CO_2Et\mathbf{L2})(CH_3COO)_2]^+$ and $[Co_2(Br\mathbf{L2})(CH_3COO)_2]^+$ B3-LYP, nonrelativistic). No complete set of parameters afforded a correct prediction of the experimental values.

The orbital mechanism (obtained from the calculations) for the magnetic exchange coupling involves interactions between the metal sites through the bridging phenolic oxygen. Most commonly a combination of Co-O σ-antibonding and π-bonding interactions as shown in Fig. 6.23.

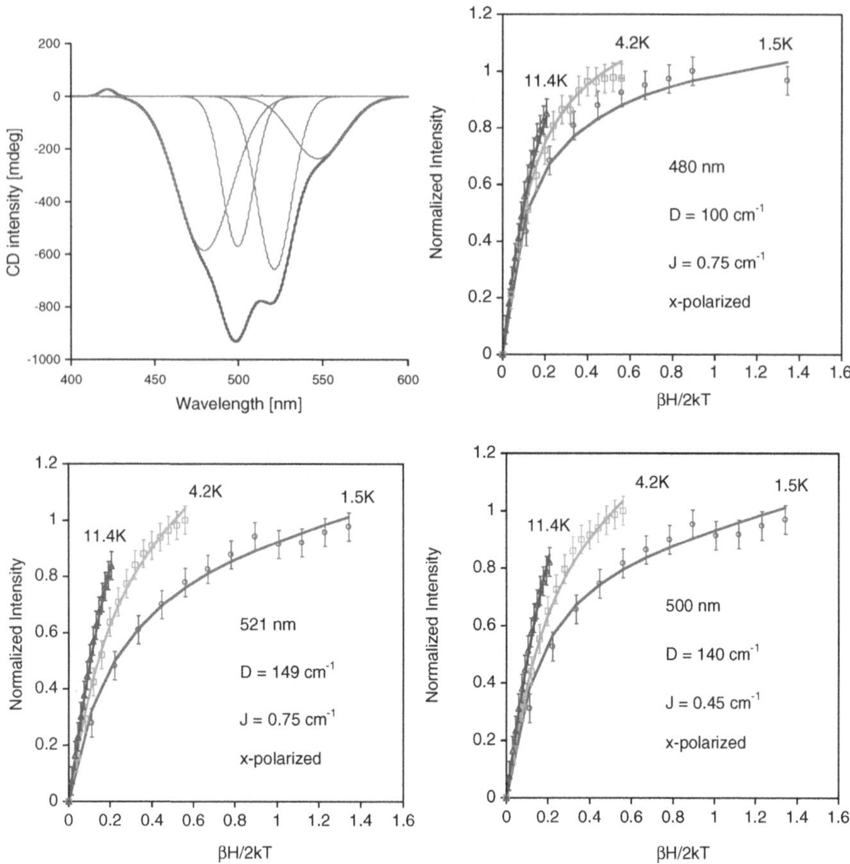

Fig. 6.22 MCD spectrum of $[Co_2(NO_2\mathbf{L2})(CH_3COO)_2]^+$ in ethanol at 1.5 K and 7.0 T (pathlength 0.62 cm, 8 mM). Gaussian deconvolved spectrum is identical with the recorded spectrum except for noise. *On the right and below* magnetization plots from VTVH analysis of the bands at 480, 500 and 521 nm

6.2.6.5 Angular Overlap Model Calculations

Spectral simulations were made using the angular overlap model (AOM) using the program AOMX [47]. Calculations were performed for $[Co_2(CO_2EtH_2\mathbf{L1})(CH_3COO)_2](PF_6)$ and $[Co_2(CO_2Et\mathbf{L2})(CH_3COO)_2](PF_6)$ based on the transitions obtained from the diffuse reflectance and MCD data. The coordinates were generated from the respective crystal structures, and each Co(II) metal site was treated separately. The calculations effectively simulated the experimental data for these two complexes; results are listed in Table 6.8. The Racah parameters C and B were fitted separately, with n in C = n B varying from 4 to 4.7. The Racah B

Table 6.6 Parameters from VTVH MCD for the $[Co_2(CO_2EtH_2\mathbf{L1})(CH_3COO)_2]PF_6$, $[Co_2(CO_2Et\mathbf{L2})(CH_3COO)_2]PF_6$, $[Co_2(CH_3\mathbf{L2})(CH_3COO)_2]PF_6$, $[Co_2(NO_2\mathbf{L2})(CH_3COO)_2]PF_6$ and $[Co_2(Br\mathbf{L2})(CH_3COO)_2]PF_6$ complexes

Complex		Transition (nm)	J[a] (cm^{-1})	D (cm^{-1})	M_{xy}	M_{xz}	M_{yz}
$[Co_2(CO_2EtH_2\mathbf{L1})(CH_3COO)_2]PF_6$	Mull	530	−0.33	100	16.91	12.61	2.97
	Mull	506	−0.33	100	10.15	15.60	3.94
	Mull	323	−0.43	100	14.32	16.17	3.18
	Ethanol	335	−0.33	100	−4.78	8.50	7.12
	Ethanol	489	−0.34	102	20.18	8.69	5.48
	Ethanol	510	−0.34	102	27.81	−4.81	5.09
$[Co_2(CO_2Et\mathbf{L2})(CH_3COO)_2]PF_6$	Ethanol	465	−0.05	108	−0.82	12.09	7.06
	Ethanol	499	−0.05	108	−1.03	43.10	2.34
	Ethanol	519	−0.05	108	0.18	31.94	0.88
	Mull	347	−0.10	100	4.26	19.74	3.16
	Mull	465	−0.10	100	0.62	27.41	6.01
	Mull	530	−0.10	100	0.72	22.83	−0.98
$[Co_2(CH_3\mathbf{L2})(CH_3COO)_2]PF_6$	Mull	240	−0.33	98	32.83	−23.22	6.11
	Mull	322	−0.33	98	21.81	14.43	5.61
	Mull	533	−0.74	98	20.34	27.11	0.73
	Ethanol	495	−0.30	100	2.91	7.57	0.84
	Ethanol	455	−0.30	100	−3.62	10.00	0.91
$[Co_2(NO_2\mathbf{L2})(CH_3COO)_2]PF_6$	Ethanol	480	0.75	100	43.01	17.79	7.37
	Ethanol	500	0.45	140	12.42	14.70	1.92
	Ethanol	510	0.75	149	18.38	16.72	1.13
$[Co_2(Br\mathbf{L2})(CH_3COO)_2]PF_6$	Ethanol	499	1.5	80	93	0	7

[a] Hamiltonian H = 2 J · S1 · S2

parameter is similar to that determined from the solution spectra, and the e_σ values are as expected in the low range of previously observed parameters; the contribution from π bonding was set to be zero to simplify calculations.

6.2.7 Binding of Phosphoesters in Solution

6.2.7.1 MCD Spectroscopy of $[Co_2(BrL2)(CH_3COO)_2]^+$ in the Presence of DPP

Upon addition of 25 equivalents of DPP, the MCD spectrum of the complex $[Co_2(Br\mathbf{L2})(CH_3COO)_2]^+$ changes (Fig. 6.24a and b). The intensities and band positions are shifted. VTVH MCD analysis of the band at ~500 nm revealed that the system is still ferromagnetically coupled in the presence of DPP; however, the coupling constant (+0.45 cm^{-1}) drops to one third of the original value (+1.5 cm^{-1}). The Co(II) ions are still hexacoordinate, and mass spectral analysis

Table 6.7 Parameters from DFT calculations for the calculation of the magnetic exchange coupling via the broken symmetry approach

Complex	Functional	Corrections	SS	E_B	ΔE	J
$[Co_2(CO_2EtH_2\mathbf{L1})(CH_3COO)_2]^+$	BP	Nonrelativistic	BS	−522.2065	−2.24	−20.8
			S = 3	−522.1833		
	B3-LYP	Nonrelativistic	BS	−611.1019	−0.31	−2.9
			S = 3	−611.0987		
	BP	Relativistic (scalar ZORA)	BS	−521.4713	−1.35	−12.5
			S = 3	−521.4573		
	B3-LYP	Relativistic (scalar ZORA)	BS	−610.2982	0.15	1.4
			S = 3	−610.2998		
$[Co_2(CO_2Et\mathbf{L2})(CH_3COO)_2]^+$	BP	Nonrelativistic	BS	−553.5710	−0.55	−5.1
			S = 3	−553.5653		
	B3-LYP	Nonrelativistic	BS		[a]	
			S = 3			
	BP	Relativistic (scalar ZORA)	BS	−552.8115	−1.20	−11.1
			S = 3	−552.7991		
	B3-LYP	Relativistic (scalar ZORA)	BS	−646.1604	−2.13	−19.8
			S = 3	−646.1383		
$[Co_2(CH_3\mathbf{L2})(CH_3COO)_2]^+$	BP	Nonrelativistic	BS	−515.0187	−5.01	−46.5
			S = 3	−514.9668		
	B3-LYP	Nonrelativistic	BS	−601.1751	−5.36	−49.8
			S = 3	−601.1195		
	BP	Relativistic (scalar ZORA)	BS	−514.2820	−3.38	−31.4
			S = 3	−514.2470		
	B3-LYP	Relativistic (scalar ZORA)	BS	−600.3664	−4.56	−42.4
			S = 3	−600.3191		

(continued)

Table 6.7 (continued)

Complex	Functional	Corrections	SS	E_B	ΔE	J
$[Co_2(NO_2\mathbf{L2})(CH_3COO)_2]^+$	BP	Nonrelativistic	BS	−514.0687	−5.10	−47.4
			S = 3	−514.0158		
	B3-LYP	Nonrelativistic	BS	−602.8913	−6.62	−61.5
			S = 3	−602.8227		
	BP	Relativistic	BS	−513.3004	1.90	17.7
		(scalar ZORA)	S = 3	−513.3201		
	B3-LYP	Relativistic	BS	−602.0422	3.00	27.9
		(scalar ZORA)	S = 3	−602.0733		
$[Co_2(Br\mathbf{L2})(CH_3COO)_2]^+$	BP	Nonrelativistic	BS	−496.4027	−1.85	−17.2
			S = 3	−496.3835		
	B3-LYP	Nonrelativistic	BS		[a]	
			S = 3			
	BP	Relativistic	BS	−495.6906	1.88	−17.5
		(scalar ZORA)	S = 3	−495.6711		
	B3-LYP	Relativistic	BS	−579.7643	0.33	3.0
		(scalar ZORA)	S = 3	−579.7677		

[a] The S = 3 calculation did not reach full convergence in several attempts, BP = Becke Perdew, ΔE = Energy difference, E_B = Bonding Energy, BS = broken symmetry state, SS = Spin State

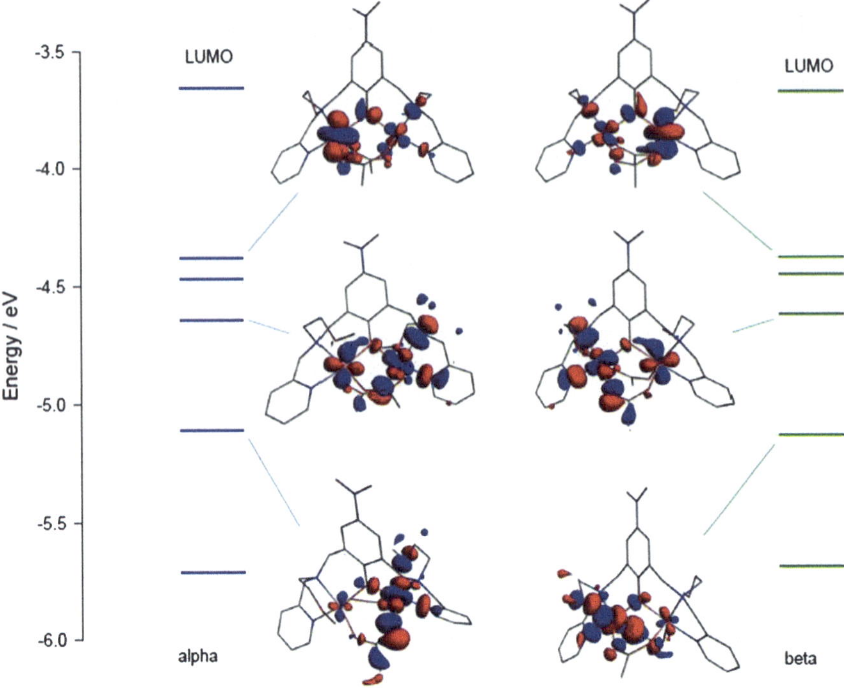

Fig. 6.23 Energy levels and spatial orbital plots for the $[Co_2(NO_2\mathbf{L2})(CH_3COO)_2]^+$ complex

revealed that two DPP molecules are bound simultaneously. The $[Co_2(Br\mathbf{L2})(CH_3COO)_2](PF_6)$ complex was chosen for the MCD experiments because its spectrum has a better signal-to-noise ratio in ethanol than the other complexes.

6.2.7.2 Mass Spectrometry of the Complexes in the Presence of DPP

The binding of phosphate esters in solution was also investigated by mass spectrometry. Figure 6.25a shows the mass spectrum of $[Co_2(CO_2EtH_2\mathbf{L1})(CH_3COO)_2]^+$ in the presence of 25 equivalents of DPP.

The base peak at *m/z* 1109.0 is proposed to arise from a species of one single negatively charged ligand, two Co(II) and two deprotonated diphenyl phosphate molecules (calc. *m/z* 1109.2 (100 %), 1110.2 (56 %)). A peak at m/z 895.0 is assigned to a species with $CO_2EtH_2\mathbf{L1}^-$, two Co(II) ions, one deprotonated DPP molecule, a water and a hydroxide molecule (calc. *m/z* 895.2 (100 %), 896.2 (43 %)). The latter peak illustrates that the acetates are readily replaced in solution by substrate and solvent molecules. In Fig. 6.25b the mass spectrum of the complex with $CO_2EtH\mathbf{L2}$ and DPP is shown. A prominent species arising from one negative charged ligand, two Co(II) ions and two deprotonated DPP molecules

Table 6.8 AOMX calculations for $[Co_2(CO_2EtH_2\mathbf{L1})(CH_3COO)_2](PF_6)$ and $[Co_2(CO_2Et\mathbf{L2})(CH_3COO)_2](PF_6)$

Complex	C	B	Transition in O_h symmetry[a]	Obs. (nm)	Calc. (nm)	ε_σ N_{tert} (cm^{-1})	ε_σ N_{py}(cm^{-1})	ε_σ O_{Phenol} (cm^{-1})	ε_σ O_{Ac} (cm^{-1})	ε_σ $O_{alcohol/ether}$ (cm^{-1})
Co(1) $[Co_2(CO_2EtH_2\mathbf{L1})$	3,390	715	$^4T_{1g} \rightarrow {}^4T_{2g}$	1,304	1,323	4,576	5,002	4,801	3,489	726
$(CH_3COO)_2](PF_6)$				1,046	992					
			$^4T_{1g} \rightarrow {}^4A_{2g}$	568	588					
			$^4T_{1g} \rightarrow {}^4T_{1g}(P)$	541	539					
				510	503					
				490	489					
			$^4T_{1g} \rightarrow$ doublet	467	468					
Co(2) $[Co_2(CO_2EtH_2\mathbf{L1})$	3,237	785	$^4T_{1g} \rightarrow {}^4T_{2g}$	1,304	1,299	3,500	3,522	4,075	3,670	999
$(CH_3COO)_2](PF_6)$				1,046	1,095					
			$^4T_{1g} \rightarrow {}^4A_{2g}$	568	560					
			$^4T_{1g} \rightarrow {}^4T_{1g}(P)$	541	532					
				510	519					
				490	487					
			$^4T_{1g} \rightarrow$ doublet	467	468					
Co(1) $[Co_2(CO_2Et\mathbf{L2})$	3,309	819	$^4T_{1g} \rightarrow {}^4T_{2g}$	1,256	1,279	3,480	4,539	3,686	3,998	980
$(CH_3COO)_2](PF_6)$				1,053	1,061					
			$^4T_{1g} \rightarrow {}^4A_{2g}$	546	543					
			$^4T_{1g} \rightarrow {}^4T_{1g}(P)$	518	518					
				500	500					
				465	466					
Co(2) $[Co_2(CO_2Et\mathbf{L2})$	3,319	808	$^4T_{1g} \rightarrow {}^4T_{2g}$	1,256	1,273	3,895	4,773	4,773	3,475	698
$(CH_3COO)_2](PF_6)$				1,053	1,047					
			$^4T_{1g} \rightarrow {}^4A_{2g}$	546	546					
			$^4T_{1g} \rightarrow {}^4T_{1g}(P)$	518	518					
				500	498					
				465	468					

[a] All near IR bands are from the diffuse reflectance spectra, all other bands are taken from MCD

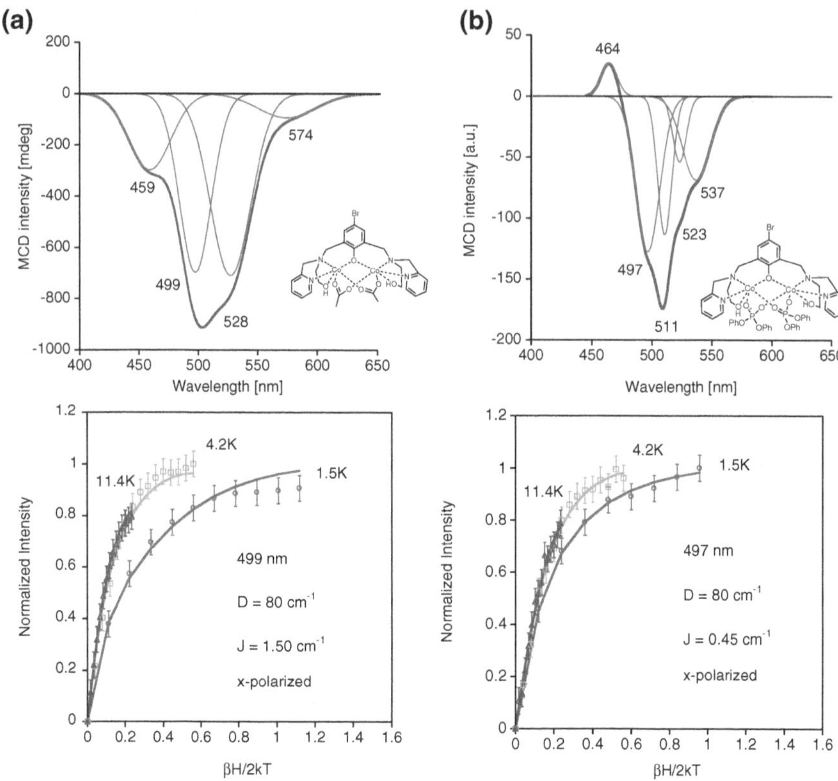

Fig. 6.24 MCD spectra of $[Co_2(Br\mathbf{L2})(CH_3COO)_2]^+$ in ethanol at 1.5 K and 7.0 T (pathlength 0.62 cm, 10 mM) in the presence and absence of 25 equivalents DPP. After addition of DPP, the solution was *left* at room temperature for 12 h prior to recording the spectra. Gaussian deconvolved spectra are identical with the recorded spectra except for noise. **a** $[Co_2(Br\mathbf{L2})(CH_3COO)_2]^+$ in ethanol including magnetization plot of the band at 499 nm below. **b** $[Co_2(Br\mathbf{L2})(CH_3COO)_2]^+$ + 10 equivalents DPP. *Below* magnetization plot from VTVH analysis of the band at 497 nm of the spectrum with DPP present

is found at *m/z* 1137.0 (calc. m/z 1137.2 (100 %), 1138.2 (59 %)). In addition a small peak from $CO_2Et\mathbf{L2}^-$, two Co(II) ions, one deprotonated DPP molecule, a hydroxide molecule and a coordinating acetonitrile is present (found *m/z* 947.0, calc *m/z* 946.2 (100 %), 947.2 (48 %)). Figure 6.26 also shows the mass spectrum of $[Co_2(Br\mathbf{L2})(CH_3COO)_2]^+$ in the presence of 25 equivalents DPP. Only one species is observed, comprised of two cobalt ions, one negatively charge ligand and two negatively charged diphenyl phosphate molecules (found *m/z* 1144.9, calc. 1143.1 (100 %), 1145.1 (97 %)).

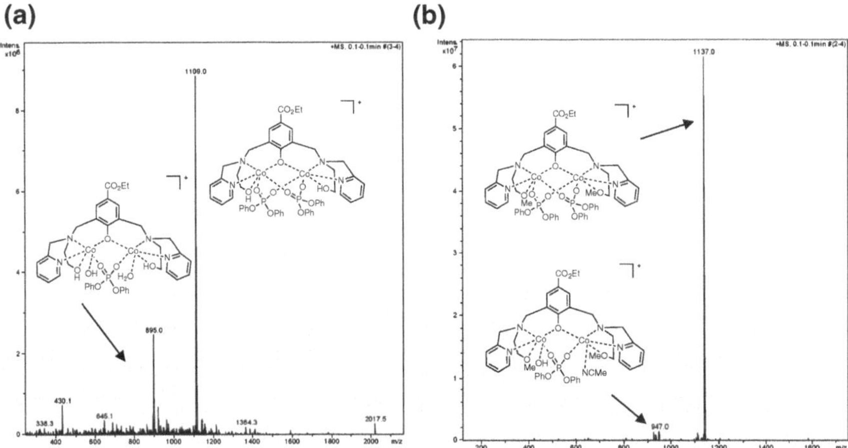

Fig. 6.25 **a** Mass spectrum of $[Co_2(CO_2EtH_2\mathbf{L1})(CH_3COO)_2]^+$ in the presence of 25 equivalents diphenyl phosphate measured in acetonitrile. $[Co_2(CO_2EtH_2\mathbf{L1})(CH_3COO)_2]^+$ was added to a solution of diphenyl phosphate in acetonitrile (0.01 mM final concentration of complex and 0.25 mM DPP) 10 min prior to spectra recording. **b** Mass spectrum of $[Co_2(CO_2Et\mathbf{L2})(CH_3COO)_2]^+$ in the presence of 25 equivalents DPP measured in acetonitrile. $[Co_2(CO_2Et\mathbf{L2})(CH_3COO)_2]^+$ was added to a solution of DPP in acetonitrile (0.01 mM final concentration of complex and 0.25 mM DPP). Spectrum recorded 5 min after mixing

6.2.8 Phosphatase-like Kinetics

The activity towards organophosphoesters using the substrate BDNPP was investigated. All complexes are good functional mimics for phosphoesterase enzymes and show one pK_a relevant for hydrolysis (Fig. 6.27a and b; Table 6.9).

The data were fit to Eq. 2.3 (see Chap. 2) derived for a monoprotic system [48]. For all **L2** ligands the kinetically relevant pK_a is in the range of 8.12–8.75 and points to a terminal water molecule bound to Co(II) as the nucleophile. For the Co(II) complex with the ligand $CO_2EtH_2\mathbf{L1}$ fitting of the data results in a pK_a of 10.54 similar to the one found in the Cd(II) complex of this ligand (Chap. 5).

Substrate dependence was measured at the pH with highest activity for each complex and followed Michaelis-Menten saturation behavior (Fig. 6.28; Table 6.9). Complex dependence was linear from 0 to 0.12 mM [complex]. The complexes are among the most active compounds known to date for the cleavage of BDNPP [32].

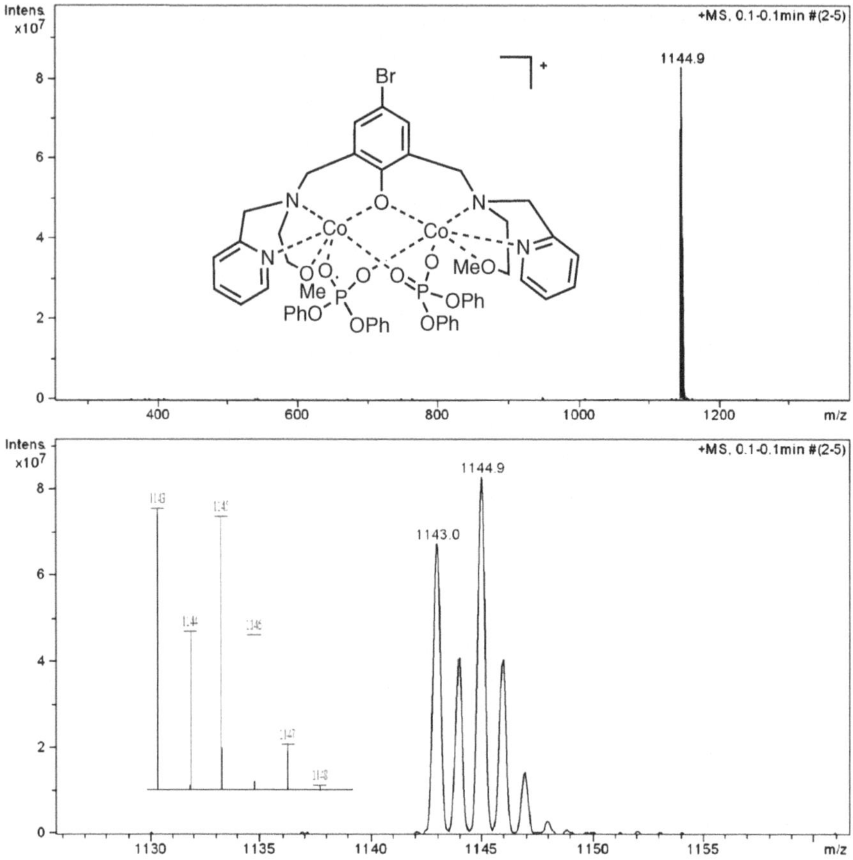

Fig. 6.26 Mass spectrum of $[Co_2(Br\mathbf{L2})(CH_3COO)_2]^+$ in the presence of 25 equivalents DPP measured in acetonitrile. The complex was added to a solution of DPP in acetonitrile (0.01 mM final concentration of complex and 0.25 mM DPP). Spectrum recorded 5 min after mixing

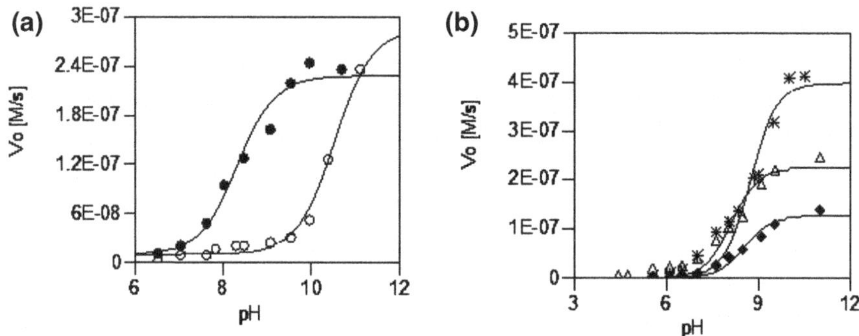

Fig. 6.27 pH dependence profiles for **a** $[Co_2(CO_2EtH_2\mathbf{L1})(CH_3COO)_2]PF_6$ (*open circle*) and $[Co_2(CO_2Et\mathbf{L2})(CH_3COO)_2]PF_6$ (*filled circle*) and **b** $[Co_2(CH_3\mathbf{L2})(CH_3COO)_2]PF_6$ (*filled diamond*), $[Co_2(NO_2\mathbf{L2})(CH_3COO)_2]PF_6$ (*open triangle*) and $[Co_2(Br\mathbf{L2})(CH_3COO)_2]PF_6$ (*asterisk*)

Table 6.9 Kinetic parameters obtained for the complexes in BDNPP hydrolysis

Complex	pK_a	pH with highest activity	v_o (M/s) $\times 10^{-7}$	k_{cat} (1/s) $\times 10^{-3}$	K_m (mM)
$[Co_2(CO_2EtH_2\mathbf{L1})(CH_3COO)_2]PF_6$	10.54 ± 0.10	10.70	3.65 ± 0.38	9.12 ± 0.95	3.26 ± 0.80
$[Co_2(CO_2Et\mathbf{L2})(CH_3COO)_2]PF_6$	8.34 ± 0.11	10.40	4.56 ± 0.44	11.40 ± 1.10	4.31 ± 0.85
$[Co_2(CH_3\mathbf{L2})(CH_3COO)_2]PF_6$	8.53 ± 0.11	11.00	2.19 ± 0.25	5.48 ± 0.63	4.48 ± 1.13
$[Co_2(NO_2\mathbf{L2})(CH_3COO)_2]PF_6$	8.12 ± 0.12	9.55	3.69 ± 0.66	9.23 ± 1.65	6.83 ± 2.48
$[Co_2(Br\mathbf{L2})(CH_3COO)_2]PF_6$	8.75 ± 0.12	10.00	7.64 ± 1.66	19.10 ± 4.15	4.62 ± 1.62

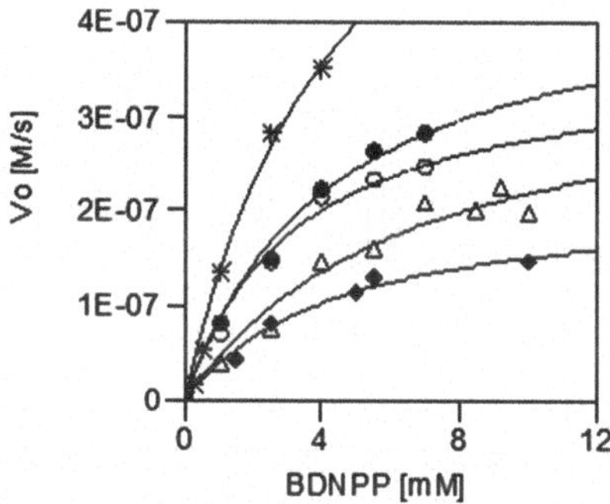

Fig. 6.28 Michaelis-Menten plots for the hydrolysis of BDNPP catalyzed by $[Co_2(CO_2EtH_2\mathbf{L1})(CH_3COO)_2]PF_6$ (*open circle*), $[Co_2(CO_2Et\mathbf{L2})(CH_3COO)_2]PF_6$ (*filled circle*), $[Co_2(CH_3\mathbf{L2})(CH_3COO)_2]PF_6$ (*filled diamond*), $[Co_2(NO_2\mathbf{L2})(CH_3COO)_2]PF_6$ (*open triangle*) and $[Co_2(Br\mathbf{L2})(CH_3COO)_2]PF_6$ (*asterisk*)

6.2.9 Binding of β-Lactams in Solution

6.2.9.1 Mass Spectra of $[Co_2(CO_2EtH_2L1)(CH_3COO)_2]^+$ and $[Co_2(CO_2EtL2)(CH_3COO)_2]^+$ in the Presence of Penicillin and Nitrocefin

Figure 6.29a shows the mass spectrum of $[Co_2(CO_2EtH_2\mathbf{L1})(CH_3COO)_2]^+$ in the presence of one equivalent nitrocefin. A peak at *m/z* 1185.0 is attributed to a species comprised of one ligand, two Co(II) ions, one acetonitrile solvent molecule and one water molecule with a hydrolyzed nitrocefin molecule bound to the complex through the alcohol arm of the ligand (calc. *m/z* 1184.2 (100 %), 1185.2 (55 %)); and the peak at *m/z* 1640.9 is assigned to a similar species as above but with an additional not hydrolyzed nitrocefin bound and no solvent molecules (calc. *m/z* 1641.18 (100.0 %), 1642.18 (79.4 %)). That nitrocefin was hydrolyzed under these conditions was also apparent from the color change from yellow to red after 1 min of mixing complex and substrate.

In Fig. 6.29b the mass spectrum of $[Co_2(CO_2Et\mathbf{L2})(CH_3COO)_2]^+$ and nitrocefin is shown. No color change was observed, indicating that the complex with $CO_2EtH\mathbf{L2}$ is not capable of hydrolyzing the substrate. This complex is, however, able to bind up to two non-hydrolyzed nitrocefin molecules as shown in the mass

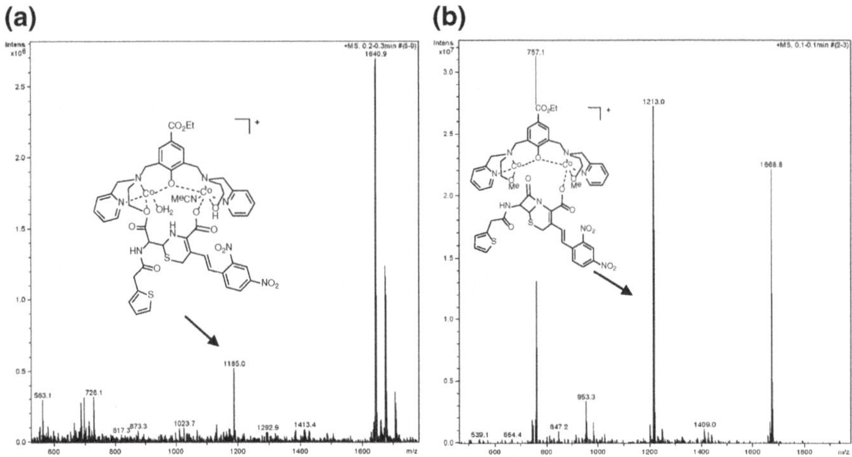

Fig. 6.29 Mass spectra of the cobalt complexes in the presence of one equivalent nitrocefin measured in acetonitrile. **a** $[Co_2(CO_2EtH_2\mathbf{L1})(CH_3COO)_2]^+$ was added to a solution of nitrocefin in acetonitrile (both 0.01 mM final concentration) and a color change from yellow to red was apparent after 1 min. The spectrum was recorded 5 min after mixing. **b** $[Co_2(CO_2Et\mathbf{L2})(CH_3COO)_2]^+$ was added to a solution of nitrocefin in acetonitrile (both 0.01 mM final concentration). No color change was observed

spectrum with the peaks at *m/z* 1213.0 (calc. *m/z* 1212.2 (100 %), 1213.2 (57 %)) and *m/z* 1668.8 (calc. *m/z* 1669.2 (100 %), 1670.2 (85 %)).

A similar experiment was conducted with the lactam antibiotic penicillin G. Figure 6.30a shows the mass spectrum of $[Co_2(CO_2EtH_2\mathbf{L1})(CH_3COO)_2]^+$ in the presence of one equivalent penicillin G. As found in the experiment with nitrocefin a peak at m/z 1003.1 is attributed to a species comprised of one ligand, two Co(II) ions, one acetonitrile molecule and one water molecule with a hydrolyzed penicillin G molecule bound to the complex through the alcohol arm of the ligand (calc. m/z 1002.23 (100.0 %), 1003.24 (49.7 %)).

In Fig. 6.30b the mass spectrum of $[Co_2(CO_2Et\mathbf{L2})(CH_3COO)_2]^+$ and penicillin G is shown. No hydrolysis of the penicillin was observed. This complex is however, also able to bind this lactam substrate as shown in the mass spectrum with the peak at *m/z* 1031.1 (calc. *m/z* 1030.3 (100 %), 1031.3 (52 %)).

6.2.9.2 In Solution Infrared Measurements of $[Co_2(CO_2EtH_2L1)(CH_3COO)_2]^+$ in the Presence of Penicillin

The in-solution IR measurements show that $[Co_2(CO_2EtH_2\mathbf{L1})(CH_3COO)_2]^+$ hydrolyzes the lactam bond in penicillin forming an ester (rather than a carboxylic acid which would occur at lower wavenumbers) as apparent in Fig. 6.31. Due to the overlapping of the two bands (1,780 and 1,760 cm^{-1}) the changes could only

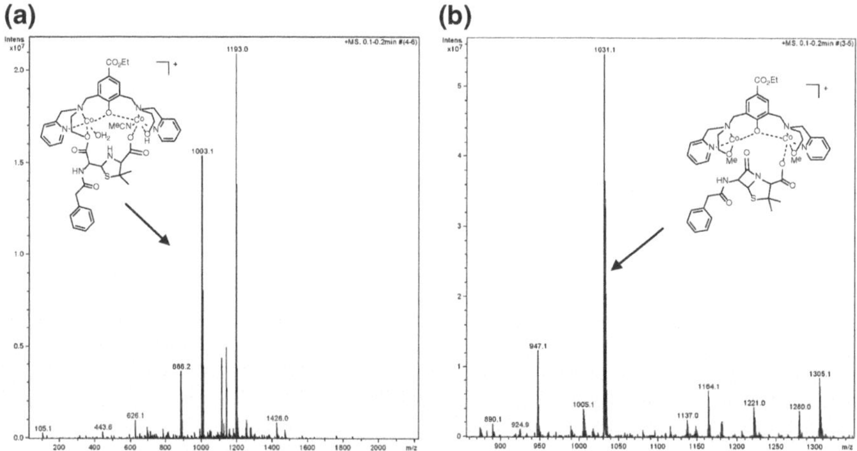

Fig. 6.30 Mass spectra of the Co(II) complexes in the presence of one equivalent penicillin G measured in acetonitrile. **a** $[Co_2(CO_2EtH_2\mathbf{L1})(CH_3COO)_2]^+$ was added to a solution of penicillin G in acetonitrile (both 0.01 mM final concentration and the spectrum was recorded 5 min after mixing. **b** $[Co_2(CO_2Et\mathbf{L2})(CH_3COO)_2]^+$ was added to a solution of penicillin G in acetonitrile (both 0.01 mM final concentration) and the spectrum was recorded 5 min after mixing

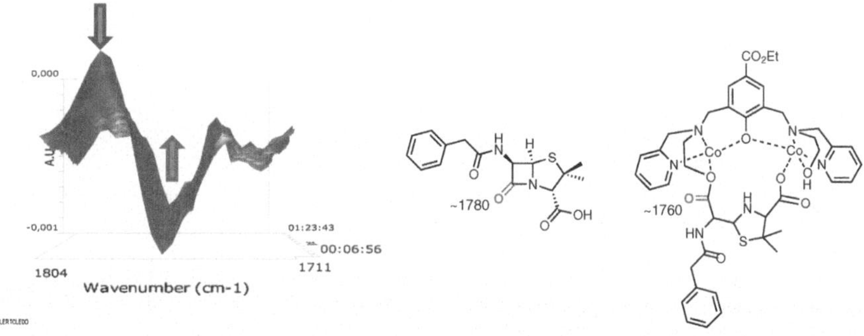

Fig. 6.31 IR spectral changes in the first derivative absorbance spectra for the hydrolysis of penicillin G by $[Co_2(CO_2EtH_2\mathbf{L1})(CH_3COO)_2]^+$ in DMSO:H_2O (9:1)

be visualized in the derivative spectra. The x-axis displays the small region of the relevant bands whereas the time course of the hydrolysis is displayed on the z-axis.

6.2.9.3 MCD Measurements of $[Co_2(CO_2EtH_2L1)(CH_3COO)_2]^+$ and $[Co_2(CO_2EtL2)(CH_3COO)_2]^+$ in the Presence of Penicillin

MCD spectra in the presence of penicillin G were recorded for the complexes $[Co_2(CO_2EtH_2\mathbf{L1})(CH_3COO)_2]^+$ and $[Co_2(CO_2Et\mathbf{L2})(CH_3COO)_2]^+$ (Figs. 6.32 and 6.34). Nitrocefin was initially used as well however, the hydrolysis product absorbs very strongly and it was thus impossible to produce a good spectrum. A color

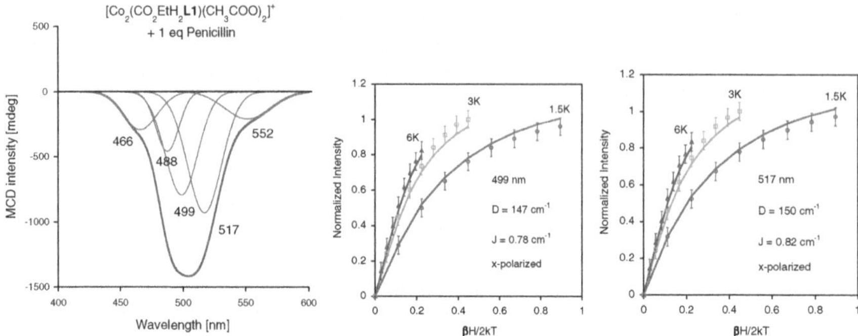

Fig. 6.32 Gaussian deconvoluted ethanol MCD spectrum of $[Co_2(CO_2EtH_2\mathbf{L1})(CH_3COO)_2]^+$ in the presence of one equivalent penicillin. The spectrum was recorded 1 h after mixing and subsequent incubation at 40 °C. *On the right* magnetization plots obtained after VTVH analysis at 1.5, 3, 6, 12, 25 and 50 K of two selected transitions (499 and 517 nm). For clarity the plots for 12, 25 and 50 K have been omitted

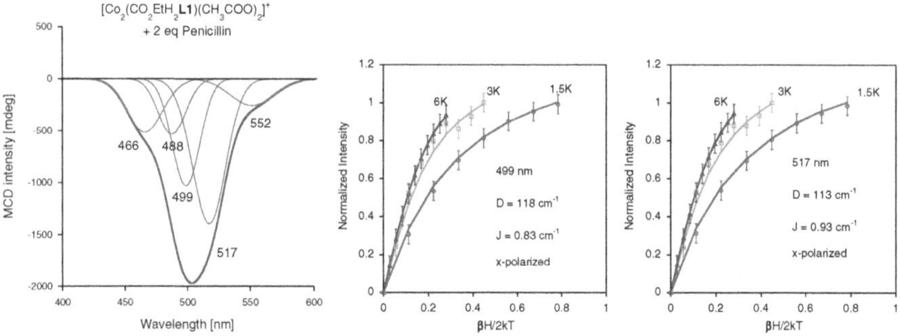

Fig. 6.33 Gaussian deconvoluted ethanol MCD spectrum of $[Co_2(CO_2EtH_2\mathbf{L1})(CH_3COO)_2]^+$ in the presence of two equivalents penicillin. The spectrum was recorded 1 h after adding the second equivalent and subsequent incubation at 40 °C. *On the right* magnetization plots obtained after VTVH analysis at 1.5, 3, 6, 12 and 50 K of two selected transitions (499 and 517 nm). For clarity the plots for 12 and 50 K have been omitted

change from pink to orange was observed for the solutions made of $[Co_2(CO_2EtH_2\mathbf{L1})(CH_3COO)_2]^+$ and one or two equivalents of penicillin suggesting hydrolysis of the β-lactam substrate (which has a yellow color when hydrolyzed). Interestingly the VTVH analysis for all complexes in the presence of penicillin showed that the two Co(II) ions are coupled ferromagnetically. The addition of a second equivalent of penicillin to the complex $[Co_2(CO_2EtH_2\mathbf{L1})(CH_3COO)_2]^+$ does not greatly affect the geometry or the magnetic coupling (Fig. 6.33). For the addition of penicillin to a solution of the complex $[Co_2(CO_2Et\mathbf{L2})(CH_3COO)_2]^+$ the second equivalent seems to influence the geometry more strongly (Fig. 6.35). A small amount of five coordinate Co(II) is detected (as apparent with the transition at 612 nm). The VTVH data of this transition were analyzed, however the spectra were

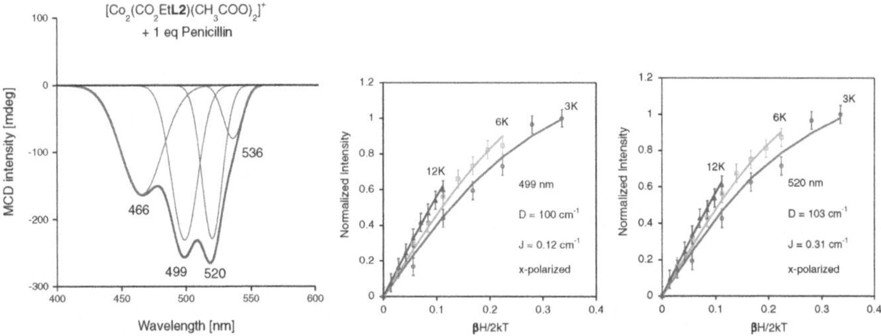

Fig. 6.34 Gaussian deconvoluted ethanol MCD spectrum of $[Co_2(CO_2Et\mathbf{L2})(CH_3COO)_2]^+$ in the presence of one equivalent penicillin. The spectrum was recorded 1 h after mixing and subsequent incubation at 40 °C. *On the right* magnetization plots obtained after VTVH analysis at 3, 6 and 12 K of two selected transitions (499 and 520 nm)

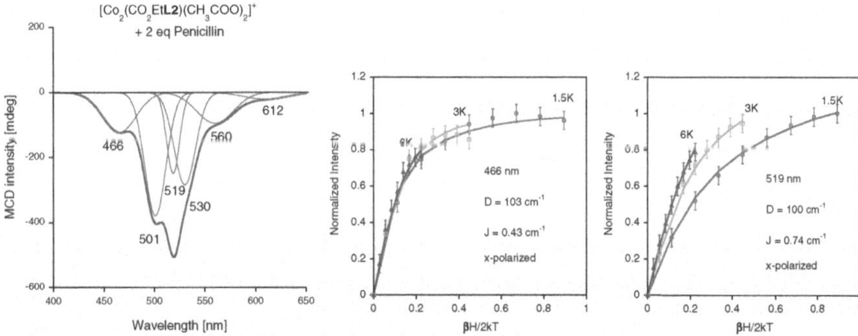

Fig. 6.35 Gaussian deconvoluted ethanol MCD spectrum of $[Co_2(CO_2Et\mathbf{L2})(CH_3COO)_2]^+$ in the presence of two equivalents penicillin. The spectrum was recorded 1 h after adding the second equivalent and subsequent incubation at 40 °C. *On the right* magnetization plots obtained after VTVH analysis at 1.5, 3, and 6 K of two selected transitions

too weak and noisy to produce a satisfactory fit. As the substrate remains intact and is not hydrolyzed by $[Co_2(CO_2Et\mathbf{L2})(CH_3COO)_2]^+$ it is competing with the ligand arms for coordination sites and is possibly partly displacing one of the only loosely coordinated methyl ether arms. For the $[Co_2(CO_2EtH_2\mathbf{L1})(CH_3COO)_2](PF_6)$ and $[Co_2(CO_2Et\mathbf{L2})(CH_3COO)_2](PF_6)$, in the absence of the substrate, magnetic susceptibility and VTVH analyses suggested that the metal ions were weakly antiferromagnetically coupled (Sect. 6.2.6.3). The result indicates that changes in the bridging ligand (hydrolyzed or non-hydrolyzed penicillin G) can influence the sign and magnitude of J, despite the major magnetic exchange occurring through the phenolic oxygen bridge.

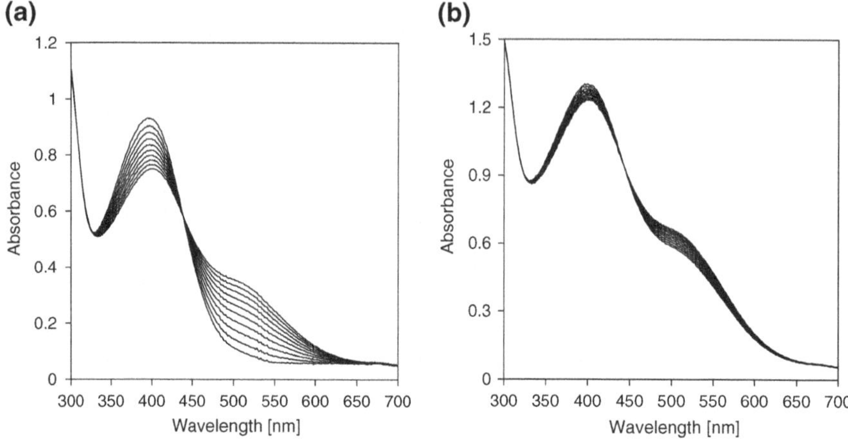

Fig. 6.36 UV-Vis spectra of the hydrolysis of successively one (**a**) and two equivalents (**b**) of nitrocefin by $[Co_2(CO_2EtH_2\mathbf{L1})(CH_3COO)_2]^+$; pH 8, buffer:MeCN 1:1, (first 10 min of the reaction are shown, initial concentration of complex and nitrocefin were 0.05 mM; spectra recorded at 1 min intervals); 37 °C

6.2.10 Lactamase-like Activity

Successively, one and two equivalents of nitrocefin were added to $[Co_2(CO_2EtH_2\mathbf{L1})(CH_3COO)_2]^+$ and it was apparent that the second equivalent of nitrocefin is hydrolyzed with rates significantly lower than those for the first equivalent (Fig. 6.36a and b), in agreement with what was found for the analogous Cd(II) complex (Refer to Chap. 5). Under the same experimental conditions $[Co_2(CO_2Et\mathbf{L2})(CH_3COO)_2]^+$ was unable to hydrolyze nitrocefin as the observed rates are consistent with autohydrolysis. It is proposed that in the complex $[Co_2(CO_2EtH_2\mathbf{L1})(CH_3COO)_2]^+$ the alkoxide of the ligand acts as the nucleophile and that the hydrolyzed nitrocefin remains bound.

The same experiment was conducted with $[Co_2(CO_2Et\mathbf{L2})(CH_3COO)_2]^+$ at pH 8, however, only autohydrolysis was detected. Figure 6.37 shows the species distribution for the hydrolysis of the first equivalent nitrocefin by $[Co_2(CO_2EtH_2\mathbf{L1})(CH_3COO)_2]^+$. The species distribution was obtained after fitting the UV-Vis data from Fig. 6.36 using the software ReactLab [49].

The data were fitted to a model where nitrocefin first binds to $[Co_2(CO_2EtH_2\mathbf{L1})(CH_3COO)_2]^+$ (k_1) and then is subsequently hydrolyzed by this complex (k_2). Analysis of the data (300–700 nm, [complex] 0.05 mM, [substrate] 0.05 mM, pH 8, 37 °C) resulted in $k_1 = 2.5 \times 10^3 \pm 1.9 \times 10^1\ M^{-1}\ min^{-1}$ and $k_2 = 1.6 \times 10^{-1} \pm 8.1 \times 10^{-4}\ min^{-1}$. It should be mentioned that the second equivalent was more slowly hydrolyzed by $[Co_2(CO_2EtH_2\mathbf{L1})(CH_3COO)_2]^+$ than with the respective Cd(II) complex (Chap. 5). It is assumed that this is due to the larger size of Cd(II) and thus, to the greater ability to accommodate and hydrolyse a second nitrocefin molecule efficiently.

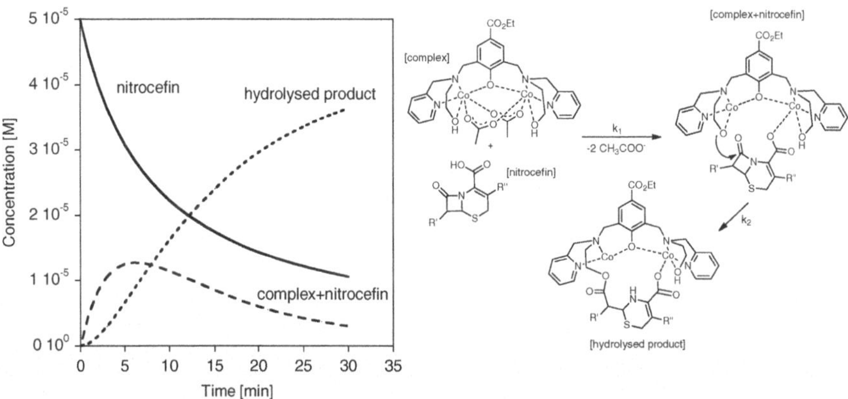

Fig. 6.37 Species distribution and proposed mechanism of nitrocefin hydrolysis by $[Co_2(CO_2EtH_2\mathbf{L1})(CH_3COO)_2]^+$

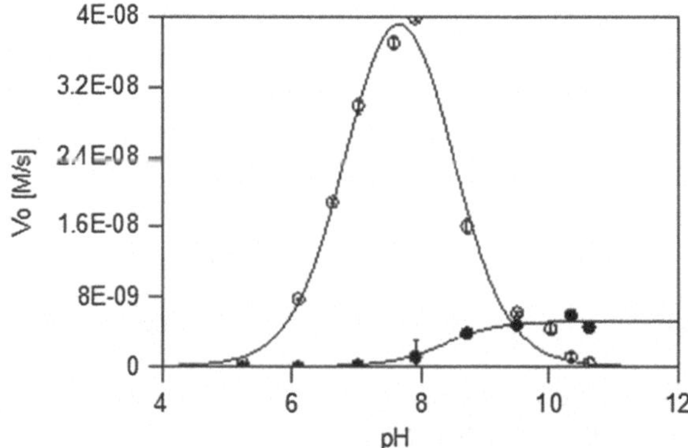

Fig. 6.38 pH rate profile for $[Co_2(CO_2EtH_2\mathbf{L1})(CH_3COO)_2]^+$ (*open square*) and $[Co_2(CO_2Et\mathbf{L2})(CH_3COO)_2]^+$ (*filled circle*); (aqueous multi-component buffer (50 mM in each MES, pH 5.50–6.50; HEPES, pH 7.00–8.50; CHES, pH 9.00–10.00; CAPS, pH 10.5–11.0, ionic strength controlled by 250 mM $LiClO_4$); 50:50 MeCN: buffer, 50 μM [complex] and [nitrocefin]

For $[Co_2(CO_2EtH_2\mathbf{L1})(CH_3COO)_2]^+$ the pH dependence with one equivalent nitrocefin resulted in a bell shaped profile with the two pK_as ($pK_{a1} = 6.88 \pm 0.74$ and $pK_{a2} = 8.45 \pm 0.68$) (Fig. 6.38). The first pK_a is attributed to the deprotonation of an alkoxide ligand arm coordinated to Co(II). The second pK_a is proposed to arise from a water molecule which, upon deprotonation, competes with the substrate for coordination, thus lowering hydrolysis rates.

For $[Co_2(CO_2Et\mathbf{L2})(CH_3COO)_2]^+$ the pH dependence with one equivalent nitrocefin resulted in a sigmoidal profile with $pK_{a1} = 8.47 \pm 0.14$ (Fig. 6.38); the pK_a is similar to that observed for organophosphate hydrolysis, and in the absence of the alkoxide the nucleophile is likely to be a terminal water molecule.

6.3 Discussion

6.3.1 Structures

The structures of the Co(II) complexes show that for the **L2** ligands the Co(II)-O ether arm distances (2.223–2.294 Å) are in general longer than with the CO_2EtH_3**L1** ligand bearing alcohol arms (2.127–2.169 Å). This is in accordance to what was found for Co(II)-methyl ether bond lengths of a similar dinucleating ligand with a phenol core [Co(bomp)$(OAc)_2$]BPh_4, with Co(II)-ether arm distances ranging from 2.160 to 2.229 Å [22]. The Co(II)-alcohol arm distances for the complex [Co(bhmp)$(OAc)_2$]BPh_4 reported by Sakiyama et al. are similar (2.070–2.178 Å) to those found for [Co_2(CO_2EtH_2**L1**)$(CH_3COO)_2$]PF_6 (2.127(8)–2.169(8) Å) [21]. A significant distortion of the coordination sphere around the Co(II) centers is apparent in all structures with the bond distances to the six donor atoms in e.g. [Co_2(CO_2EtH_2**L1**)$(CH_3COO)_2$]PF_6 ranging from 1.989 Å (Co(1)-O(1)) to 2.187 Å (Co(1)-O(6)). The Co(1)•••Co(2) separation is 3.394 Å in [Co_2(CO_2EtH_2**L1**) $(CH_3COO)_2$]PF_6, 3.372 Å for [Co_2(CO_2Et**L2**)$(CH_3COO)_2$]PF_6, 3.346 Å [Co_2(CH_3**L2**)$(CH_3COO)_2$]PF_6 and 3.363 Å for [Co_2(Br**L2**)$(CH_3COO)_2$]PF_6. Similar separation was found in the cobalt complexes [Co(bhmp)$(OAc)_2$]BPh_4 and [Co(bomp)$(OAc)_2$]BPh_4 (3.356 and 3.336 Å) [21, 22]. Interestingly, the calculated separation of the cobalt atoms in the ligand [Co_2(NO_2**L2**)$(CH_3COO)_2$]PF_6 differs significantly from the other structures with a large separation of 3.455 Å. For all other complexes calculated bond lengths and angles are in good agreement with the experimental values.

6.3.2 Spectroscopy and Magnetism

The magnetic susceptibility measurements showed that the complexes exhibit weak antiferromagnetic (J = −0.66, −0.10 and −0.67 cm^{-1} at 500 Oe for [Co_2 (CO_2EtH_2**L1**)$(CH_3COO)_2$]PF_6, [Co_2(CO_2Et**L2**)$(CH_3COO)_2$]PF_6 and [Co_2(CH_3**L2**) $(CH_3COO)_2$](PF_6), respectively, or ferromagnetic coupling (J = 2.38 and 3.09 at 500 Oe for [Co_2(NO_2**L2**)$(CH_3COO)_2$]PF_6 and [Co_2(Br**L2**)$(CH_3COO)_2$]PF_6) suggesting that the substituent in *para*-position of the phenolic oxygen influences the magnetic exchange coupling between the two Co(II) centers. The zero field splitting parameter D moreover is typical for six coordinate Co(II) and ranges from 95 to 146 cm^{-1} for the complexes. ZFS parameters of similar complexes have been reported to range from 69 to 144 cm^{-1} [21, 22]. Other dinuclear Co(II) complexes have exhibited D values as low as 30 and 44 cm^{-1} [14, 19]. DFT calculations suggested the main exchange pathway proceeds via the bridging phenol, however, a linear correlation between the Hammett parameter of the *para*-substituents, the Co-O-Co distance or angle could not be found. The MCD spectra of the complexes in ethanol were dominated by three negative bands (approximately around 490, 515,

540 nm) arising from the $^4T_{1g} \rightarrow {}^4T_{1g}(P)$ transition in high spin hexacoordinate Co(II). A weak positive band around 400 nm present in some of the spectra ($[Co_2(CO_2EtH_2\mathbf{L1})(CH_3COO)_2]PF_6$ and $[Co_2(CO_2Et\mathbf{L2})(CH_3COO)_2]PF_6$ in ethanol) was tentatively assigned to a transition arising from a paramagnetic ground state ($^4T_{1g} \rightarrow$ doublet Co(II)). The excellent agreement between the (AOM) calculated and experimental band positions (Table 6.8) gives confidence that assigning these bands to d–d transitions is correct and that they are not caused by charge transfer transitions. This is further supported by comparing relative intensities of the solution spectra (or diffuse reflectance spectra) with the MCD spectra. Charge transfer bands are relatively intense in absorption spectra and relatively weak in MCD spectra; whereas the opposite is true for d–d transitions [12]. VTVH analysis of the MCD spectra showed that the obtained fits were in remarkably good agreement with the data obtained from magnetic susceptibility measurements.

6.3.3 Mechanism of Phosphodiester Hydrolysis

There are surprisingly few studies of Co(II) complexes as phosphoesterase mimics. For the complexes in the present study the activity towards organophosphoesters using BDNPP a commonly used model substrate, was investigated. All complexes are good functional mimics for phosphodiesterases and show one pK_a relevant for hydrolysis. For the complexes with the methyl-ether donor, the absence of an alkoxide nucleophile and the kinetically relevant pK_a in the range 8.12–8.75 suggests that a terminal water molecule bound to Co(II) is the active nucleophile. For $[Co_2(CO_2EtH_2\mathbf{L1})(CH_3COO)_2](PF_6)$, fitting of the data resulted in a pK_a of 10.54. For this complex, as with previous studies with this type of ligand (Chap. 5), the possibility exists that the alkoxide may be the active nucleophile. Previous studies have suggested that the coordinated alcohol is deprotonated at or below the same pH as a coordinated water molecule and that a coordinated alcohol is a stronger nucleophile than a coordinated hydroxide [50, 51]. However, the pK_a of 10.5 for this complex is still in the range of that for a Co(II)-OH_2, and hence the identity of the nucleophile is uncertain [52]. What is apparent is that the catalytic activity of the Co(II) complexes towards the substrate BDNPP is similar to that displayed by the analogous Zn(II) complexes although it does not approach the efficiency of $Co(II)_2$-GpdQ ($k_{cat} = 1.62\ s^{-1}$, $k_{cat}/K_m = 1.16\ mM^{-1}\ s^{-1}$) [15].

6.3.4 Mechanism of β-Lactam Hydrolysis

It is proposed that at pH 8, the alkoxide deprotonates below the pK_a of a putative Co-OH_2/OH species and is a more reactive nucleophile. Thus, $[Co_2(CO_2EtH_2\mathbf{L1})(CH_3COO)_2]^+$ is active and $[Co_2(CO_2Et\mathbf{L2})(CH_3COO)_2]^+$ is inactive at this pH. The kinetic data at pH 8 support this proposition. There is precedence for this

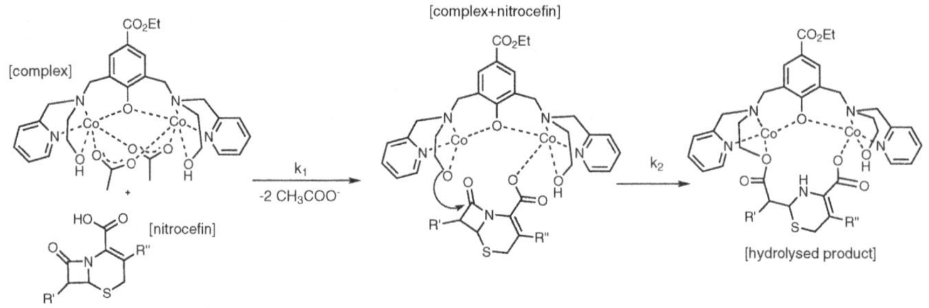

Fig. 6.39 Proposed mechanism of nitrocefin hydrolysis by $[Co_2(CO_2EtH_2\mathbf{L1})(CH_3COO)_2]^+$ at pH 8

observation in Zn(II) complexes where there is considerable evidence both that a coordinated alcohol is deprotonated at or below the same pH as a coordinated water molecule [50, 51]. It is also proposed that a coordinated alcohol is a stronger nucleophile than a coordinated hydroxide, Thus, for $[Co_2(CO_2EtH_2\mathbf{L1})(CH_3COO)_2]^+$ the proposed mechanism involves similar steps to those described for the Zn(II) complexes, the difference being that following coordination of the nitrocefin to one Co(II) site through the carboxylate, the deprotonated alkoxide of the ligand acts as the nucleophilic agent at the carbonyl of the β-lactam. The ring-opened hydrolyzed product remains bound to the Co(II) sites, preventing the active site from being regenerated (Fig. 6.39). Binding and subsequent hydrolysis of a second molecule of the substrate is possible, but not facile, most likely due to steric crowding around the di-Co(II) site.

6.4 Conclusion

The Co(II) complexes of the five phenol-based ligands $CO_2EtH_3\mathbf{L1}$, $CO_2EtH\mathbf{L2}$, $CH_3H\mathbf{L2}$, $BrH\mathbf{L2}$, and $NO_2H\mathbf{L2}$ have been prepared. Attempts have been made to correlate structural parameters with the strength of the magnetic coupling after analyzing the magnetic and spectroscopic properties of the complexes. Computational studies have also been conducted to verify the experimental magnetic coupling constants. Bonding analysis suggests that for all the complexes in this study the most important orbital mechanism for interaction between the metal sites involves the phenoxo-bridge; the differences in magnetic behavior are most likely attributable to minor electronic and structural effects. Mass spectral and MCD studies in the presence of the slow-reacting substrate DPP showed that the complexes can bind up to two substrate molecules and that the geometry of the active site, as well as the magnetic coupling, changes upon substrate binding. Kinetic analysis with the activated substrate BDNPP suggested that, for the complexes derived from **L2** ligands, terminal water is the nucleophile with a kinetically relevant pK_a in the range 8.12–8.75. For the complex with $CO_2EtH_3\mathbf{L1}$, however, the possibility that an alkoxide ligand arm is the active nucleophile cannot be discounted. The complexes in this study are good functional models for phosphatase enzyme systems. The complexes $[Co_2(CO_2EtH_2\mathbf{L1})(CH_3COO)_2]^+$ and $[Co_2(CO_2Et\mathbf{L2})(CH_3COO)_2]^+$ were also analyzed for their ability to hydrolyze lactam substrates and differ distinctly in their reactivity towards nitrocefin at pH 8. The $[Co_2(CO_2EtH_2\mathbf{L1})(CH_3COO)_2]^+$ complex is reactive towards nitrocefin whereas $[Co_2(CO_2Et\mathbf{L2})(CH_3COO)_2]^+$ is unreactive. The difference between the two complexes is due to the presence of an alkoxide in the former and a combination of kinetic, MS, UV-vis and IR results suggest that, at the pH of the experiment, this is the nucleophilic agent.

The next chapter will introduce two asymmetric ligands for the generation of both functional and structural models of the enzyme GpdQ.

References

1. N. Wiberg, A.F. Holleman, E. Wiberg, *Inorganic Chemistry* (Academic Press, NY, 2001)
2. J.J.R.F. da Silva, R.J.P. Williams, *The Biological Chemistry of the Elements: The Inorganic Chemistry of Life* (Oxford University Press, NY, 2001)
3. J. Webb, P. Morris, J. Chem. Educ. **52**, 53 (1975)
4. W.M.B.L. Vallee, in *Methods Enzymol*, vol 226, ed. by F.R. James, L.V. Bert (Academic Press, New York, 1993), pp. 52–71
5. J.A. Larrabee, C.M. Alessi, E.T. Asiedu, J.O. Cook, K.R. Hoerning, L.J. Klingler, G.S. Okin, S.G. Santee, T.L. Volkert, J. Am. Chem. Soc. **119**, 4182–4196 (1997)
6. K.S. Hadler, N. Mitic, S.H. Yip, L.R. Gahan, D.L. Ollis, G. Schenk, J.A. Larrabee, Inorg. Chem. **49**, 2727–2734 (2010)
7. G. Schenk, F. Ely, K.S. Hadler, N. Mitić, L.R. Gahan, D.L. Ollis, N.M. Plugis, M.T. Russo, J.A. Larrabee, J. Biol. Inorg. Chem. **16**, 777–787 (2011)
8. E.I. Solomon, Inorg. Chem. **40**, 3656–3669 (2001)
9. E.I. Solomon, K.O. Hodgson, *Spectroscopic Methods in Bioinorganic Chemistry* (American Chemical Society, New York, 1998)
10. W.R. Mason, *A Practical Guide to Magnetic Circular Dichroism Spectroscopy* (Wiley-Interscience, Hoboken, 2007)
11. F. Neese, Power Point Presentation, Max Planck Insitut für Bioanorganische Chemie Mülheim an der Ruhr (2004)
12. J.A. Larrabee, W.R. Johnson, A.S. Volwiler, Inorg. Chem. **48**, 8822–8829 (2009)
13. E.I. Solomon, A. Decker, N. Lehnert, Proc. Natl. Acad. Sci. U. S. A. **100**, 3589–3594 (2003)
14. F.B. Johansson, A.D. Bond, U.G. Nielsen, B. Moubaraki, K.S. Murray, K.J. Berry, J.A. Larrabee, C.J. McKenzie, Inorg. Chem. **47**, 5079–5092 (2008)
15. K.S. Hadler, E.A. Tanifum, S.H. Yip, N. Mitić, L.W. Guddat, C.J. Jackson, L.R. Gahan, K. Nguyen, P.D. Carr, D.L. Ollis, A.C. Hengge, J.A. Larrabee, G. Schenk, J. Am. Chem. Soc. **130**, 14129–14138 (2008)
16. F. Ely, K.S. Hadler, N. Mitić, L.R. Gahan, D.L. Ollis, N.M. Plugis, M.T. Russo, J.A. Larrabee, G. Schenk, J. Biol. Inorg. Chem. **16**, 777–787 (2011)
17. J.A. Larrabee, S.A. Chyun, A.S. Volwiler, Inorg. Chem. **47**, 10499–10508 (2008)
18. B.N. Figgis, *Ligand Field Theory and Its Applications* (Wiley, NY, 2000)
19. S.M. Ostrovsky, R. Werner, D.A. Brown, W. Haase, Chem. Phys. Lett. **353**, 290–294 (2002)
20. H. Sakiyama, Inorg. Chim. Acta **360**, 715–716 (2007)

21. M.J. Hossain, M. Yamasaki, M. Mikuriya, A. Kuribayashi, H. Sakiyama, Inorg. Chem. **41**, 4058–4062 (2002)
22. H. Sakiyama, R. Ito, H. Kumagai, K. Inoue, M. Sakamoto, Y. Nishida, M. Yamasaki, Eur. J. Inorg. Chem. **2001**, 2027–2032 (2001)
23. H. Sakiyama, J. Chem. Software **7**, 171–177 (2001)
24. L. Noodleman, J. Chem. Phys. **74**, 5737–5743 (1981)
25. E. Ruiz, J. Cano, S. Alvarez, P. Alemany, J. Comput. Chem. **20**, 1391–1400 (1999)
26. G. Cavigliasso, R. Stranger, Inorg. Chem. **47**, 3072–3083 (2008)
27. D.R. Jones, L.F. Lindoy, A.M. Sargeson, J. Am. Chem. Soc. **106**, 7807–7819 (1984)
28. R.E. Mirams, S.J. Smith, K.S. Hadler, D.L. Ollis, G. Schenk, L.R. Gahan, J. Biol. Inorg. Chem. **13**, 1065–1072 (2008)
29. P. Karsten, A. Neves, A.J. Bortoluzzi, M. Lanznaster, V. Drago, Inorg. Chem. **41**, 4624–4626 (2002)
30. A. Neves, M. Lanznaster, A.J. Bortoluzzi, R.A. Perlata, A. Casellato, E.E. Castellano, P. Herrald, M.J. Riley, G. Schenk, J. Am. Chem. Soc. **129**, 7486–7487 (2007)
31. R.R. Buchholz, M.E. Etienne, A. Dorgelo, R.E. Mirams, S.J. Smith, S.Y. Chow, L.R. Hanton, G.B. Jameson, G. Schenk, L.R. Gahan, Dalton Trans. **43**, 6045–6054 (2008)
32. L.R. Gahan, S.J. Smith, A. Neves, G. Schenk, Eur. J. Inorg. Chem. **19**, 2745–2758 (2009)
33. R.A. Peralta, A.J. Bortoluzzi, B. de Souza, R. Jovito, F.R. Xavier, R.A.A. Couto, A. Casellato, F. Nome, A. Dick, L.R. Gahan, G. Schenk, G.R. Hanson, F.C.S. de Paula, E.C. Pereira-Maia, S.d.P. Machado, P.C. Severino, C. Pich, T. Bortolotto, H. Terenzi, E.E. Castellano, A. Neves, M.J. Riley, Inorg. Chem. **49**, 11421–11438 (2010)
34. B. Bauer-Siebenlist, F. Meyer, E. Farkas, D. Vidovic, S. Dechert, Chem. Eur. J. **11**, 4349–4360 (2005)
35. S.C. Batista, A. Neves, A.J. Bortoluzzi, I. Vencato, R.A. Peralta, B. Szpoganicz, V.V.E. Aires, H. Terenzi, P.C. Severino, Inorg. Chem. Commun. **6**, 1161–1165 (2003)
36. M. Lanznaster, A. Neves, A.J. Bortoluzzi, B. Szpoganicz, E. Schwingel, Inorg. Chem. **41**, 5641–5643 (2002)
37. J.S. Seo, N.-D. Sung, R.C. Hynes, J. Chin, Inorg. Chem. **35**, 7472–7473 (1996)
38. S.J. Smith, R.A. Peralta, R. Jovito, A. Horn, A.J. Bortoluzzi, C.J. Noble, G.R. Hanson, R. Stranger, V. Jayaratne, G. Cavigliasso, L.R. Gahan, G. Schenk, O.R. Nascimento, A. Cavalett, T. Bortolotto, G. Razzera, H. Terenzi, A. Neves, M.J. Riley, Inorg. Chem. **51**, 2065–2078 (2012)
39. G. te Velde, F.M. Bickelhaupt, E.J. Baerends, C. Fonseca Guerra, S.J.A. van Gisbergen, J.G. Snijders, T. Ziegler, J. Comput. Chem. **22**, 931–967 (2001)
40. K. Nakamoto, *Infrared and Raman Spectra of Inorganic and Coordination Compounds* (Wiley, NY, 1978)
41. F.A. Cotton, *Advanced Inorganic Chemistry* (Wiley, NY, 1999)
42. J.-L. Tian, W. Gu, S.-P. Yan, D.-Z. Liao, Z.-H. Jiang, Z. Anorg, Allg. Chem. **634**, 1775–1779 (2008)
43. D.F. Evans, J. Chem. Soc. 2003 (1959)
44. Thermo-Scientific, Grams/AI 9.0 Software
45. M.J. Riley, VTVH 2.1.1 Program for the simulation and fitting variable temperature—variable field MCD spectra (2008)
46. C. Fonseca Guerra, J.G. Snijders, G. te Velde, E.J. Baerends, Theor. Chim. Acta **99**, 391–403 (1998)
47. H. Adamsky, AOMX Program (1996)
48. I.H. Segel, *Enzyme Kinetics: Behavior and Analysis of Rapid Equilibrium and Steady-State Enzyme Systems* (Wiley, NY, 1975)
49. M. Maeder, ReactLab KINETICS
50. M. Livieri, F. Mancin, U. Tonellato, J. Chin, Chem. Commun. (24), 2862–2863 (2004)
51. J. Xia, Y.B. Shi, Y. Zhang, Q. Miao, W.X. Tang, Inorg. Chem. **42**, 70–77 (2003)
52. J. Burgess, *Metal Ions in Solution* (Halsted Press, Chichester, 1978)

Chapter 7
Asymmetric Zn(II) Complexes as Structural and Functional Models for GpdQ

7.1 Introduction

7.1.1 Asymmetric Ligand Environments

In order to generate more structurally relevant biomimetics for dinuclear metallohydrolases much effort has been devoted to the synthesis of asymmetric ligands. These ligands are considered to be more suitable models for the asymmetric coordination environment found in enzymatic systems. Nordlander et al. proposed that asymmetric complexes are not only more appropriate functional models for the active site of phosphoesterase enzymes, but also that they exhibit enhanced catalytic rates compared with their symmetric counterparts [1–3]. A selection of ligands used to generate purple acid phosphatase [1, 4, 5, 6–10], phosphoesterase [11], urease [12, 13], catechol oxidase [14] and manganese catalase biomimetics [15, 16] is displayed in Fig. 7.1.

Some of these ligands generate a hard and a soft coordination site for the generation of heterodinuclear model complexes for purple acid phosphatase metalloenzymes [6, 8]. In some cases the ligands have one structural variation in one arm [6], in others one donor arm has been omitted [1, 14]. Often the vacant coordination site is found to be occupied by water or solvent molecules in the complex [15, 16].

7.1.2 Aim and Relevance

The aim of this chapter is to design asymmetric ligands which enforce a 5,6-mixed coordination sphere similar to the active site in GpdQ.[1] Two ligands will be discussed. One ligand, with a phenol, two pyridine donors, one tertiary and one

[1] Parts of this chapter have appeared in L.J. Daumann et al., "*Asymmetric Zinc(II) Complexes as Functional and Structural Models for Phosphoesterases*" Dalton Trans. **2013**; 42, 9574–9584—Reproduced by permission of The Royal Society of Chemistry.

L. J. Daumann, *Spectroscopic and Mechanistic Studies of Dinuclear Metallohydrolases and Their Biomimetic Complexes*, Springer Theses, DOI: 10.1007/978-3-319-06629-5_7,

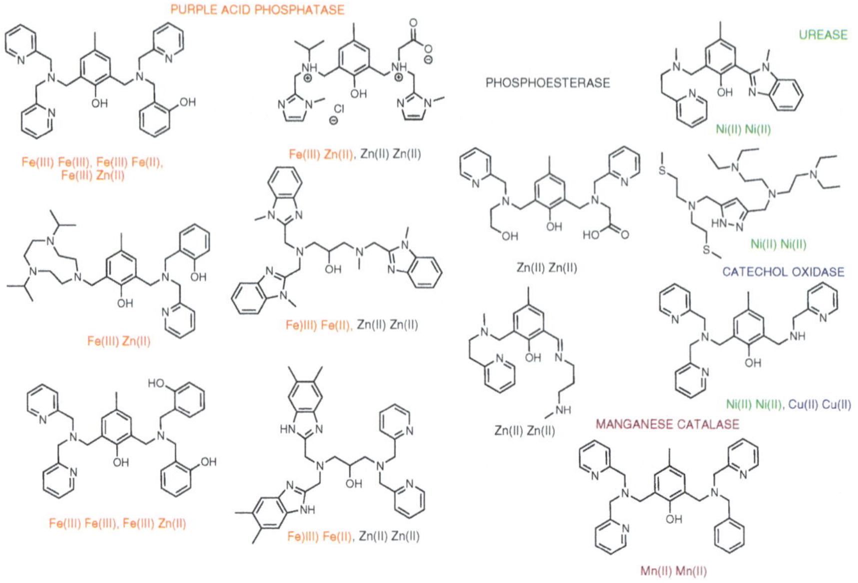

Fig. 7.1 Asymmetric ligands used in biomimetic studies

secondary nitrogen atom and one methyl ether arm (CH_3H**L4**), the second ligand will feature the same donor atom set but the secondary nitrogen is reacted with vinyl benzyl chloride to shield this coordination site (CH_3H**L5**). It is proposed that these ligands will reproduce the active site of GpdQ more accurately and leave sites open to coordination of solvent molecules. The Zn(II) complexes of both ligands will be synthesized and characterized; in addition, the Co(II) derivative of CH_3H**L4** will be investigated with MCD in the absence and presence of a substrate mimic to shed light on the coordination geometry in solution. The activity of the Zn(II) complexes towards BDNPP will be investigated.

7.2 Results

7.2.1 Ligand and Complex Synthesis

The CH_3H**L4** ligand synthesis was achieved using a protocol similar to that published by Boudalis et al. [17]. Addition of 2-methoxy-N-(pyridin-2-ylmethyl)aminoethanol to the 3-(chloromethyl)-2-hydroxy-5-methylbenzaldehyde precursor, subsequent Schiff base condensation with 2-aminomethylpyridine and reduction with sodium borohydride gave the desired ligand CH_3H**L4** as a yellow oil in moderate yield. Treatment of CH_3H**L4** with vinyl benzyl chloride in the presence of potassium carbonate readily converted this ligand into CH_3H**L5** (Fig. 7.2).

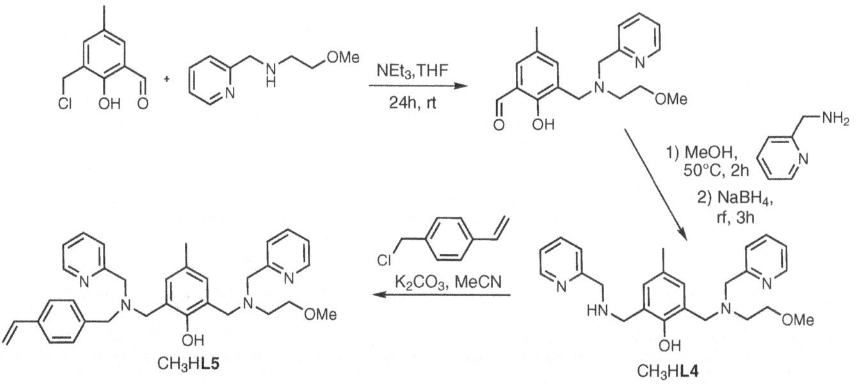

Fig. 7.2 Synthetic route to the ligands CH_3HL4 and CH_3HL5

The syntheses of the respective Zn(II) complexes proved to be more challenging than with the symmetric ligands. For the Zn(II) complex with CH_3HL4, crystals suitable for X-ray crystallography were obtained after multiple crystallizations using ether diffusion into a methanolic solution of the complex. Crystals of $[Zn_2(CH_3L5)(CH_3COO)_2]BPh_4$ were obtained after slow diffusion (3 weeks) of hexane into an acetone solution of the complex. The crystals desiccated upon removal from the mother liquor, making low temperature X-ray structure analysis necessary. The remaining analysis of this complex was undertaken with the $[Zn_2(CH_3L5)(CH_3COO)_2]PF_6$ derivative. Initially the Co(II) complexes of the two ligands were also investigated, they were however prone to (partial) oxidation making a thorough analysis difficult.

7.2.2 *Crystal Structures of the Zn(II) Complexes with CH_3HL4 and CH_3HL5*

The Zn(II) complex of CH_3L4^- crystallized in the triclinic space group P-1 as an asymmetric complex with one six-coordinate Zn(II) and one Zn(II) in a five-coordinate (trigonal bipyramidal) environment (Fig. 7.3, Tables 7.1 and 7.2). Figure 7.3a shows the ORTEP representation of the asymmetric unit which contains two $[Zn_2(CH_3L4)–(CH_3COO)_2]PF_6$ complex entities. Two acetate anions, but no additional water or solvent molecules, are present. Interestingly, the crystal structure shows the presence of two isomers, one present as a minor component in one of the two asymmetric units. In the major isomer, the pyridine nitrogen atoms are *trans* (*anti*) to each other while in the minor isomer the pyridines are *cis* (*syn*) in respect of the phenolic oxygen (See also Chap. 4). The $[Zn_2(CH_3L5)–(CH_3COO)_2]BPh_4$ complex crystallized in the P − 1 space group with one Zn(II) ion six coordinate and the other in a five coordinate environment. The vinylbenzyl group is arranged in such

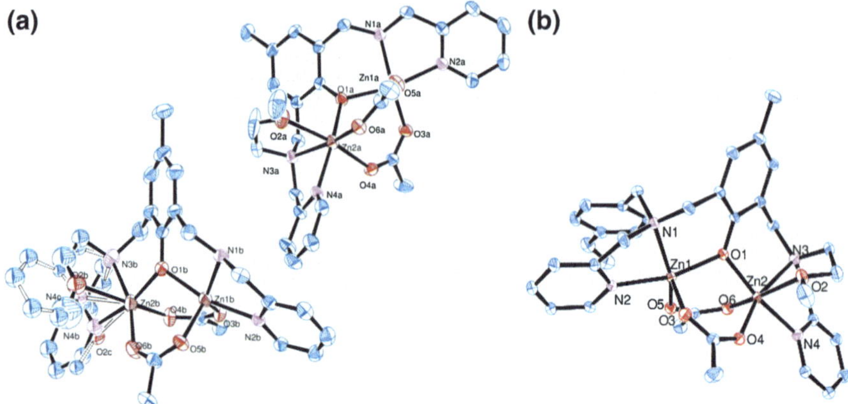

Fig. 7.3 ORTEP plots of the two molecules in the asymmetric unit of $[Zn_2(CH_3\mathbf{L4})(CH_3COO)_2]PF_6$ (*left*), with the pyridine and methyl-ether arm positions of the minor isomer represented in dashed bonds and $[Zn_2(CH_3\mathbf{L5})(CH_3COO)_2]BPh_4$ with 25 % (*right*). Counterions, solvent molecules and hydrogen atom were omitted for clarity

Table 7.1 Crystal data and structure refinement parameters for $[Zn_2(CH_3\mathbf{L4})(CH_3COO)_2]PF_6$ and $[Zn_2(CH_3\mathbf{L5})(CH_3COO)_2]BPh_4$

	$[Zn_2(CH_3\mathbf{L4})(CH_3COO)_2]PF_6$	$[Zn_2(CH_3\mathbf{L5})(CH_3COO)_2]BPh_4$
Empirical formula	$C_{28}H_{35}N_4O_6Zn_2$, P_1F_6	$C_{37}H_{39}N_4O_6Zn_2$, $C_{24}H_{20}B$, C_3H_8O
Formula weight	799.31	1145.77
Temperature [K]	293(2)	150(2)
Wavelength [Å]	0.71073 (MoK_α)	1.5418 (CuK_α)
Crystal system	triclinic	triclinic
Space group	P − 1	P − 1
a [Å]	13.137(5)	13.3209(6)
b [Å]	13.669(6)	15.0129(7)
c [Å]	20.822(8)	16.4893(6)
α [°]	85.047(3)	91.810(3)
β [°]	85.468(3)	108.350(4)
γ [°]	66.511(4)	110.363(4)
Volume [$Å^3$]	3412(2)	2897.2(2)
Z	4	2
ρ [mg/m^3]	1.556	1.313
μ [mm^{-1}]	1.530	1.466
F(000)	1,632	1,199.8
Θ range for data collection [°]	1.63 to 24.14	2.86 − 61.63
Reflections collected	33,208	26,492
Independent reflections (R_{int})	10,796 (0.0290)	8,953 (0.0277)
GOOF on F^2	1.032	1.035
Final R indices [$I > 2\sigma(I)$]	R1 = 0.0661, wR2 = 0.1831	R1 = 0.0480, wR2 = 0.1321
R indices (all data)	R1 = 0.0779, wR2 = 0.1940	R1 = 0.0557, wR2 = 0.1401

Table 7.2 Selected bond lengths [Å] and angles [°] for $[Zn_2(CH_3\mathbf{L4})(CH_3COO)_2]PF_6$

	$[Zn_2(CH_3\mathbf{L4})–(CH_3COO)_2]PF_6^*$	$[Zn_2(CH_3\mathbf{L5})–(CH_3COO)_2]BPh_4$
N1-Zn1	2.133(5)	2.192(3)
N2-Zn1	2.121(4)	2.108(3)
N3-Zn2	2.182(4)	2.184(3)
N4-Zn2	2.121(4)	2.131(3)
O1-Zn1	2.026(4)	2.016(2)
O1-Zn2	2.014(4)	2.014(2)
O2-Zn2	2.320(4)	2.370(2)
O3-Zn1	1.984(4)	2.020(2)
O4-Zn2	2.112(4)	2.029(2)
O5-Zn1	1.998 (4)	1.986(2)
O6-Zn2	2.015(4)	2.072(2)
Zn1-O1-Zn2	108.17(16)	106.98(9)
Zn1–Zn2	3.2714(12)	3.2391(6)

* Only the bond lengths and angles of the zinc complex labeled with 'a' (e.g. Zn1a) are displayed

way that it shields one site on the second Zn(II) ion leaving this site almost square pyramidal (Fig. 7.3b).

7.2.3 Infrared Spectroscopy

The $[Zn_2(CH_3\mathbf{L4})(CH_3COO)_2]PF_6$ complex exhibited stretches typical for bridging acetate (1,596 and 1,422 cm^{-1}) and for the PF_6^- counter ion (832, 556 cm^{-1}) in addition to the C–O stretching vibration of the methyl ether at 1,022 cm^{-1} and the N–H stretch at 3,307 cm^{-1} [18]. For the $[Zn_2(CH_3\mathbf{L5})(CH_3COO)_2]PF_6$ complex the stretches at 1,594 and 1,429 cm^{-1} also suggest the presence of bidentate bridging acetate [18]. Moreover C–H (2,925, 2,842 cm^{-1}), C–O (1,022 cm^{-1}) and bands typical for aromatic functionalities (766, 662 cm^{-1}) are similar to those observed for the complex $[Zn_2(CH_3\mathbf{L4})(CH_3COO)_2]PF_6$ (C–H 2,925, aromatic 768 and 663 cm^{-1}). It can thus be assumed that a similar coordination environment with two bridging acetates for $[Zn_2(CH_3\mathbf{L5})(CH_3COO)_2]PF_6$ is found in the solid state.

7.2.4 Mass Spectrometry

7.2.4.1 Mass Spectrometry of the Complexes in Different Solvents

The electrospray mass spectra of the two Zn(II) complexes were recorded in MeOH and acetonitrile, respectively. The spectrum of $[Zn_2(CH_3\mathbf{L4})(CH_3COO)_2]PF_6$ measured in MeOH is displayed in Fig. 7.4a and shows a peak at *m/z* 655.1 (100 %) (calc. *m/z* = 653.1 (100 %), 655.1 (96 %), $[C_{28}H_{35}N_4O_6Zn_2]^+$) corresponding to

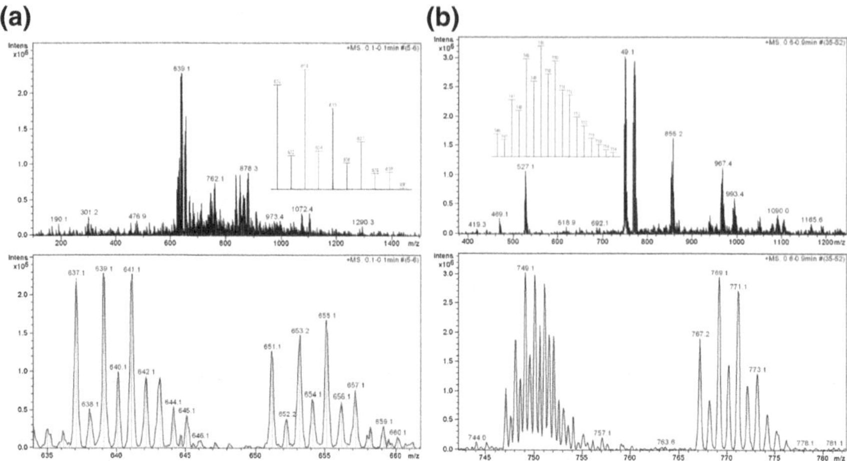

Fig. 7.4 Mass spectra of $[Zn_2(CH_3\mathbf{L4})(CH_3COO)_2]PF_6$ in methanol (**a**) and acetonitrile (**b**). Inset with *red* numbers shows the calculated isotope pattern for the major (identified) ion peak

one negatively charged ligand two Zn(II) ions and two acetates. The major ion peak at *m/z* 641.1 is assigned to $[Zn_2(CH_3\mathbf{L4})(CH_3COO)(HCOO)]^+$ (calc. *m/z* = 641.1 (100 %), 639.1 (99 %), 637.1 (82 %), $[C_{27}H_{33}N_4O_6Zn_2]^+$) The spectrum recorded in acetonitrile showed multiple doubly charged species, the major ion peak at 749.1 is assigned to a complex with two $CH_3\mathbf{L4}^-$, four Zn(II), four acetates, six water molecules and two acetonitrile molecules (calc. *m/z* 749.2 (100 %), 748.2 (79 %), $[C_{60}H_{88}N_{10}O_{18}Zn_4]^{2+}$, Fig. 7.4b).

The major ion in the mass spectrum of the $[Zn_2(CH_3\mathbf{L5})(CH_3COO)_2]PF_6$ complex measured in methanol corresponds to a $[Zn_2(CH_3\mathbf{L5})(MeO)_2(H_2O)_2]^+$ species (calc. for $C_{35}H_{47}N_4O_6Zn_2^+$ m/z = 749.2 (100 %), 751.2 (96.1 %), found *m/z* = 751.1 (100 %), 749.1 (95 %), Fig. 7.5a). Higher molecular weight species *m/z* > 800 are also found and are most likely due to coordination of additional solvent and anionic components. A peak due to free ligand is also found at *m/z* = 523.2 ($CH_3H\mathbf{L5}$ + H). The mass spectrum for $[Zn_2(CH_3\mathbf{L5})(CH_3COO)_2]PF_6$ in acetonitrile shows at least four major species: (i) *m/z* = 851.3 assigned to $[Zn_2(CH_3\mathbf{L5})(CH_3COO)_2(\text{acetonitrile})_2]^+$ (calc. 851.23 (100 %), 849.23 (85.5 %), $[C_{41}H_{49}N_6O_6Zn_2]^+$); (ii) *m/z* = 769.1 corresponds to a $[Zn_2(CH_3\mathbf{L5})(CH_3COO)_2]^+$ species (calc. 769.17 (100 %), 767.18 (87 %), $[C_{37}H_{43}N_4O_6Zn_2]^+$; (iii) a peak at *m/z* = 723.1 is due to the loss of acetates and addition of water (calc. 721.17 (100 %), 732.17 (96.6 %), $[C_{33}H_{43}N_4O_6Zn_2]^+$), and (iv) the peak at *m/z* = 653.1 can be attributed to the loss of the vinylbenzyl group (calc. *m/z* 653.11 (100 %), 655.11 (96.2 %), $[C_{28}H_{35}N_4O_6Zn_2]^+$).

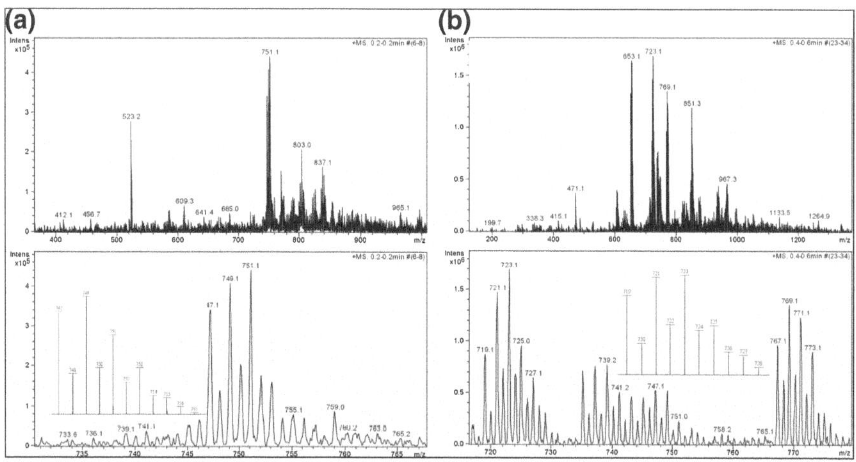

Fig. 7.5 Mass spectra of $[Zn_2(CH_3\mathbf{L5})(CH_3COO)_2]PF_6$ in methanol (**a**) and acetonitrile (**b**). Inset with red numbers shows the calculated isotope pattern for the major peak

7.2.4.2 Mass Spectrometry of $[Zn_2(CH_3L4)(CH_3COO)_2]PF_6$ and $[Zn_2(CH_3L5)–(CH_3COO)_2]PF_6$ in the Presence of the Substrate Mimic DPP

To investigate relevant species occurring during phosphate ester hydrolysis, the mass spectra of both complexes were measured in acetonitrile in the presence of different amounts of DPP. The $[Zn_2(CH_3\mathbf{L4})(CH_3COO)_2]PF_6$ complex readily loses acetates under mass spectral conditions as shown in the section above and forms, even in the presence of only one equivalent DPP (Fig. 7.6a), a complex with two DPP bound (100 %), illustrating the high affinity of phosphoester substrates. In addition, a small peak ($\sim$25 %) is observed, arising from the complex with one DPP, one hydroxide and an acetonitrile solvent molecule, a structure being possibly more relevant for the initial stages of phosphoester hydrolysis. For $[Zn_2(CH_3\mathbf{L5})(CH_3COO)_2]PF_6$ the mass spectrum in the presence of one equivalent DPP is different. The base peak arises from one species with one DPP, a hydroxide and acetonitrile (100 %). Two smaller peaks from intact complex with two acetates (29 %) or two DPP molecules (47 %) bound are present. It is proposed that $[Zn_2\text{-}(CH_3\mathbf{L4})(CH_3COO)_2]PF_6$ has a high affinity (under mass spectrometry conditions) for DPP while $[Zn_2(CH_3\mathbf{L5})(CH_3COO)_2]PF_6$, due to the bulky pendant vinylbenzyl group, needs an excess of substrate to form species with one or two DPP bound.

Figure 7.7b illustrates how a large excess of DPP changes the species distribution for $[Zn_2(CH_3\mathbf{L5})(CH_3COO)_2]PF_6$, the main species now corresponding to $[Zn_2(CH_3\mathbf{L5})(DPP)_2]^+$ (Fig. 7.7b). The mass spectrometric experiments demonstrate that the complexes can bind one or two phosphate substrate molecules readily and show that the acetates are not inhibiting substrate binding.

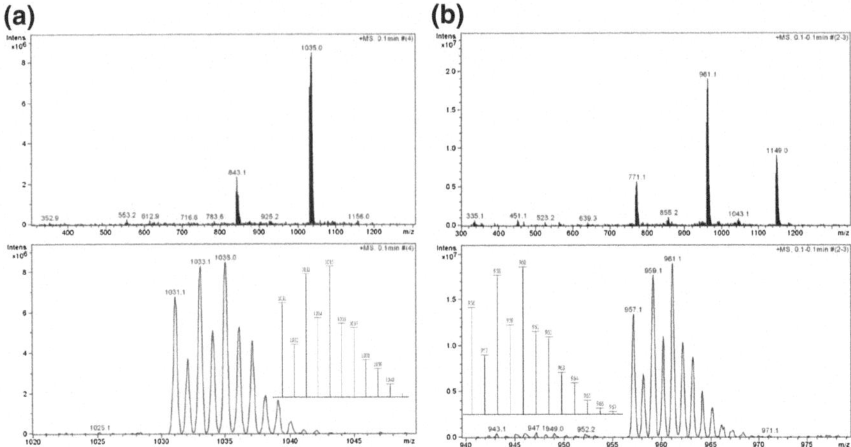

Fig. 7.6 **a** Mass spectrum of $[Zn_2(CH_3\mathbf{L4})(CH_3COO)_2]PF_6$ in the presence of 1 eq. DPP in acetonitrile. **b** Mass spectrum of $[Zn_2(CH_3\mathbf{L5})(CH_3COO)_2]PF_6$ in the presence of 1 eq. DPP in acetonitrile. Final concentration of complexes and substrate were both 10 μM

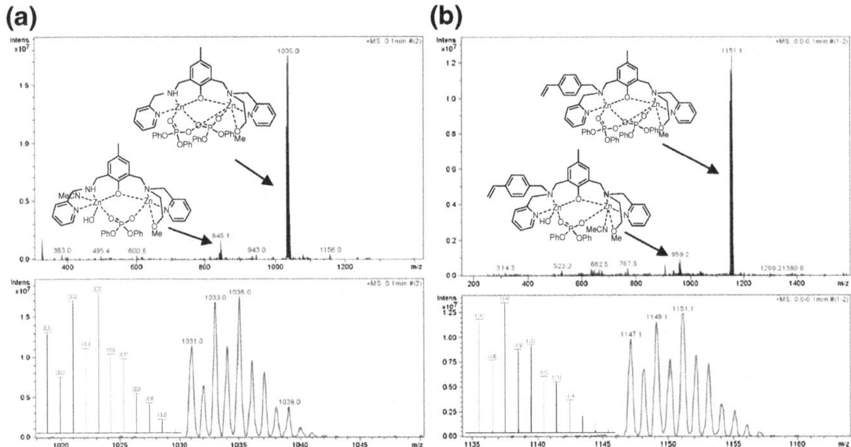

Fig. 7.7 **a** Mass spectrum of $[Zn_2(CH_3\mathbf{L4})(CH_3COO)_2]PF_6$ in the presence of excess (>25 eq) DPP in acetonitrile. **b** Mass spectrum of $[Zn_2(CH_3\mathbf{L5})(CH_3COO)_2]PF_6$ in the presence of excess (>25 eq) DPP in acetonitrile. Final concentration of complexes and substrate were 10 and 250 μM, respectively

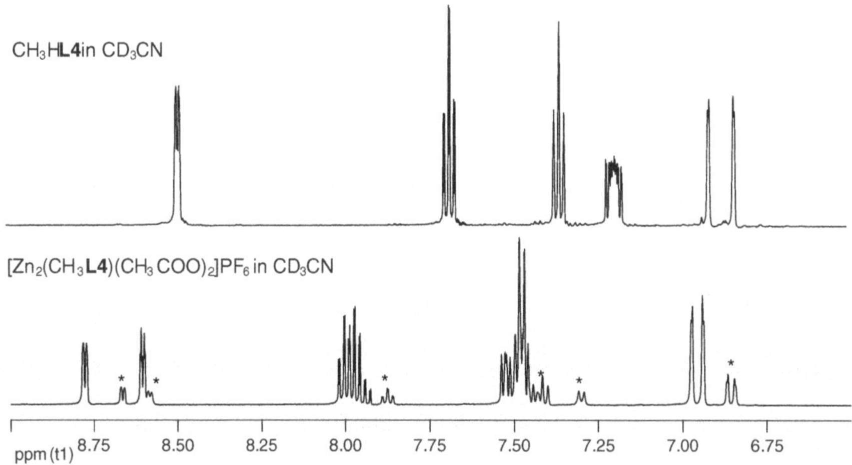

Fig. 7.8 Aromatic region of CH_3H**L4** and its Zn(II) complex

7.2.5 NMR Spectroscopy

7.2.5.1 NMR Spectroscopy of CH_3HL4 and CH_3HL5 and Their Zn(II) Complexes

Given the diamagnetic nature of Zn(II), ^1H, ^{13}C and ^{31}P NMR spectroscopy are excellent methods to investigate the structure of Zn(II) complexes in solution. NMR spectra were recorded for both complexes in a range of solvents. Figures 7.8 and 7.9 show a comparison between the free CH_3H**L4** ligand and its Zn(II) complex. A low intensity structure, not stemming from free ligand, is present (marked in Fig. 7.8 with an asterisk). It is possible that these resonances represent either the minor isomer observed in the X-ray crystal structure, or is due to a complex with the ether arm not coordinated. A combination of both is possible.

The region assigned to the aliphatic protons of CH_3H**L4** and its Zn(II) complex is shown in Fig. 7.9 and demonstrates how the protons of the CH_2-groups in the free ligand become inequivalent upon metal ion coordination. This is in accordance with the lack of a C_2-symmetry axis in the complex.

It should be noted that the ratio of low intensity isomer (marked with * in Fig. 7.8) to the major isomer depended strongly on the water content in the mixture. In an experiment where aliquots of D_2O were titrated into a solution of the complex in deuterated acetonitrile, the proportion of the low intensity species first reduced and then another set of low intensity resonances appeared at higher field (Fig. 7.10).

To monitor how readily CH_3H**L5** would assemble a catalytically competent binuclear Zn(II) center a ^1H NMR titration experiment was conducted. Figure 7.11 shows the spectral changes after successive addition of one and two equivalents of

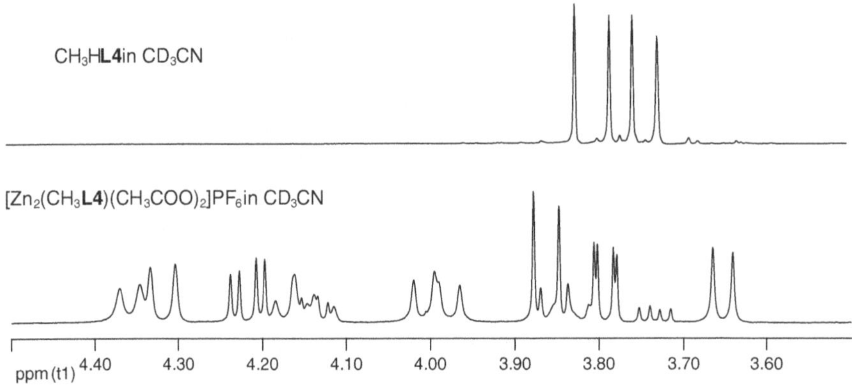

Fig. 7.9 Aliphatic region of CH_3HL4 and its zinc complex

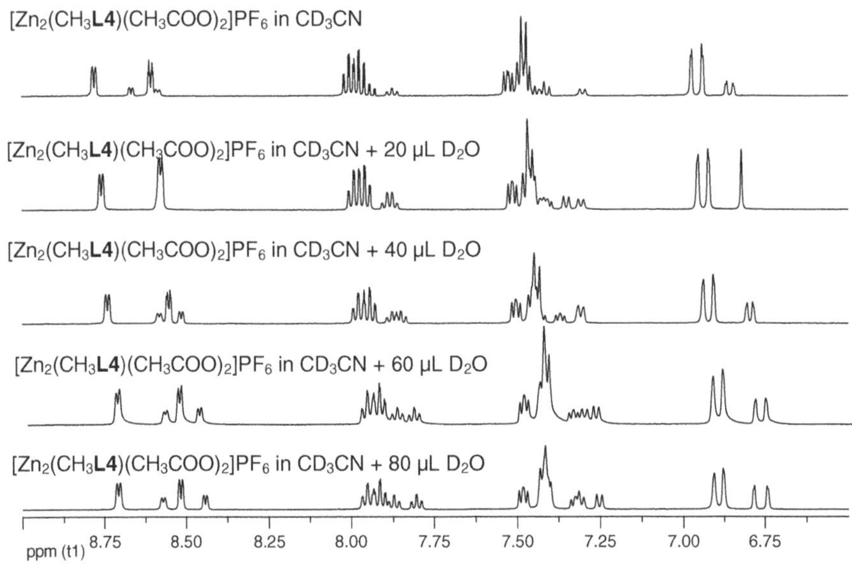

Fig. 7.10 D_2O titration experiment for the complex $[Zn_2(CH_3\mathbf{L4})(CH_3COO)_2]PF_6$

zinc(II) acetate to a solution of $CH_3H\mathbf{L5}$ in deuterated acetonitrile. After each addition the solvent in the NMR tube was brought to reflux for a couple of minutes prior to spectra recording. After addition of one equivalent of zinc(II) acetate the resonances of the vinyl protons had experienced minimal shifts and it is thus concluded that the ligands binds Zn(II) ions in a stepwise manner, occupying first the less sterically hindered site, without the pendant vinyl benzyl group. It is also apparent that after the addition of two equivalents Zn(II) acetate, the solution still contains partially uncomplexed ligand, and thus another equivalent of Zn(II)

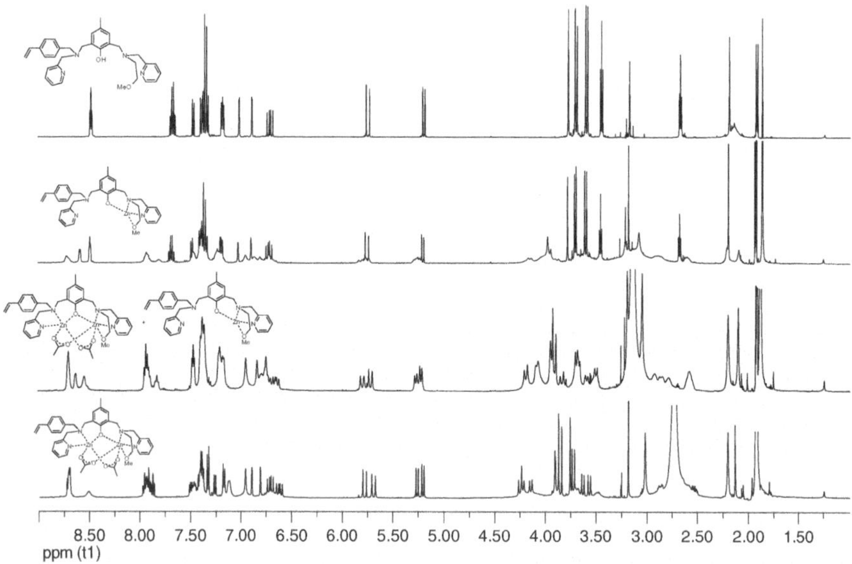

Fig. 7.11 ^{1}H-NMR titration of zinc(II) acetate into the CH_3H**L5** ligand in CD_3CN and proposed species

acetate was added. Intriguingly a set of two independent resonances for the vinyl-protons were detected in the spectrum of the fully assembled binuclear complex, suggesting two isomeric forms in solution. The vinyl group might adapt two possible conformations one pointing towards the Zn(II) center and one away from it.

7.2.5.2 NMR Spectra of the Complexes in the Presence of the Substrate Mimic DPP

Substrate binding was investigated with ^{31}P NMR spectroscopy. The addition of one equivalent DPP to a solution of [Zn_2(CH_3**L4**)(CH_3COO)$_2$]PF_6 in deuterated acetonitrile shows multiple phosphate ester species but no resonance from free DPP in accordance with the electrospray mass spectrometry experiments which showed high affinity for this substrate (Fig. 7.12a). Given the observed species in the mass spectrum with complex having one or two species bound, the different signals observed in the ^{31}P-spectrum are probably due to a mix of species with one DPP monodentately or bidentately bound (−8.0 and −8.3 ppm), and two DPP molecules also mono- or bidentately bound (−9.1 and −9.5 ppm). Addition of a second equivalent of DPP leads to an increase in intensity of the two resonances at −9.5 and −9.1 ppm (Fig. 7.12b). This underlines the assignment of having two DPP bound to the zinc complex in either a either mono- or bidentate fashion. Further addition of DPP gives rise to a signal at −12.2 ppm attributed to free DPP (Fig. 7.12c).

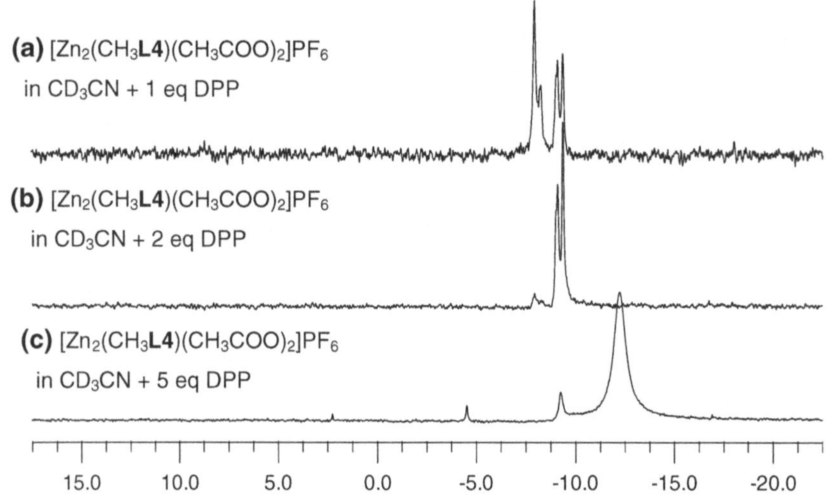

Fig. 7.12 ^{31}P NMR spectra of $[Zn_2(CH_3\mathbf{L4})(CH_3COO)_2]PF_6$ measured in CD_3CN in the presence of **a** one equivalent DPP **b** two equivalents DPP and (**c**) five equivalents DPP

In the ^{1}H-spectrum of $[Zn_2(CH_3\mathbf{L5})(CH_3OO)_2]^+$ with one and two equivalents substrate (Fig. 7.13c and d) the signals are found broadened suggesting a bridging DPP conformation [19]. Interestingly one of the two sets of vinyl proton resonances loses intensity upon the addition of one or two DPP molecules to the NMR mixture (Fig. 7.13c–e). Given the bulky nature of the vinyl benzyl moiety these results support the notion that in solution the vinyl benzyl group exists in two conformations one directed towards the Zn(II) center and one pointing in the opposite direction, the latter not being influenced by substrate binding.

Moreover, the vinyl benzyl group displays, after the addition of five equivalents DPP (Fig. 7.13e) ^{1}H-chemical shifts similar to those found in the spectrum of the free ligand (Fig. 7.13a). This implies that the vinyl benzyl group points away (and is not further influenced) from the Zn(II) ions when substrate is bound. When comparing the ^{13}C NMR spectra of free complex to that in the presence of five equivalents of DPP two things are apparent: (i) the former spectrum has two sets of vinyl resonances at 114.9 and 115.2 ppm while (ii) the latter exhibits only one resonance at 115.2 ppm. Moreover, the carbonyl$_{OAc}$ resonance shifts, upon addition of DPP to the complex, from 179.0 ppm (bridging acetate) to 173.1 ppm (free acetate). The methyl$_{OAc}$ resonance also exhibits a shift (from 23.1 to 20.8 ppm). This clearly shows that the acetates are not inhibiting substrate binding and that DPP binds strongly to the complex, in addition, the vinyl benzyl group is orientated away from the Zn(II) ions. The ^{31}P NMR spectrum of the $[Zn_2(CH_3\mathbf{L5})(CH_3OO)_2]^+$ complex, in the presence of one equivalent DPP shows one broad resonance. This resonance arises from multiple species such as monodentately or bidentately bridging bound and also free DPP in the mixture (see Gaussian deconvolution in Fig. 7.14a). Upon

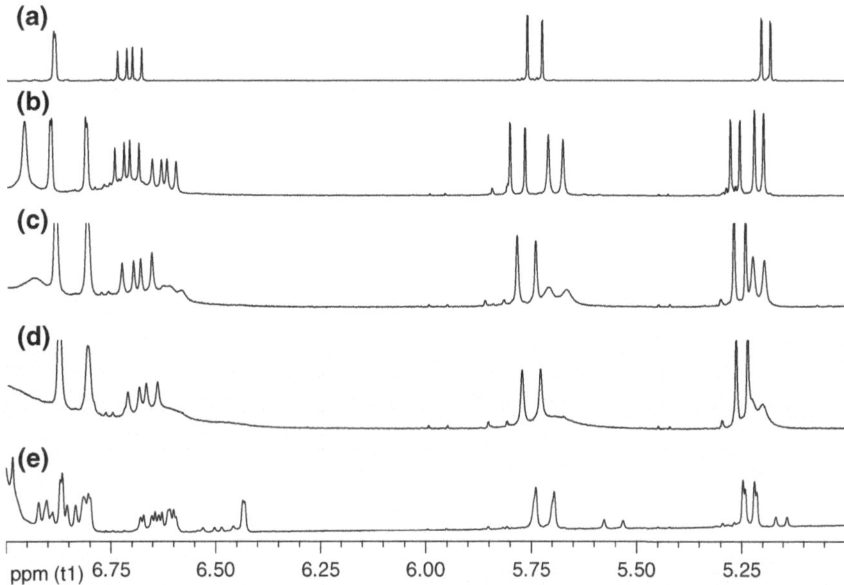

Fig. 7.13 The ^{1}H NMR spectrum of **a** CH_3H**L5**; **b** $[Zn_2(CH_3$**L5**$)(CH_3OO)_2]^+$ and **c** $[Zn_2(CH_3$**L5**$)(CH_3OO)_2]^+$ + 1 equivalent DPP; **d** $[Zn_2(CH_3$**L5**$)(CH_3OO)_2]^+$ + 2 equivalents DPP; **e** $[Zn_2(CH_3$**L5**$)(CH_3OO)_2]^+$ + 5 equivalents DPP

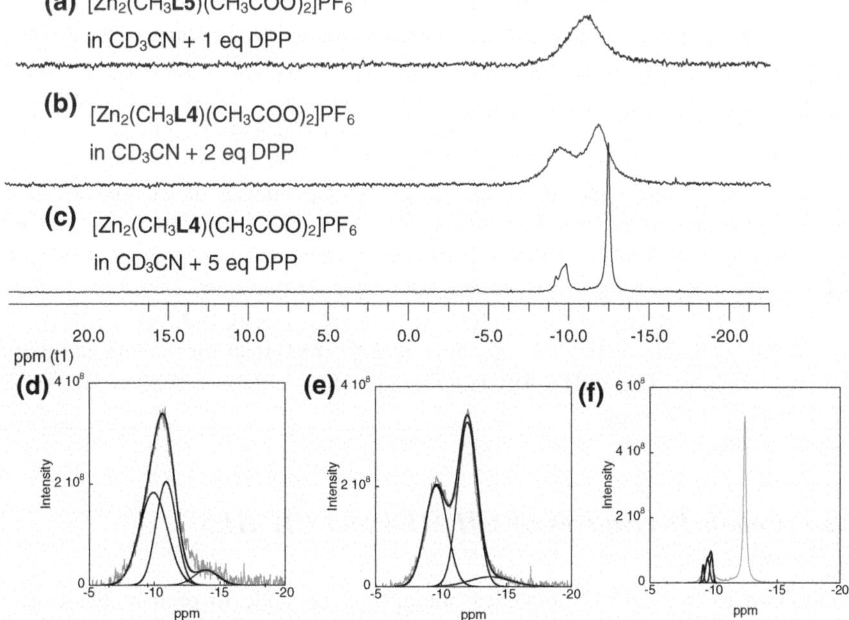

Fig. 7.14 ^{31}P NMR spectra of $[Zn_2(CH_3$**L5**$)(CH_3OO)_2]^+$ in CD_3CN the presence of **a** 1 equivalent DPP. **b** 2 equivalents DPP. **c** 5 equivalents DPP at room temperature. Gaussian deconvolution of the phosphorous signal of spectrum with **d** one equivalent DPP. **e** two equivalents DPP and **f** five equivalents DPP

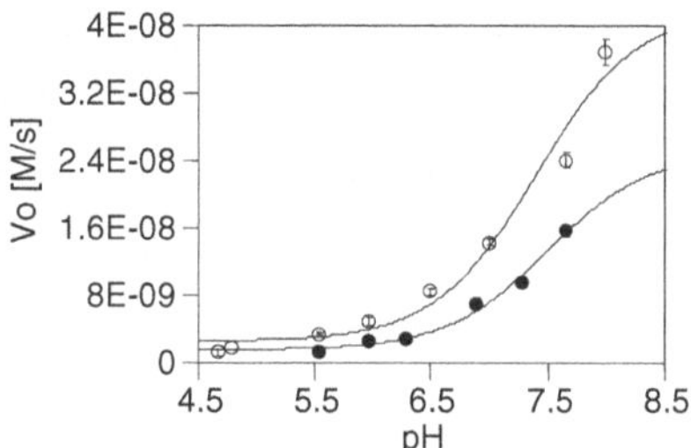

Fig. 7.15 pH dependence of the rate of BDNPP hydrolysis by $[Zn_2(CH_3\mathbf{L4})(CH_3COO)_2]^+$ (*open circle*) and $[Zn_2(CH_3\mathbf{L5})(CH_3COO)_2]^+$ (*filled circle*)

addition of a further equivalent of DPP to the mixture the resonance arising from free DPP becomes more clearly resolved (Fig. 7.14b). This resonance becomes the major species upon addition of excess DPP (see spectrum c in Fig. 7.14).

7.2.6 Phosphoesterase Activity

Similar to the other complexes reported in this thesis, the phosphoesterase activity of the Zn(II) complexes of ligand $CH_3H\mathbf{L4}$ and $CH_3H\mathbf{L5}$ was investigated. The pH dependence of BDNPP hydrolysis was measured from pH 4.75 to 10 and is shown in Fig. 7.15 for both complexes. The activity for both complexes dropped dramatically above pH 8 and the data became irreproducible. In some cases a white precipitate was observed, presumed to be zinc hydroxide. The leaching of zinc at pH values higher than 8 had been previously reported by Meyer et al. [20]. The pK_a values determined for $[Zn_2(CH_3\mathbf{L4})(CH_3COO)_2]^+$ and $[Zn_2(CH_3\mathbf{L5})(CH_3COO)_2]^+$ (7.39 ± 0.19 and 7.50 ± 0.19, respectively) are typical for a zinc bound water molecule [21]. The dependence of BDNPP concentration on the rate of hydrolysis by $[Zn_2(CH_3\mathbf{L4})(CH_3COO)_2]^+$ and $[Zn_2(CH_3\mathbf{L5})(CH_3COO)_2]^+$ followed Michaelis-Menten kinetics and is displayed in Fig. 7.16. These studies were conducted at the pH optimum 7.66.

The rate of activity is lower for the complex $[Zn_2(CH_3\mathbf{L5})(CH_3COO)_2]^+$ ($k_{cat} = 0.97 \pm 0.21 \times 10^{-3}\ s^{-1}$, K_m 7.01 ± 2.57 mM) than for the less sterically hindered complex $[Zn_2(CH_3\mathbf{L4})(CH_3COO)_2]^+$ ($k_{cat} = 2.45 \pm 0.27 \times 10^{-3}\ s^{-1}$, $K_m = 9.48 \pm 1.74$ mM).

7.2.7 Cobalt Derivatives of CH_3HL4 and CH_3HL5

In addition to the Zn(II) complexes attempts were made to prepare the Co(II) derivatives of $CH_3H\mathbf{L4}$ and $CH_3H\mathbf{L5}$. It was anticipated that the Co(II) complexes would permit a more detailed spectroscopic analysis and, by analogy to the

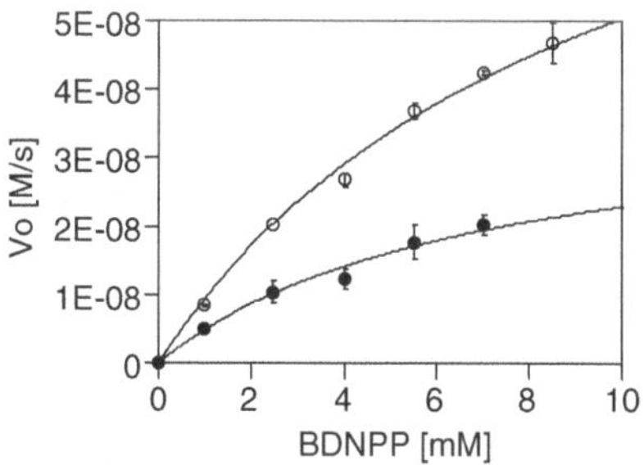

Fig. 7.16 Substrate concentration dependence on the catalytic rate for $[Zn_2(CH_3\mathbf{L4})(CH_3COO)_2]^+$ (*open circle*) and $[Zn_2(CH_3\mathbf{L5})(CH_3COO)_2]^+$ (*filled circle*)

corresponding Zn(II) complexes, provide more information concerning the structures of the complexes in solution and in the presence of substrates. It was found however, that the Co(II) complexes were sensitive to oxidation and were thus more difficult to handle than the cobalt(II) complexes reported earlier. The lack of one oxygen donor from the ligand arm, as opposed to the **L2**-ligand (Chaps. 2, 4 and 5) series, with a subsequent shift of the redox potential in favor of Co(III) is suspected to be the reason for the sensitivity to oxidation. An attempt to generate the Co(III)Co(III) complex using sodium hexanitrocobaltate(III) was not successful with a mixture of yellow, brown and green compounds formed. It is proposed while one site favors Co(III) the other site (with the methyl ether) still favors Co(II) and so direct synthesis of a Co(III)Co(III) complex is not favored. This result is in accord with the experiments that were conducted for the symmetrical complex with $CH_3\mathbf{L2}^-$ where a Co(III) species could not be produced electrochemically.

The complex $[Co(II)Co(III)(CH_3\mathbf{L4})(CH_3CO_2)_2](PF_6)_2$ was isolated as a brown solid after allowing a solution of the complex to evaporate in air over a period of one week. Microanalysis confirmed the presence of two sodium hexafluorophosphate counter ions, in accordance with one Co(II) being oxidized to Co(III). Mass spectral analysis shortly after synthesis of the complex (data not shown) identified both Co(II)Co(II) and Co(II)Co(III) species, based on the observed m/z values and charges of the ions. The infrared spectrum of this complex showed bridging acetate (1,601 and 1,415 cm^{-1}), the $\Delta\nu_{asym\text{-}sym} = 185\ cm^{-1}$ suggesting the presence of a bridging bidentate acetate. $[Co(II)Co(III)(CH_3\mathbf{L5})(CH_3CO_2)_2](PF_6)_2$ was isolated when the cobalt complex was synthesized from Co(II) acetate. The IR spectrum also displays acetate stretches with the same $\Delta\nu_{asym\text{-}sym} = 182\ cm^{-1}$ which is attributed to the more ionic character/higher charge of one of the cobalt atoms [18].

7.2.8 VTVH-MCD of the Cobalt Derivative of CH_3HL4

Due to their sensitivity to oxygen for the MCD measurements, the complexes were prepared in situ in degassed ethanol and immediately analyzed (the complexes

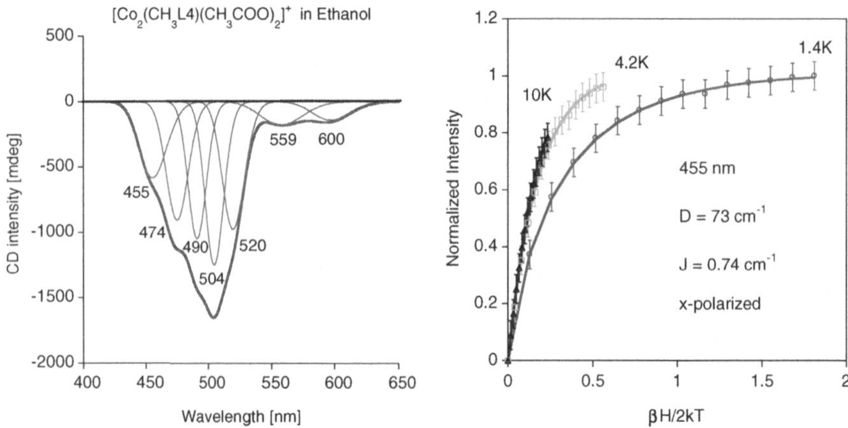

Fig. 7.17 MCD spectrum of $[Co_2(CH_3\mathbf{L4})(CH_3COO)_2]^+$ in ethanol at 1.5 K and 7.0 T (pathlength 0.62 cm, 8 mM). On the *right* magnetization plot from VTVH analysis of the band at 455 nm

oxidized completely in the MCD cell over three days at 1.5 K as shown be the absence of any Co(II) signals after this time). Figure 7.17 shows the Gaussian deconvoluted spectrum of $[Co_2(CH_3\mathbf{L4})(CH_3COO)_2]^+$ in ethanol. As can be seen from the spectrum, only a small amount of five coordinate Co(II) is present (peaks at ~559 and 600 nm). It is proposed that in solution this complex exists mainly as a dinuclear complex with two octahedral Co(II) ions as apparent from the peaks in the spectrum arising from six coordinate Co(II) (~455, 474, 490, 504 and 520 nm).

VTVH analysis yielded a zero field splitting parameter of 73 cm^{-1}, typical for hexacoordinate Co(II) [22]. The magnetic exchange coupling between the two metal centers suggests weak ferromagnetic coupling (+ 0.74 cm^{-1}). To the in situ generated Co(II) complex were added 25 equivalents of DPP (Fig. 7.18). In the presence of this substrate mimic the bands arising from transitions in six coordinate Co(II) change intensity and were shifted to lower wavelengths while the bands from five coordinate cobalt were shifted to higher wavelengths (Fig. 7.18).

The magnetic exchange coupling is not influenced upon substrate binding, the ZFS parameter D, obtained with VTVH fitting of the data, is however found at higher energy (124 cm^{-1}). This demonstrates that upon substrate binding the hexadentate coordination sphere and active site geometry is retained.

7.2.8.1 MCD Spectrum of $[Co_2(CH_3L5)(CH_3COO)_2]^+$

The MCD spectrum of $[Co_2(CH_3\mathbf{L5})(CH_3COO)_2]^+$ in ethanol is shown in Fig. 7.19. As was observed with the Co(II) complex of the ligand $CH_3H\mathbf{L4}$, six-coordinate Co(II) transitions dominate the spectrum. It is intriguing that despite the

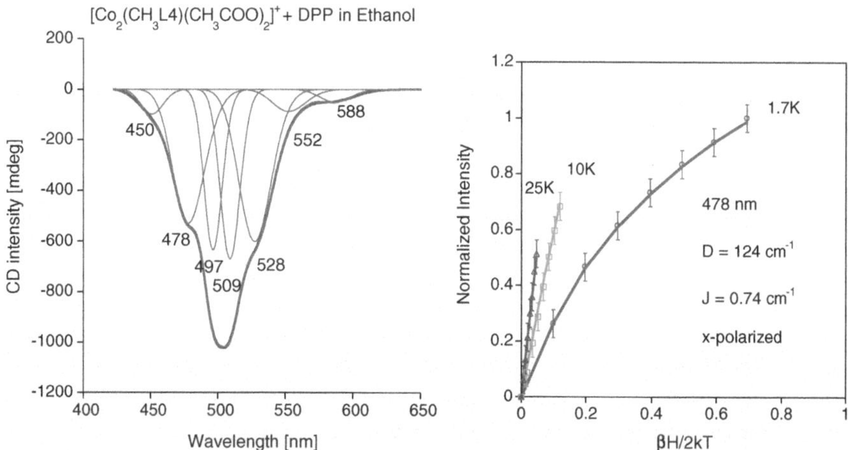

Fig. 7.18 MCD spectrum of $[Co_2(CH_3\mathbf{L4})(CH_3COO)_2]^+$ in ethanol at 1.5 K and 7.0 T (pathlength 0.62 cm, 8 mM) in the presence of 25 equivalents DPP. On the *right* magnetization plot from VTVH analysis of the band at 478 nm

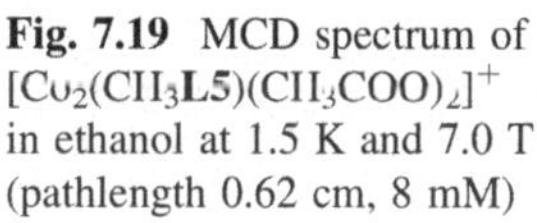
Fig. 7.19 MCD spectrum of $[Co_2(CH_3\mathbf{L5})(CH_3COO)_2]^+$ in ethanol at 1.5 K and 7.0 T (pathlength 0.62 cm, 8 mM)

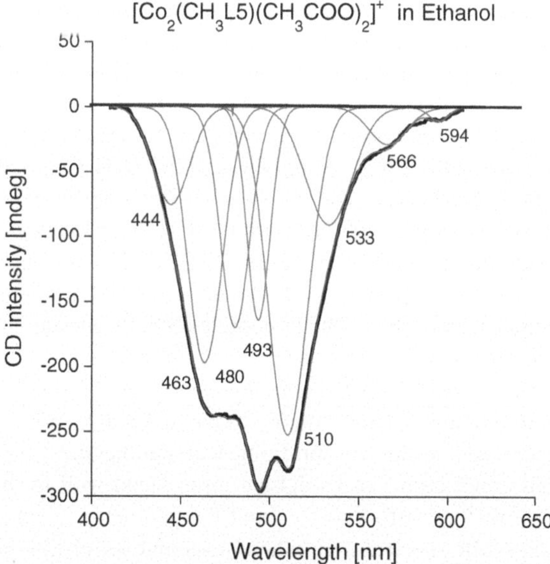

bulky vinylbenzyl group, a hexadentate coordination sphere is favored in solution for complexes derived from the $CH_3H\mathbf{L5}$ ligand. VTVH analysis was not conducted.

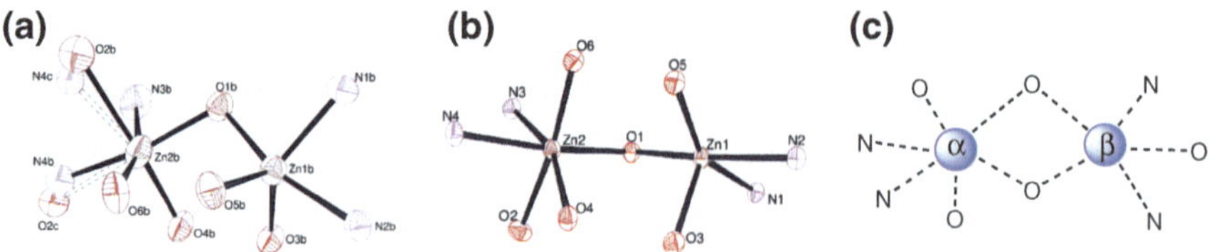

Fig. 7.20 First coordination sphere of the two Zn(II) ions in the $[Zn_2(CH_3\mathbf{L4})(CH_3COO)_2]PF_6$ complex (**a**), $[Zn_2(CH_3\mathbf{L5})(CH_3COO)_2]BPh_4$ (**b**) and in GpdQ (**c**)

7.3 Discussion

7.3.1 Structures in the Solid State and in Solution

X-ray crystal structures showed that the ligands enforced a 5,6-mixed coordination sphere in the solid state. For $[Zn_2(CH_3\mathbf{L4})\text{-}(CH_3COO)_2]PF_6$ two isomers were present in the crystal structure. The comparison of the bond lengths to the symmetrical counterpart $[Zn_2(CH_3\mathbf{L2})\text{–}(CH_3COO)_2]PF_6$ (Chap. 4) show that the O_{Phenol}-Zn bonds are similar with a mean of 2.024 Å for the asymmetric and 2.015 Å for the symmetric counterpart. The methyl ether distance is shorter (2.320 Å) than in the symmetric complex (mean 2.432 Å). N_{tert}–Zn (2.124 Å) and N_{py}–Zn (2.150 Å) distances similar to $[Zn_2(CH_3\mathbf{L2})(CH_3COO)_2]PF_6$ (2.186 and 2.1355 Å). The $O_{acetate}$–Zn bond lengths in $[Zn_2(CH_3\mathbf{L4})(CH_3COO)_2]PF_6$ vary from 1.971 to 2.058 Å while those from $[Zn_2(CH_3\mathbf{L2})(CH_3COO)_2]PF_6$ encompass a range from 1.992 to 2.076 Å. The Zn–O_{Phenol}–Zn angle is reduced to 108.17° in the asymmetric complex (109.70° with the $CH_3H\mathbf{L2}$ ligand). Zn–Zn distances are reduced in the asymmetric complexes (3.222, 3.271 Å) opposed to 3.296 Å in the symmetric counterpart. Overall, the ligands $CH_3H\mathbf{L4}$ and $CH_3H\mathbf{L5}$ enforce a geometry which makes their Zn(II) complexes appropriate structural mimics of phosphoesterase enzymes such as GpdQ. A close up view of the first coordination sphere of the two Zn(II) ions in the complexes is dis-played in Fig. 7.20 along with the first coordination sphere donor atoms in GpdQ. While the acetate ligands do not accurately mimic the bridging water molecules, the general structure, geometry and donor-atoms of GpdQ are reproduced well in this crystal structure.

Structural information about the geometry in solution was gathered using inter alia NMR spectroscopy where a second set of low intensity signals was detected. In the case of $[Zn_2(CH_3\mathbf{L4})(CH_3COO)_2]PF_6$ this is attributed to two isomers, previously also detected in the solid state structure of this complex. For the second complex with the vinylbenzyl group, NMR studies reveal that the position of this group in solution is influenced by the presence of substrate. If substrate is present it is proposed that the vinyl benzyl group shifts from a position shielding one Zn(II) site to a conformation pointing away from it. MCD studies with the rather unstable Co(II) derivatives of $CH_3H\mathbf{L4}$ and $CH_3H\mathbf{L5}$ revealed that in solution the

5,6-coordinate ligand environment is not retained. Additional coordination of solvent molecules to the vacant site complete a second, octahedral, coordination sphere.

7.3.2 Substrate Binding and Mechanism of Hydrolysis

The less sterically hindered complex $[Zn_2(CH_3\mathbf{L4})(CH_3COO)_2]^+$ binds substrate molecules more readily as shown by NMR and mass spectrometry experiments. This is unsurprising since the vinyl benzyl group of the $CH_3H\mathbf{L5}$-ligand was shown to shield one Zn(II) ion. Only in the presence of excess substrate is the binding facile. The nucleophile in organophosphate hydrolysis is proposed to be a terminal water molecule. The pK_a values determined for $[Zn_2(CH_3\mathbf{L4})(CH_3COO)_2]^+$ and $[Zn_2(CH_3\mathbf{L5})(CH_3COO)_2]^+$ (7.39 ± 0.19 and 7.50 ± 0.19, respectively) are typical for a Zn(II)-bound water molecule. They are also in the range of the pK_a values found for the symmetric Zn(II) complexes (Chap. 4). The efficiency of BDNPP hydrolysis for the $[Zn_2(CH_3\mathbf{L5})(CH_3COO)_2]^+$ complex ($0.14 \times 10^{-3}\ s^{-1}\ mM^{-1}$) approaches only half of the efficiency of $[Zn_2(CH_3\mathbf{L4})(CH_3COO)_2]^+$ ($k_{cat}/K_m = 0.26 \times 10^{-3}\ s^{-1}\ mM^{-1}$). Compared to the catalytic efficiency of Zn_2-GpdQ ($0.05\ s^{-1}mM^{-1}$) the complexes are ~200 and 360 times less efficient than the enzyme. The substrate affinity for $[Zn_2(CH_3\mathbf{L4})(CH_3COO)_2]^+$ was found to be slightly higher than for the vinyl benzyl complex derivative although the error for K_m values is quite large. These findings are in accordance with the interpretations from NMR and mass spectrometry experiments. Surprisingly, both asymmetric complexes are less active than the symmetric counterparts. It had been previously proposed by Nordlander et al. that asymmetric complexes have an advantage over their symmetric counterparts [1–3]. That this did not apply for the systems studies herein might be due to the fact that both Zn(II) complexes leached $Zn(OH)_2$ to some extent. This type of metal ion loss was previously observed by Meyer et al. for a related Zn(II) complex based on a pyrazolate ligand system [23].

7.4 Conclusion

The crystal structure of $[Zn_2(CH_3\mathbf{L4})(CH_3COO)_2]PF_6$ shows one vacant site for coordination of nucleophiles and solvent molecules which is readily occupied with solvent molecules and potential nucleophiles in solution as shown by mass spectrometry and also by MCD measurements of the Co(II) derivative. The asymmetric ligand sphere did not seem to have a beneficial effect on hydrolytic activity. The omitted ligand donor arm seemed in fact to cause instability of the complexes at pH values higher than 8. The complex $[Zn_2(CH_3\mathbf{L5})(CH_3COO)_2]PF_6$ with the more bulky ligand has an impaired substrate affinity as shown by mass spectrometry and ^{31}P NMR. Also the diminished hydrolytic activity is attributed to steric effects of the vinylbenzyl group. Nevertheless, both complexes are competent mimics of the phosphoesterase activity of GpdQ and also good structural models for this enzyme.

The next chapter will present an overview on how model complexes can, potentially, be used in applications for water decontamination by attaching them to insoluble polymers, for example, Merrifield resin. In addition, GpdQ will be immobilized onto dendrimer-modified magnetite nanoparticles to generate a potentially recyclable organophosphate bioremediation system.

References

1. H. Carlsson, E. Nordlander, M. Jarenmark, C. R. Chim. **10**, 433–462 (2007)
2. M. Jarenmark, E. Csapo, J. Singh, S. Wockel, E. Farkas, F. Meyer, M. Haukka, E. Nordlander, J. Chem. Soc., Dalton Trans. **39**, 8183–8194 (2010)
3. M. Jarenmark, S. Kappen, M. Haukka, E. Nordlander, Dalton Trans. **8**, 993–996 (2008)
4. S. Albedyhl, M.T. Averbuch-Pouchot, C. Belle, B. Krebs, J.L. Pierre, E. Saint-Aman, S. Torelli, Eur. J. Inorg. Chem. (6), 1457–1464 (2001)
5. M. Lanznaster, A. Neves, A.J. Bortoluzzi, B. Szpoganicz, E. Schwingel, Inorg. Chem. **41**, 5641–5643 (2002)
6. E. Lambert, B. Chabut, S. Chardon-Noblat, A. Deronzier, G. Chottard, A. Bousseksou, J.-P. Tuchagues, J. Laugier, M. Bardet, J.-M. Latour, J. Am. Chem. Soc. **119**, 9424–9437 (1997)
7. C. Belle, I. Gautier-Luneau, L. Karmazin, J.-L. Pierre, S. Albedyhl, B. Krebs, M. Bonin, Eur. J. Inorg. Chem. **2002**, 3087–3090 (2002)
8. C. Belle, I. Gautier-Luneau, J.L. Pierre, C. Scheer, E. SaintAman, Inorg. Chem. **35**, 3706–3708 (1996)
9. S. Albedyhl, D. Schnieders, A. Jancso, T. Gajda, B. Krebs, Eur. J. Inorg. Chem. **2002**, 1400–1409 (2002)
10. J.H. Satcher, M.W. Droege, M.M. Olmstead, A.L. Balch, Inorg. Chem. **40**, 1454–1458 (2001)
11. R.R. Buchholz, M.E. Etienne, A. Dorgelo, R.E. Mirams, S.J. Smith, S.Y. Chow, L.R. Hanton, G.B. Jameson, G. Schenk, L.R. Gahan, Dalton Trans. **43**, 6045–6054 (2008)
12. F. Meyer, E. Kaifer, P. Kircher, K. Heinze, H. Pritzkow, Chem.—Eur. J. **5**, 1617–1630 (1999)
13. H. Adams, S. Clunas, D. E. Fenton, D. N. Towers, J. Chem. Soc., Dalton Trans. 3933–3935 (2002)
14. I.A. Koval, D. Pursche, A.F. Stassen, P. Gamez, B. Krebs, J. Reedijk, Eur. J. Inorg. Chem. **2003**, 1669–1674 (2003)
15. L. Dubois, R. Caspar, L. Jacquamet, P.-E. Petit, M.-F. Charlot, C. Baffert, M.-N. Collomb, A. Deronzier, J.-M. Latour, Inorg. Chem. **42**, 4817–4827 (2003)
16. L. Dubois, D.-F. Xiang, X.-S. Tan, J. Pécaut, P. Jones, S. Baudron, L. Le Pape, J.-M. Latour, C. Baffert, S. Chardon-Noblat, M.-N. Collomb, A. Deronzier, Inorg. Chem. **42**, 750–760 (2003)
17. A.K. Boudalis, R.E. Aston, S.J. Smith, R.E. Mirams, M.J. Riley, G. Schenk, A.G. Blackman, L.R. Hanton, L.R. Gahan, Dalton Trans. (44), 5132–5139 (2007)
18. K. Nakamoto, *Infrared and Raman Spectra of Inorganic and Coordination Compounds* (Wiley, New York, 1978)
19. K. Selmeczi, C. Michel, A. Milet, I. Gautier-Luneau, C. Philouze, J.-L. Pierre, D. Schnieders, A. Rompel, C. Belle, Chem.—Eur. J. **13**, 9093–9106 (2007)
20. B. Bauer-Siebenlist, F. Meyer, E. Farkas, D. Vidovic, J.A. Cuesta-Seijo, R. Herbst-Irmer, H. Pritzkow, Inorg. Chem. **43**, 4189–4202 (2004)

21. J. Burgess, *Metal Ions in Solution* (Halsted Press, Chichester, 1978)
22. J.A. Larrabee, C.M. Alessi, E.T. Asiedu, J.O. Cook, K.R. Hoerning, L.J. Klingler, G.S. Okin, S.G. Santee, T.L. Volkert, J. Am. Chem. Soc. **119**, 4182–4196 (1997)
23. B. Bauer-Siebenlist, F. Meyer, E. Farkas, D. Vidovic, J.A. Cuesta-Seijo, R. Herbst-Irmer, H. Pritzkow, Inorg. Chem. **43**, 4189–4202 (2004)

Chapter 8
Immobilization of GpdQ and Related Models for Bioremedial Applications

8.1 Introduction

8.1.1 Strategies to Immobilize Enzymes and Catalysts

Dendrimer modified and plain magnetite nanoparticles (MNP) have been extensively studied for their potential application in environmental decontamination, medicine and biology [1–4]. For example, Grüttner et al. reported MNP functionalized with chelating groups on the surface to bind lanthanides and actinides in order to remove radionuclides from nuclear wastes [3]. MNPs also have potential in transport of pharmaceuticals by coating them with, for example, streptavidin (SA) which displays an extremely high affinity for biotin [2]. Biotin itself can be easily conjugated to many different diagnostic reagents [2]. Magnetically assisted processes have been reported to purify water [4] and MNP have been shown to be useful in protein immobilization and purification, and DNA-extraction [1, 5]. A biomimetic catalyst, comprised of aspartate and histidine residues coated on magnetite, has been reported for the facile removal of paraoxon (phosphoester) and 4-nitrophenyl acetate (carboxylic ester) from water [6]. The nanoparticles converted 77 % of 10 equivalents paraoxon within 48 h to diethyl phosphate, significantly more than in the uncatalyzed reaction (1 %). The main advantage of magnetite nanoparticles is that they can be easily recovered from the reaction media by magnetic separation.

In an approach to utilize enzymes to degrade pesticides recombinant organophosphate hydrolase (OPH) was anchored to the surface of *E.coli* cells and packed in a bioreactor [7]. A similar approach used *E.coli* cells with OPH on the surface, immobilized on glass beads [7], silica beads [8] or nylon [9]. In 2005 a group reported the immobilization of the pesticide degrading enzyme OPH on a cellulose matrix. The enzyme was attached to the matrix via a cellulose binding domain [10]. When packed in a column bioreactor the immobilized OPH was able to degrade a 0.2 mM solution of the pesticide coumaphos rapidly and without loss of activity over 2 months [11]. Another possibility to use enzymes in a solid form to catalyze

L. J. Daumann, *Spectroscopic and Mechanistic Studies of Dinuclear Metallohydrolases and Their Biomimetic Complexes*, Springer Theses, DOI: 10.1007/978-3-319-06629-5_8,

reactions are CLEAs (cross-linked enzyme aggregates) [12–14]. CLEAs have been generated for a large number of enzymes (laccases, lipases, esterases etc.), after addition of a precipitation agent e.g. ethanol, ammonium sulfate, the aggregated enzyme is permanently cross-linked with glutaraldehyde. The cross-linking has the effect of stabilization and also the CLEAs are immiscible with water which makes it easy to remove them from the reaction media and reuse [13, 14].

The immobilization of metal complexes onto inorganic or organic supports is also a well-known concept. Modified silica, organic resins such as Tentagel or Merrifield resin and other polymers have served as supports. Merrifield resin-attached complexes have been reported to be useful for various applications such as Ni(II) catalyzed cross coupling reactions [15], the oxidative DNA cleavage catalyzed by an immobilized cyclen Cu(II) [16], Pd(II) catalyzed Suzuki-Miyaura cross-coupling reactions [17], salen Mn(II) complexes for epoxidation reactions [18] or the Co(II) catalyzed polymerization of vinyl acetate [19]. The advantage is that the complex can be easily separated from the reaction mixture and the complex loaded resin can be recycled. A Fe(III)Zn(II) purple acid phosphatase model complex was reported by Neves et al. that had been immobilized on 3-aminopropyl silica [20]. The reaction rates of the immobilized systems were similar to those found in the reaction with free complex and the immobilized catalyst could be reused. A PAP model with pendant olefinic arms suitable for copolymerization and thus incorporation into a polymer has also been reported recently [21].

8.1.2 Aims of this Chapter

The aims of this chapter are to generate immobilized model complexes and enzymes for potential water treatment purposes.[1] A protocol will be established to immobilize ligands on MNP and Merrifield Resin (MR) and their Zn(II) complexes will be generated. The immobilized systems will be further analyzed for their ability to degrade BDNPP. Moreover GpdQ will be attached to PAMAM dendrimer modified MNP and the functionality will be tested in a specifically designed assay. Not only kinetic characterization of the particles will be conducted but also a thorough characterization using IR, microanalysis, transmission electron microscopy (TEM) and X-ray photoelectron spectroscopy (XPS).

---

[1] Reprinted from "L.J. Daumann et al., "*Immobilization of the enzyme GpdQ on magnetite nanoparticles for organophosphate pesticide bioremediation*" J. Inorg. Biochem. **2014**, 131, 1–7", Copyright (2014), with permission from Elsevier:

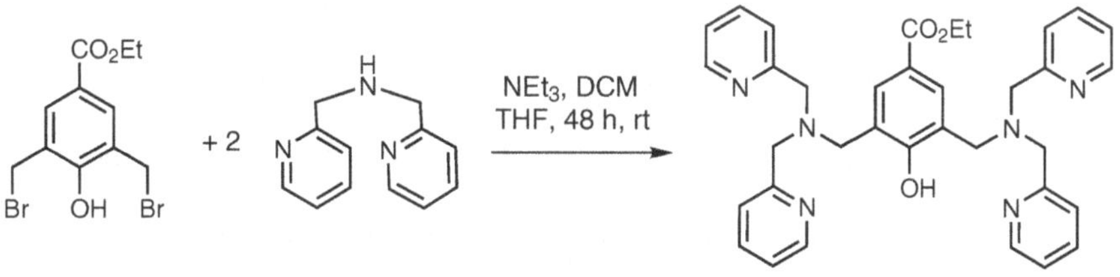

Fig. 8.1 Ligand synthesized to investigate ester hydrolysis in the absence of alcohol arms

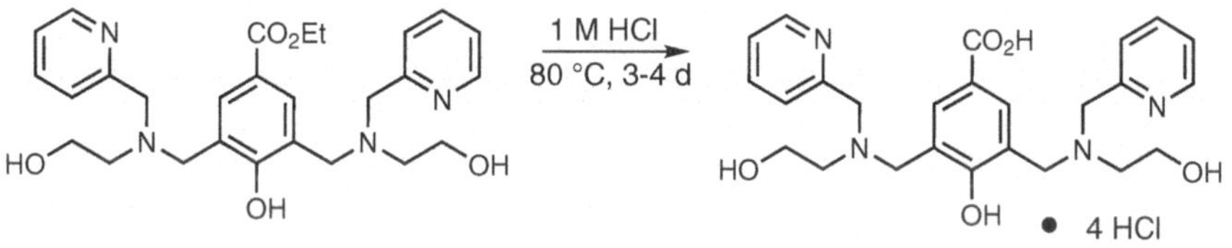

Fig. 8.2 Successful hydrolysis of the ester moiety of $EtCO_2H_3$**L1**

8.2 Early Attempts to Immobilize/Polymerize Model Complexes

The first attempt to immobilize the model systems was to cleave the ester of CO_2EtH_3**L1** to generate a pendant carboxylic acid function which could be coupled to an amine from a resin using peptide coupling methods. A similar approach has been reported by Scrimin et al., who had used Tentagel resin as support (pendant NH_2 groups) [22]. Attempts to hydrolyze the ester group of CO_2EtH_3**L1** with NaOH in ethanolic solution were unsuccessful although this had been described in the literature for the similar ligand methyl 3,5-bis [9-bis(2-pyridylmethyl)amino)methyl]-hydroxybenzoate [23]. Initially it was thought that the alcohol arms were causing the problems during the alkaline hydrolysis. Hence the ligand shown in Fig. 8.1 was synthesized and the cleavage of the ester group attempted as a 'proof of principle' of the literature reaction. This resulted in only very low yields of the desired product. After a close examination of the supporting information it was clear that this type of reaction was not very clean and thus other methods had to be evaluated [23].

As base hydrolysis was deemed unsuitable, acid hydrolysis was investigated. The ethyl ester ligand was dissolved in 1 M HCl and stirred at 80 °C for 4 days (Fig. 8.2). After removal of the solvent, NMR analysis showed that the desired carboxylic acid compound was obtained as hydrochloride in acceptable yield (50 %) with all nitrogen atoms protonated.

Subsequently, attempts were made to couple the ligand CO_2HH_3**L1** to commercially available Tentagel S-NH_2 and Tentagel MAP-4 resin. The latter resin had been treated previously with piperidine in DMF to cleave the FMOC

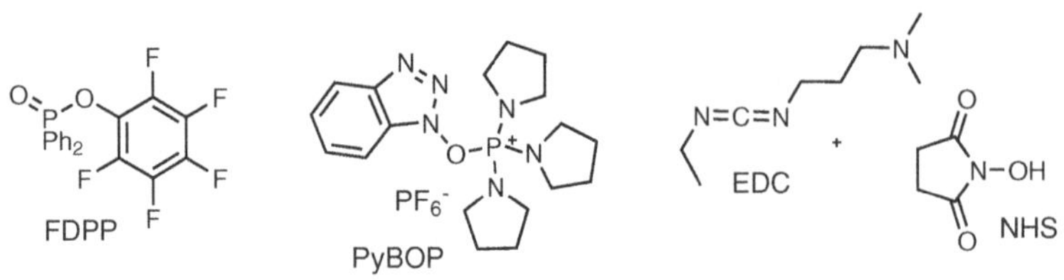

Fig. 8.3 Peptide coupling reagents tested for the immobilization of CO_2HH_3**L1**

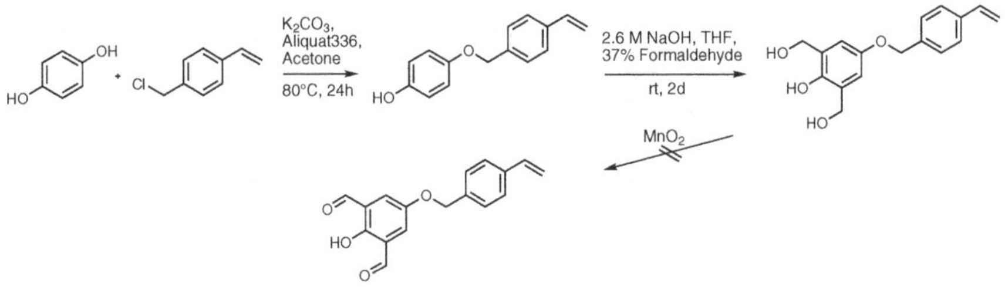

Fig. 8.4 Failed attempt to generate a polymerizable ligand

(fluorenylmethyloxycarbonyl) protecting group. PyBOP (benzotriazol-1-yl-oxy-tripyrrolidinophosphonium hexafluoro-phosphate) with NMO (n-methylmorpholine), FDPP (pentafluorophenyl diphenyl phosphinate) with DIPEA (diispropylethyl amine) and EDC (1-ethyl-3-(3-dimethylamino-propyl)carbodiimide) with NHS (N-hydroxysuccinimide) were utilized as coupling reagents (Fig. 8.3).

However, none of those coupling reagents facilitated the immobilization of CO_2HH_3**L1** on the Tentagel resins; this pathway of immobilization via peptide coupling was thus not continued. Immobilization via co-polymerisation was investigated as this had been described previously by Striegler et al. [24]. A ligand featuring a pendant vinyl group at the pyridines had been described by Zee et al. [21]. however a suitable polymerization procedure was not found, the failure attributed to the bulkiness of the ligand [21]. Thus, a different approach with a ligand bearing a pendant vinyl benzyl group in the *para*-position to the phenolic oxygen was considered. Step one and two of this ligand synthesis were conducted as described in the literature (Fig. 8.4, together with Miss Harriet McAtee) [25, 26]. The following step proved to be difficult, since a functionalization of the hydroxymethyl groups with hydrogen bromide or chloride could not be pursued due to the reactivity of the vinyl benzyl group towards those reagents. Manganese dioxide was used to convert the OH groups into aldehydes which could then be potentially further functionalized with amino methyl pyridine to build up the ligand arms.

The reaction did, however, not proceed cleanly to give the desired product but rather a mixture of three compounds which could not be separated.

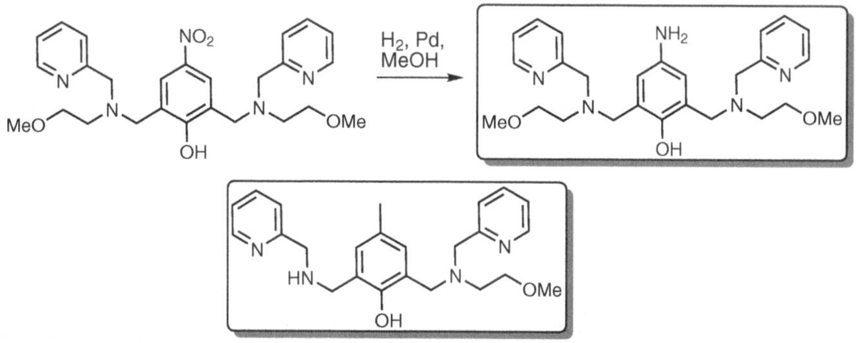

Fig. 8.5 The two ligands NH_2H**L2** and CH_3H_2**L4** with NH-functionalities that were used for immobilization studies

Inspired by the failed attempts to couple or polymerize the ligands described above it was decided that a functionalization via amine groups would be pursued where NH_2 and NH functionalities of the ligands would be reacted with chloro-functionalised resins or condensed with pendant aldehyde residues. The reduction of the NO_2H**L2** ligand (Chaps. 2, 4 and 6) was straightforward and accomplished using an H-Cube equipped with a palladium/charcoal cartridge [27]. The asymmetric ligand described in Chap. 7 also features a NH-moiety, hence this ligand was also used in the immobilization studies (Fig. 8.5).

8.3 Immobilization of Model Complexes on Merrifield Resin

8.3.1 Synthesis of Merrifield Resin Immobilized Zn(II) Complexes

Merrifield resin (MR) was chosen due to its ready availability and high degree of substitution (3.5–4.5 mmol/g Cl^- loading, Aldrich Chemical Company). 3-Chloropropyl functionalized silica gel was also considered initially, but the degree of immobilization on this support was shown to be low. The respective ligands NH_2H**L2** and CH_3H_2**L4** were stirred for five days at room temperature with MR in the presence of dry, powdered potassium carbonate in acetonitrile. After this time the resin was filtered off, washed with copious amounts of water to remove potassium carbonate and then methanol to remove any unbound ligand. To generate the immobilized Zn(II) complexes, zinc(II) acetate was added to a suspension of the MR with the respective immobilized ligand in methanol (Fig. 8.6) After 30 min reflux, the resins were filtered, washed with methanol and dried under high vacuum.

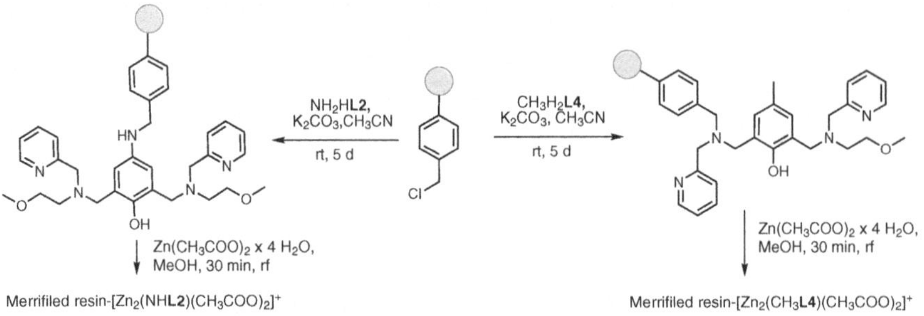

NH2HL2,
K2CO3,CH3CN
rt, 5 d
CH3H2L4,
K2CO3, CH3CN
rt, 5 d
HN
OH
Cl
Zn(CH3COO)2 x 4 H2O,
MeOH, 30 min, rf
Zn(CH3COO)2 x 4 H2O,
MeOH, 30 min, rf
Merrifiled resin-[Zn2(NHL2)(CH3COO)2]+
Merrifiled resin-[Zn2(CH3L4)(CH3COO)2]+

Fig. 8.6 Immobilization of ligands NH_2HL2 and CH_3H_2L4 and generation of the Zn(II) complexes

(a)
(b)
%T
Wavenumber [cm-1]
Merrifield Resin
Merrifield Resin-[Zn2(NHL2)(CH3COO)2]+
Merrifield Resin-[Zn2(CH3L4)(CH3COO)2]+

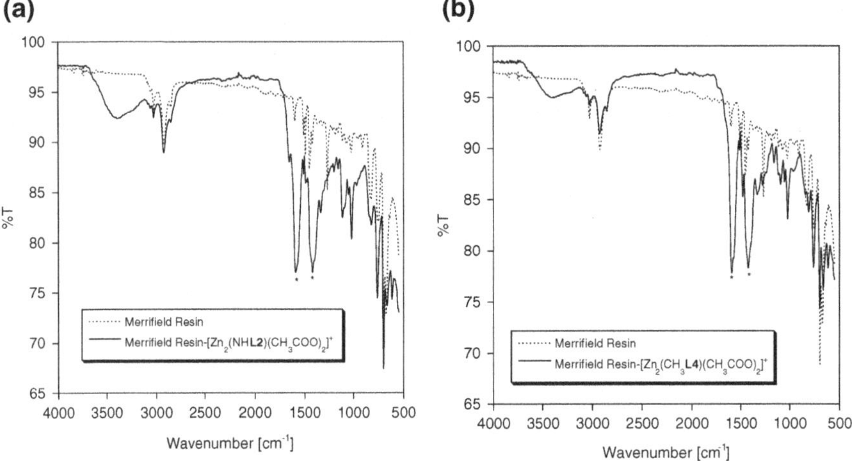

Fig. 8.7 IR spectra of **a** plain MR and MR-$[Zn_2(NH\mathbf{L2})(CH_3COO)_2]^+$ and **b** pain MR and MR-$[Zn_2(CH_3\mathbf{L4})(CH_3COO)_2]^+$

To investigate whether MR would take up Zn(II) ions, which could potentially catalyze organophosphate hydrolysis, unsubstituted resin was also subjected to the aforementioned treatment.

8.3.2 Infrared Spectroscopy of the Complexes on MR

Figure 8.7 shows the IR spectra of unsubstituted MR (dotted line) and the resins with the zinc(II) complexes MR-$[Zn_2(NH\mathbf{L2})(CH_3COO)_2]^+$ (a) and MR-$[Zn_2(CH_3\mathbf{L4})(CH_3COO)_2]^+$ (b). Clearly visible are the two (asymmetric and symmetric) acetate stretches (marked with *). Also broad bands at 3,300 cm^{-1},

Table 8.1 Microanalyses of the different resins

Resin type	%C	%H	%N
MR	80.18	6.64	0
MR treated with $Zn(CH_3COO)_2$	80.39	6.76	0
MR-NHH**L2**	78.41	7.35	5.23
MR-$[Zn_2$(NH**L2**)$(CH_3COO)_2]^+$	64.56	6.43	3.84
MR-CH_3H**L4**	77.39	7.22	4.86
MR-$[Zn_2(CH_3$**L4**$)(CH_3COO)_2]^+$	64.42	6.18	3.58

indicate OH moieties present, probably in the form of Zn(II) bound water or methanol.

A $\Delta\nu_{asym-sym}$ of 175 cm^{-1}, found for both complexes, is indicative for successful complex formation, with acetate bound in a bidentate bridging fashion and is virtually identical to the values in the free complexes (e.g. Chap. 7) [28]. IR bands attributed to acetate stretches were not observed when unsubstituted resin was treated with zinc(II) acetate. It was thus concluded that MR alone would not bind large amounts of zinc(II) acetate that could be relevant for activity.

8.3.3 *Elemental Analysis of the Zn(II) Complexes on MR*

Successful immobilization was also confirmed with elemental analysis. Table 8.1 lists the values obtained and the results show that the C, H and N values for MR treated with zinc(II) acetate, differ only slightly from commercially available untreated MR. This supports the results obtained in the IR spectroscopic measurements that unsubstituted resin does not bind Zn(II) ions. After immobilization of the ligands, the percentage of nitrogen increased substantially which, after generation of the Zn(II) complexes then decreases again as expected when adding non nitrogen containing compounds to the resin (Table 8.1).

8.3.4 *XPS of MR and Immobilized Complexes*

The XPS of commercially available MR showed, in addition to the expected peaks from carbon and oxygen large amounts of tin (Fig. 8.8a) [29–31]. A closer inspection of the oxygen peak lead to the conclusion that it was tin oxide, presumably from the manufacturing process of the resin [30].

The supplier of the MR (Sigma-Aldrich) did not disclose any information whether tin oxide was used in the manufacturing process of this resin so the origin of this contamination is currently unknown. The resin treated with zinc(II) acetate showed only a very small amount of zinc (Fig. 8.8b) and no peaks from nitrogen. The XPS spectra of the immobilized ligands show nitrogen peaks in addition to the peaks from

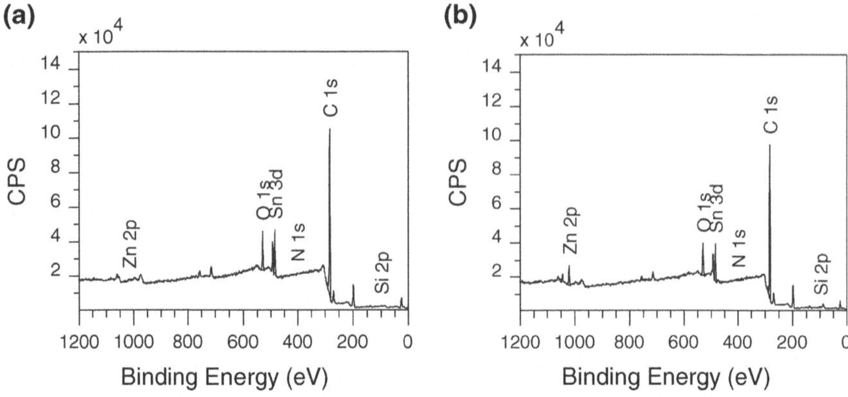

Fig. 8.8 XPS survey spectra of MR (**a**) and MR treated with $Zn(CH_3COO)_2$ (**b**)

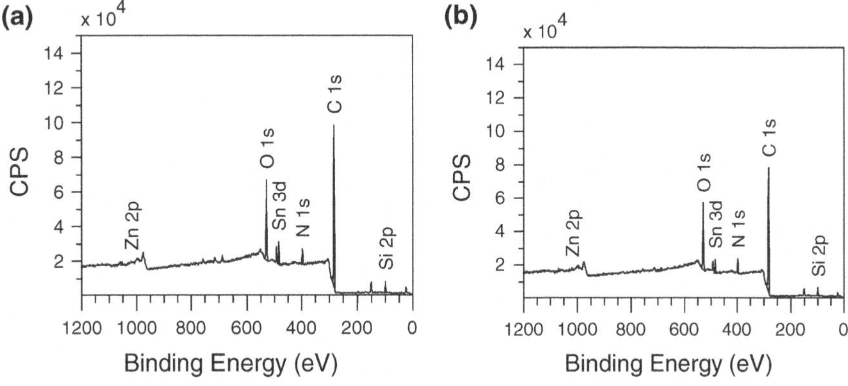

Fig. 8.9 XPS survey spectra of MR-CH_3H**L4** (**a**) and MR-NHH**L2** (**b**)

MR as expected. Also a greater degree of oxygen relative to unsubstituted resin is observed (Fig. 8.9a and b). Moreover Si 2p peaks are present in the immobilized ligand MR [31]. This is due to the fact that the resins were dried on a high vacuum oil pump and silicone oils are a known cause for this contamination [32].

Finally the XPS spectra of the immobilized complexes showed the two typical peaks of the Zn 2p electrons (Fig. 8.10a and b) confirming the uptake of Zn(II) ions into the resin/ligand [31].

A very small peak for chlorine (Cl 2p) was found in all spectra (not labeled) and a detailed analysis suggested the presence of both ionic chloride and carbon bound chloride (Fig. 8.11) [31]. The former is suggested to arise after replacing the Cl with the ligand amine and the latter is due to unreacted Cl-residues on the resin surface. As expected, two peaks are observed for the Cl 2p electrons due to spin orbital splitting [33]. Only one type of nitrogen is observed/resolved, the binding energy typical for tertiary nitrogen (pyridine, tertiary amine). The peak for the Zn

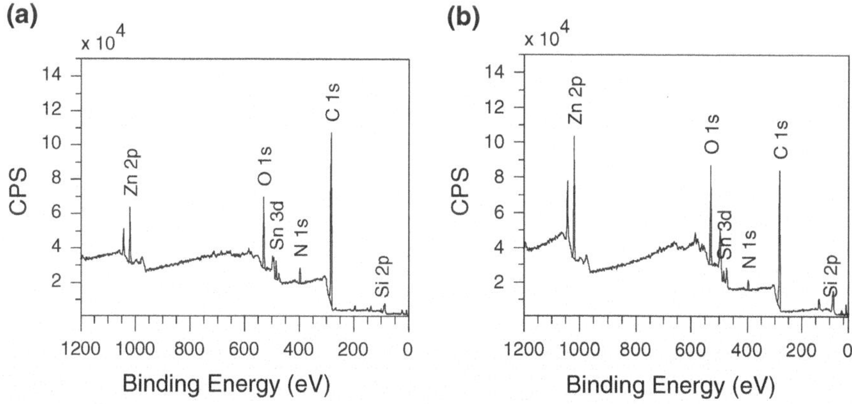

Fig. 8.10 XPS survey spectra of MR-[$Zn_2(CH_3$**L4**$)(CH_3COO)_2]^+$ (**a**) and MR-[Zn_2(NH**L2**)$(CH_3COO)_2]^+$ (**b**)

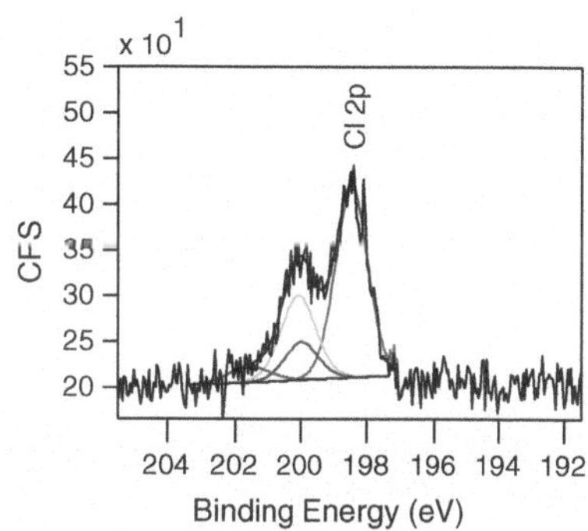

Fig. 8.11 Gaussian deconvoluted Cl 2p spectrum

2p electrons displays the typical pattern with a second smaller peak at higher energy (Fig. 8.12c) [31, 33].

The XPS spectra of the MR-[Zn_2(NH**L2**)$(CH_3COO)_2]^+$ compound were similar and are shown in Fig. 8.13. Figure 8.13a shows the peaks arising from the Zn(II) ions in the complex while (b) demonstrates that oxygen is present in at least three different chemical environments in the sample: C–O (ether and phenol), C=O (acetates) and one peak arising from the contaminant tin oxide (e). Two nitrogen species were observed accounting for the tertiary nitrogen in the ligand (N_{py} and $N_{tert,}$ 400.09 eV) and one from the secondary amine at 399.41 eV (Fig. 8.13d) [34].

8.3.5 *Kinetic Properties of the Immobilized Zn(II) Complexes*

For the kinetic assays a method had to be established which would allow determination of the kinetic parameters for BDNPP hydrolysis. In order to obtain initial

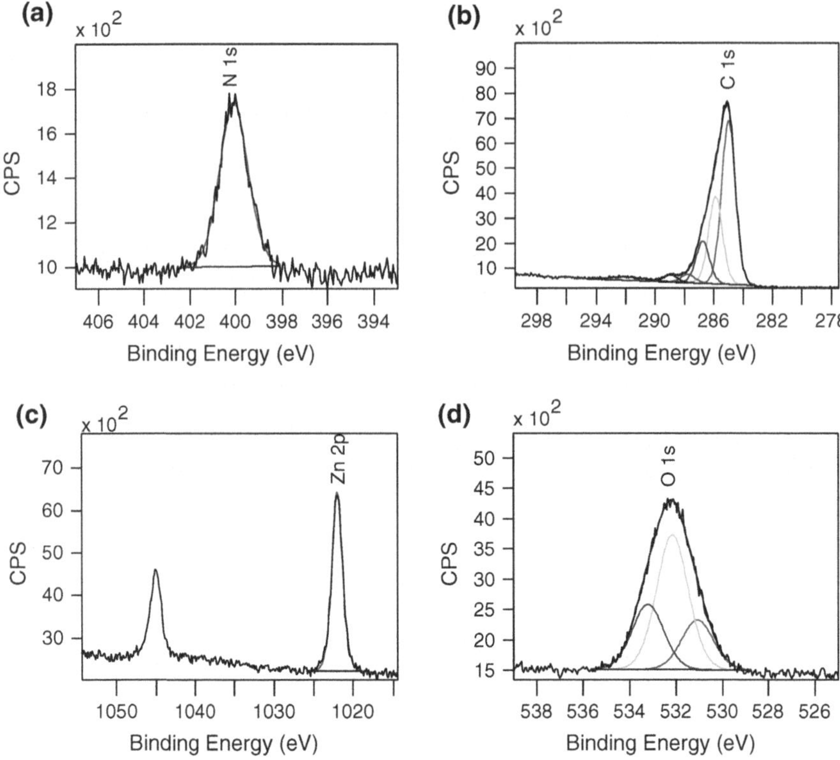

Fig. 8.12 Gaussian deconvoluted (**a**) N 1 s (**b**) C 1 s (**c**) Zn 2p and (**d**) O 1 s spectra of MR-$[Zn_2(CH_3\mathbf{L4})(CH_3COO)_2]^+$

rates, the absorbance of a solution of multicomponent buffer, acetonitrile and BDNPP was determined at the beginning of the experiment (T = 0) and then 0.5 mg of resin were added. After 15 min, 1 mL of the suspension was transferred to an Eppendorf tube and centrifuged for 10 s. The absorbance of the supernatant was then measured again (T = 15) and the Abs/min determined. For every value this was done in triplicate, however, due to swelling properties of the resin and possibly inhomogeneous complex loading on the Merrifield beads the errors were larger than for the free complexes [35]. Figure 8.14 shows the pH dependence of BDNPP hydrolysis for MR-$[Zn_2(CH_3\mathbf{L4})(CH_3COO)_2]^+$ and MR-$[Zn_2(NH_2\mathbf{L2})(CH_3COO)_2]^+$.

For the complex MR-$[Zn_2(CH_3\mathbf{L4})(CH_3COO)_2]^+$ no data above pH 8 were collected as the resin leached Zn(II) at alkaline pH [36]. This is in accordance to the findings in Chap. 7 where the free asymmetric Zn(II) complexes also showed this leaching of Zn(II) (hydroxide). As Fig. 8.14 demonstrates, the pH dependence of both complexes is similar to the free Zn(II) complexes (Chaps. 4 and 7). The pK_a values were 7.61 ± 0.71 and 6.41 ± 0.31 for MR-$[Zn_2(CH_3\mathbf{L4})$

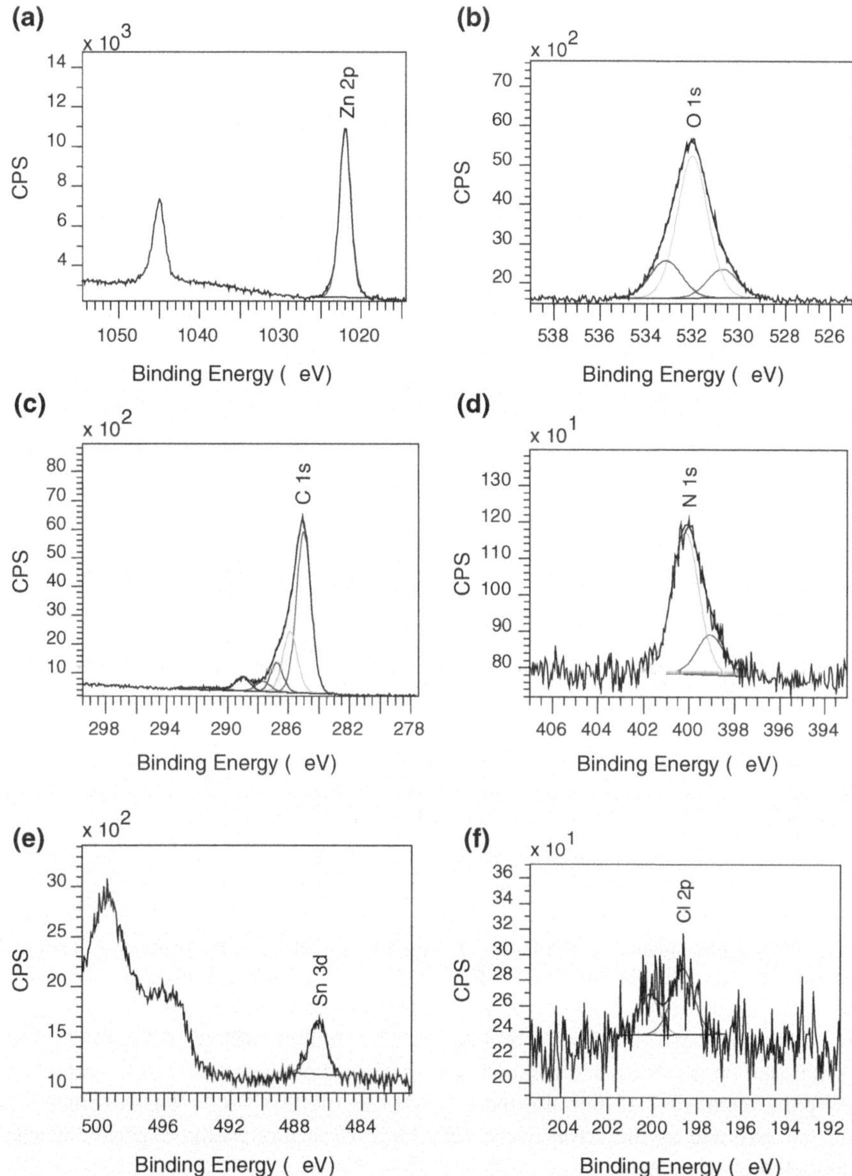

Fig. 8.13 Gaussian deconvoluted (**a**) Zn 2p (**b**) O 1 s (**c**) C 1 s (**d**) N 1 s (**e**) Sn 3d and (**f**) Cl 2p spectra of MR-[Zn_2(NH**L2**)($CH_3COO)_2]^+$

$(CH_3COO)_2]^+$ and MR-[$Zn_2(NH_2$**L2**)($CH_3COO)_2]^+$, respectively. In general the errors were larger than for the free complexes and the fits shown above have to interpreted with caution. For the latter complex, a reduction in activity due to a

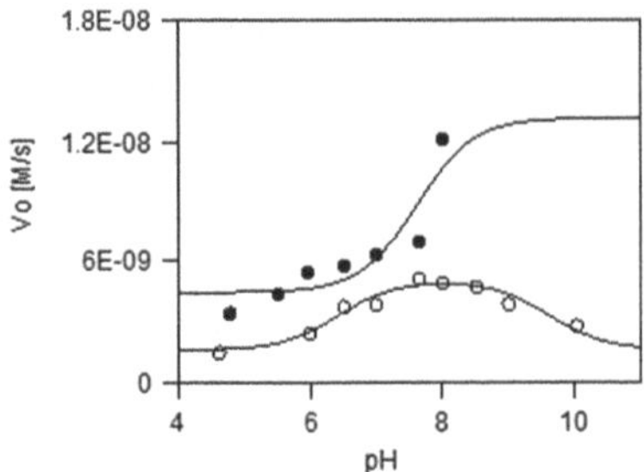

Fig. 8.14 pH dependence of BDNPP hydrolysis by MR-$[Zn_2(CH_3\mathbf{L4})(CH_3COO)_2]^+$ (*filled circle*) and MR-$[Zn_2(NH_2\mathbf{L2})(CH_3COO)_2]^+$ (*open circle*)

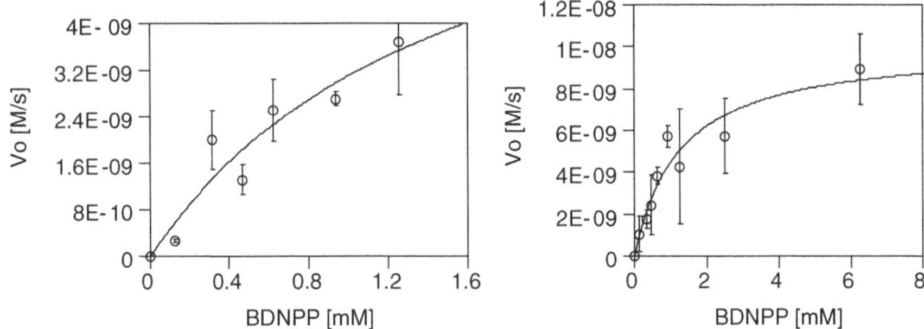

Fig. 8.15 Michaelis-Menten profile for BDNPP hydrolysis by (**a**) MR-$[Zn_2(CH_3\mathbf{L4})(CH_3COO)_2]^+$ at pH 8 and (**b**) by MR-$[Zn_2(NH_2\mathbf{L2})(CH_3COO)_2]^+$ at pH 8

deprotonation of a terminal water molecule, which is competing with the substrate for coordination space, is proposed. A Michaelis-Menten type profile was found for the complexes at pH 8 with BDNPP (Fig. 8.15). The same method was used for the pH dependence. For MR-$[Zn_2(CH_3\mathbf{L4})(CH_3COO)_2]^+$ a $V_{max} = 7.97$ $10^{-9} \pm 4.4 \times 10^{-9}$ M s^{-1} and $K_m = 1.57 \pm 1.34$ mM were found (Fig. 8. 15a). Based on the estimated catalyst loading (based on the amount of ligand taken up by the resin) of 1.45 mmol/g the k_{cat} is $11 \pm 6 \times 10^{-6}$ s^{-1}. These values are only an estimate as the errors were very large even though the experiment was conducted in triplicate.

The symmetric complex MR-$[Zn_2(NH_2\mathbf{L2})(CH_3COO)_2]^+$ yielded a V_{max} of 1.01 $10^{-8} \pm 1.20$ 10^{-9} M s^{-1} and a $K_m = 1.24 \pm 0.36$ mM. Based on the amount of ligand immobilized the catalyst loading was estimated to be ~0.92 mmol/g and thus $k_{cat} = 8.82 \pm 0.26 \times 10^{-5}$ s^{-1}.

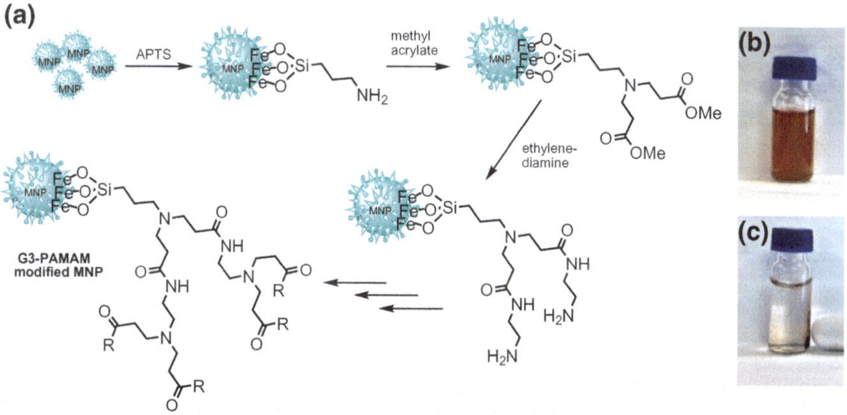

Fig. 8.16 **a** Synthesis of the magnetite nanoparticles (MNP) functionalized with G3-PAMAM dendrimer. MNP suspended in water before (**b**) and after magnetic separation (**c**)

8.4 Immobilization of Model Complexes and GpdQ on Magnetite Nanoparticles

8.4.1 Synthesis and Characterisation of G3-PAMAM MNP

Generation 3 PAMAM dendrimer modified magnetite nanoparticles were synthesized after a modification of previously published methods (Fig. 8.16a) [1, 2]. In brief, aqueous solutions of ferric chloride and ferrous sulfate were combined under vigorous stirring and the magnetite subsequently co-precipitated by adding ammonia solution.

After anchoring APTS on the surface of the magnetite, the PAMAM dendrimer was build up stepwise by alternating addition of methyl acrylate and ethylenediamine until a generation 3 (G3) dendrimer had formed on the surface. Figure 8.16 also demonstrates the appearance of the suspended nanoparticles in water after sonication (b) and after bringing a simple stirring magnet in close proximity (c).

8.4.2 Infrared Spectroscopy of G3-PAMAM MNP

Infrared analysis of unsubstituted magnetite nanoparticles showed the stretches from Fe-O at 560 cm^{-1}. Once amino propyl trimethyl silane (APTS) was attached to the surface, additionally C–H stretches at (2949, 2866 cm^{-1}) and Si–O stretches at 1015 cm^{-1} were found. In the final G3-PAMAM MNP N-H and $C{=}O_{Amide}$ stretches were apparent (3314 and 1625, 1551 cm^{-1})

Table 8.2 Elemental Analyses from the MNP at the different stages of modification

	%C	%H	%N	%S
MNP-plain	0.24	0.35	0.05	0.75
MNP-APTS	1.69	0.64	0.47	0.69
MNP-G3 PAMAM	3.53	0.75	1.44	0.00

8.4.3 Elemental Analysis of G3-MNP

Elemental analyses were conducted to confirm successful synthesis of the G3-PAMAM MNP and to monitor the increasing nitrogen content of the G0-G3 dendrimer (Table 8.2). Some sulfur contamination is apparent (presumably from the ferrous sulfate used in the MNP synthesis) which is, however, washed out during the repeated sonication/washing processes during subsequent PAMAM dendrimer modification.

8.4.4 XPS of G3-PAMAM MNP

Figure 8.17 shows the survey spectra of the nanoparticles reflecting the different degrees of functionalization. In Fig. 8.17a the XPS of unsubstituted magnetite nanoparticles is shown. While this material consists mainly of Fe_3O_4, a C 1 s peak is observed [37]. It is suggested that ethanol from the washing process remains bound to the surface, and is not removed by the vacuum drying. Figure 8.17b shows that after the immobilization of APTS and a G1-PAMAM dendrimer, the surface analysis now also shows nitrogen and small amount of silicon present [34]. The nitrogen peak increases further as the PAMAM dendrimer generation advances (Fig. 8.17c).

A closer evaluation of the G3-PAMAM spectrum reveals that the core material is indeed Fe_3O_4 [31, 37]. This was concluded form the binding energies of the Fe 2p peak ($\sim$710.0 eV) and also the O 1 s peak $\sim$529.5 eV confirms the core material being Fe_3O_4 (Fig. 8.18a and b) [37, 38]. Moreover oxygen peaks in two different environments could be assigned to the following species: C=O, C–O [34, 39].

8.4.5 Transmission Electron Microscopy of the Nanoparticles

Transmission electron microscopy was conducted on the nanoparticles to investigate the size, shape and clustering behavior. Prior to the measurements $\sim$20 mg of the nanoparticles were diluted with 1 mL of water and sonicated for 30 min.

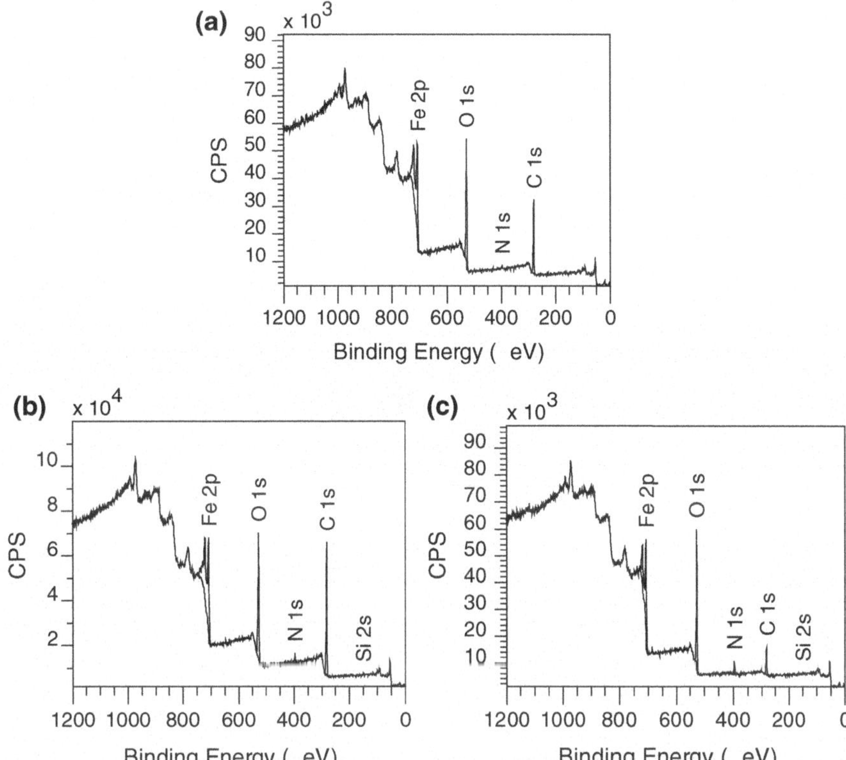

Fig. 8.17 XPS survey spectra of plain MNP (**a**), G1-PAMAM MNP (**b**) and G3-PAMAM MNP (**c**)

The suspension was then applied onto 600 mesh carbon grids and dried in air for several minutes. Figure 8.19 demonstrates that particle sizes from 5–20 nm are observed, confirming the 'nanoparticle nature' of the magnetite. Figure 8.19a also demonstrates that the MNP appear close to spherical, although TEM shows only a 2D picture. Moiré fringes are also observed from interference pattern by the atomic layers of the iron oxide [40]. Moreover the magnetic behavior is demonstrated by the clustering of the nanoparticles as demonstrated in Fig. 8.19b.

8.4.6 Immobilization of $[Zn_2(NH_2L2)(CH_3COO)_2]^+$ on G3-MNP

The nanoparticles were further functionalized with glutaraldehyde, washed with methanol and subsequently a methanolic solution of the NH_2H**L2** ligand was added (Fig. 8.20). After standing for 5 days at room temperature, the nanoparticles

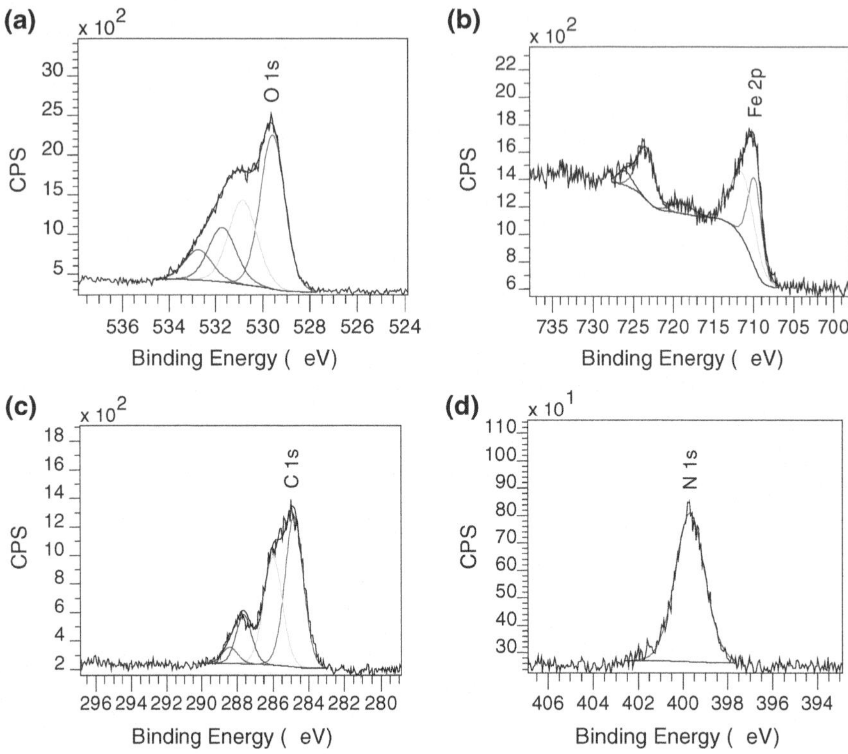

Fig. 8.18 Gaussian deconvoluted (**a**) O 1 s (**b**) Fe 2p (**c**) C 1 s and (**d**) N 1 s spectra of G3-PAMAM MNP

were washed until the washing solution was colorless. To generate the Zn(II) complex the nanoparticles were stirred with zinc(II) acetate for 30 min and then washed and dried.

8.4.6.1 Infrared Spectroscopy of G3- MNP-$[Zn_2(NH_2L2)(CH_3COO)_2]^+$

The IR spectrum of G3-MNP-NHH**L2** showed in addition to the phenol O–H stretch at 3359 bands at 1111 cm^{-1} (C–O–C) from the methyl ether arm and typical bands arising from aromatic functionalities at 818, 757 and 699 cm^{-1} (Ar-H/Py-H). For its Zn(II) complex the two characteristic stretches from bridging acetate were observed at 1599 and 1414 cm^{-1} [28]. It should be noted that G3-MNP that had been treated with Zn(II) acetate also showed bands typical for zinc(II) acetate. It was thus concluded that the PAMAM dendrimer took up zinc(II) acetate which could potentially act as catalyst as well. Since this would create

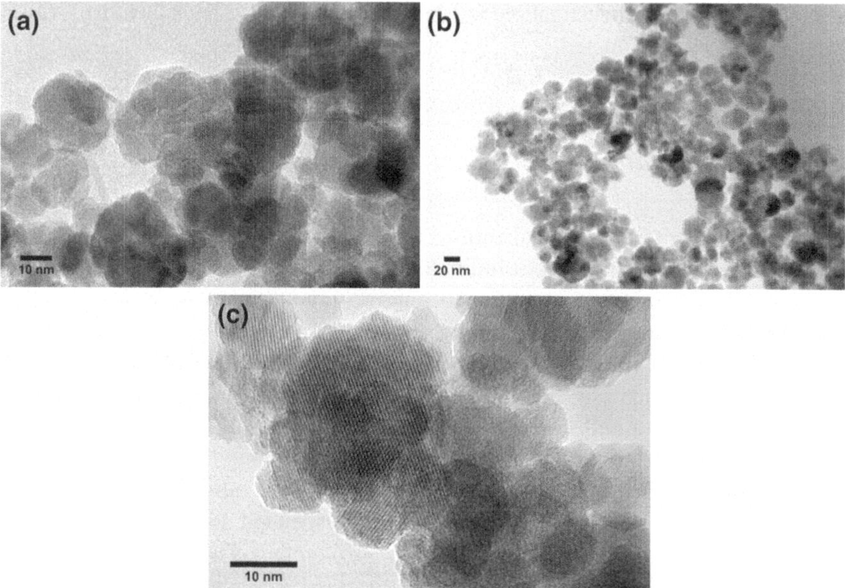

Fig. 8.19 TEM images of plain magnetite nanoparticles at two magnifications (50 k (**a**) and 200 k (**b**)) and (**c**) TEM image of magnetite nanoparticles with APTS anchored at the surface at a magnification of 400 k

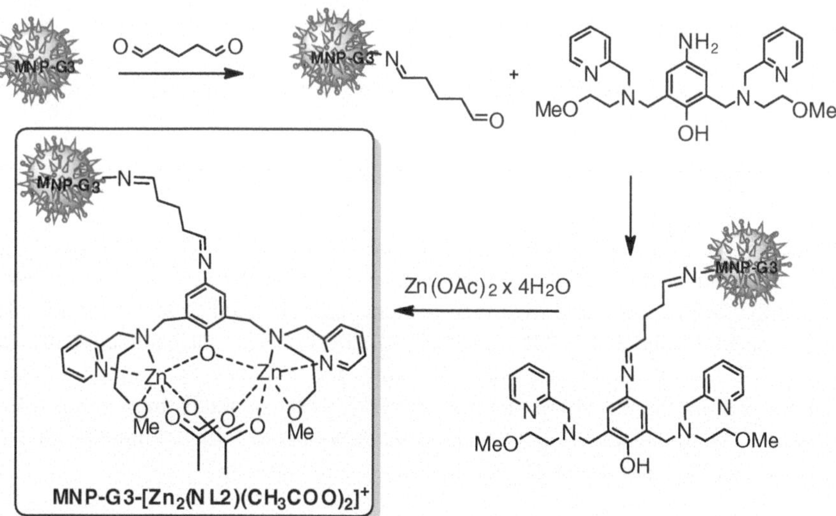

Fig. 8.20 Functionalization of the G3-MNP with the model complex

problems, a further investigation of the G3- MNP-$[Zn_2(NH_2\mathbf{L2})(CH_3COO)_2]^+$ complex was not conducted.

8.4.7 Immobilization of the Ser127Ala Mutant of GpdQ

The G3-MNP, after functionalization with glutaraldehyde, were added to a Ser127Ala GpdQ mutant stock solution and after 5 days at 4 °C the nanoparticles were washed thoroughly with buffer to remove any unbound enzyme [41]. After this time the amount of enzyme bound to the nanoparticles via amide formation between the lysine residues of the enzyme and the pendant aldehydes on the particles did not increase any further. This was confirmed by the decrease of absorbance (by up to half of the initial value) at 280 nm of the supernatant. The degree of immobilization fluctuated depending on how long the nanoparticles had been stirred with glutaraldehyde previously and also the ratio of nanoparticles to enzyme. G3-MNP-GpdQ were stored at 4 °C suspended in a small amount of buffer.

8.4.7.1 XPS of G3-MNP-GpdQ

G3-MNP-GpdQ were dried in high vacuum prior to XPS analysis. The survey spectrum shows that the elements present on the surface are mainly iron, oxygen, nitrogen and carbon (Fig. 8.21a) [34]. While the O 1 s and C 1 s peaks could be resolved into multiple species the assignment for nitrogen is less clear (Fig. 8.21c, d and e). In general the survey spectrum displays more noise as the others previously reported in this work. This is attributed to the multitude of trace elements that are bound to the enzyme and thus appear in the XPS spectrum.

8.4.7.2 Elemental Analysis

The elemental analysis of the immobilized enzyme on the nanoparticles demonstrated that both carbon and hydrogen content were elevated (4.66 %C, 1.13 %H) with respect to the G3-MNP (3.53 %C, 0.75 %H). The nitrogen content was lowered (1.22 %N) which is unsurprising since the G3-NMP have a high nitrogen content due to the PAMAM dendrimer (1.55 %N). A small amount of sulfur (0.04 %S) is found after GpdQ had been immobilized which might be due to cysteine residues and traces of sulfate bound to the enzyme.

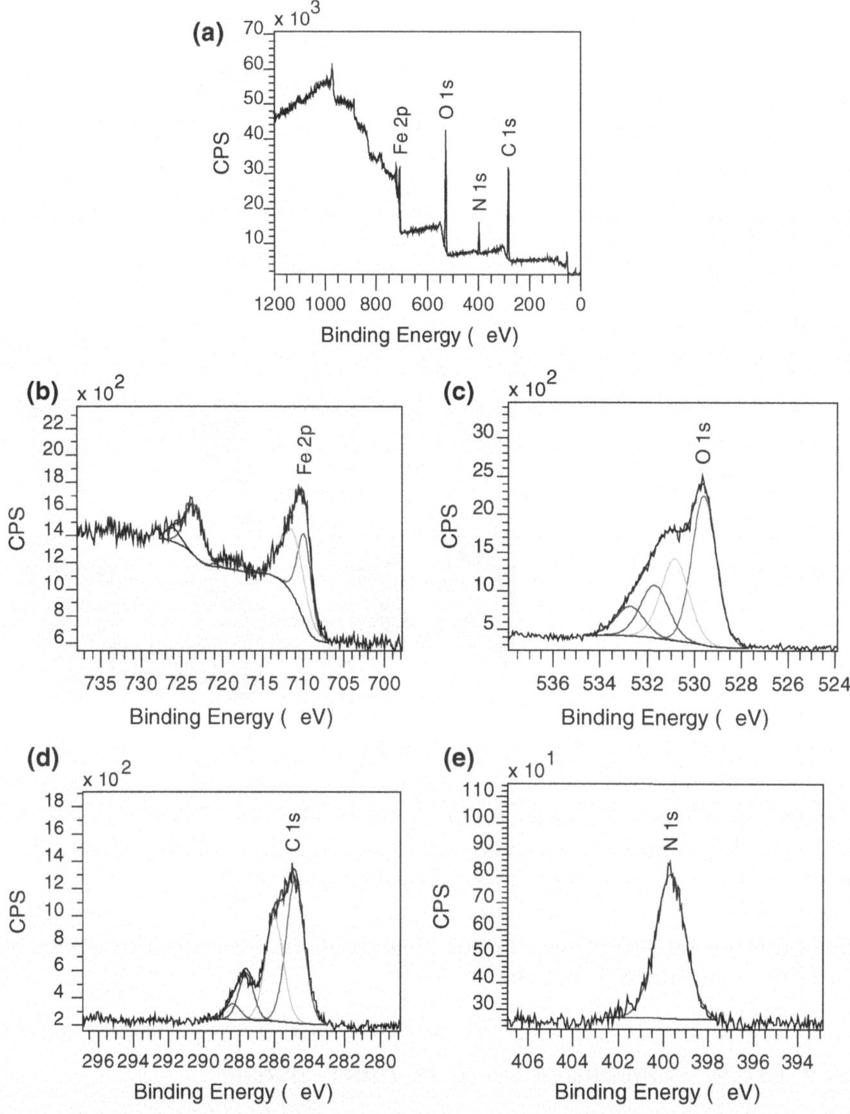

Fig. 8.21 Gaussian deconvoluted (**a**), Fe 2p (**b**), O 1 s (**c**), C 1 s (**d**) and N 1 s (**e**) spectra of G3-MNP-GpdQ

8.4.8 Transmission Electron Microscopy of GpdQ bound to G3-MNP

The TEM sample preparation of the immobilized enzyme was difficult since only water could be used to suspend the nanoparticles (usually a more volatile solvent is

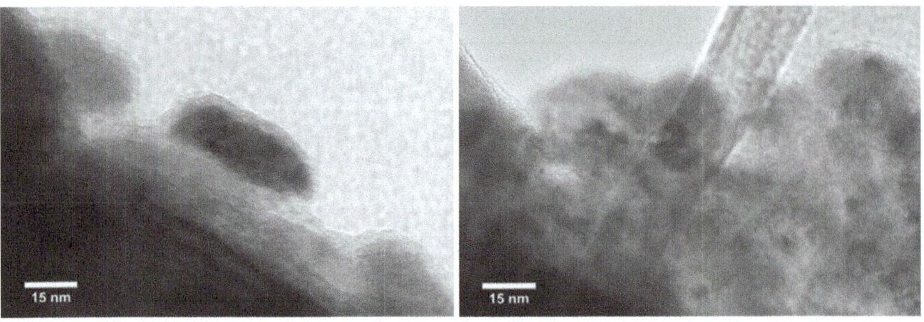

Fig. 8.22 TEM images of GpdQ immobilized on G3-MNP

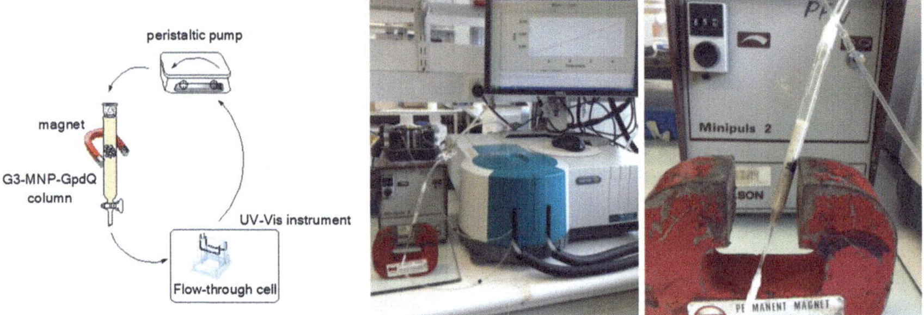

Fig. 8.23 Experimental setup to test activity of the GpdQ enzyme immobilized on G3-MNP

favoured) and prolonged sonication times were avoided since this could cause the enzyme to detach from the nanoparticles. However, a range of images was taken and the two best resulting TEM images are shown in Fig. 8.22.

The images demonstrate that the nanoparticles are surrounded by an organic matter which is presumably the enzyme. Analysis of the diffraction patterns of the darker spots confirmed them to be Fe_3O_4.

8.4.9 Use of Immobilized GpdQ in Filter Systems

To test whether the immobilized enzyme was still active towards organophosphates an experimental setup had to be designed, mimicking water filtration conditions. The nanoparticles were embedded between two layers of sand in a Pasteur pipette and were held in place with a conventional magnet. The bottom of the column was connected with a small tube to a flow through UV-Vis cuvette (Fig. 8.23). A buffered solution of the organophosphate substrate BPNPP was circled thought the column and UV-vis cell with a peristaltic pump. The hydrolysis of the substrate was monitored by the formation of the nitrophenolate ion at 405 nm.

After cycling the BPNPP solution through the column for 3 h, the column was detached from the system and washed with buffer solution and stored at 4 °C. After one week the experiment was repeated. The activity towards BPNPP was shown to be unaffected by the storage.

8.5 Discussion

8.5.1 Immobilization of Biomimetics on Merrifield Resin

After initial difficulties to find a suitable method to attach model complexes on inorganic or organic supports, the two amine ligands, CH_3**HL4** and NH_2**HL2**, were found to be the most promising systems. Both ligands were shown to react readily with the chloromethyl groups of Merrifield resin in the presence of base. To confirm that both ligands assemble competent dinuclear Zn(II) complexes, IR proved to be a useful tool. The two (symmetric and asymmetric) acetate stretches were present in the spectra after stirring the ligand-containing resins with zinc(II) acetate. The $\Delta v_{asym-sym}$ of 175 cm^{-1} showed that acetate is bound in a bidentate bridging fashion [28]. The characteristic acetate stretches were not observed when unsubstituted resin was treated with zinc(II) acetate. MR alone does not bind large amounts of zinc(II) acetate that could be relevant for catalysis. Elemental analysis also confirmed that only MR substituted with ligand is able to bind Zn(II). Moreover, the nitrogen content determined by this method confirmed not only the successful immobilization of the ligands but also, indirectly, the uptake of Zn(II). An attempt to dissolve the MR with the complexes in hot concentrated nitric acid or mixtures of nitric acid and hydrochloric acid was unsuccessful even after a prolonged heating time. Thus no atomic absorption analysis for the zinc content could be conducted. However the exact amount of ligand that was taken up by the resin was recorded by weighing the ligand remaining after immobilization. With 500 mg unsubstituted Merrifield resin, 294 mg out of 320 mg CH_3**HL4** were found to be immobilized and for NH_2**HL2** the resin (150 mg) took up 64 mg out of 100 mg ligand. For the respective Zn(II) complexes it was assumed that all ligands would form a hydrolytically active, competent dinuclear center since Zn(II) acetate was added in a large excess. To further analyze the composition of elements on the resin surface, XPS was extremely useful. XPS analysis showed that the commercially available resin was contaminated with tin oxide [30]. Also a contamination with silicon was detected. The latter is present only in the samples that had been dried with a high vacuum pump and thus it is concluded that this is the source of contamination [32]. Nonetheless the changes of the peak areas of the C, O, N and Zn peaks proved to be very useful to gain information about chemical environments and to confirm effective complex immobilization on the resin. As expected, immobilization of the ligands on MR gave rise to a N 1 s peak. For CH_3**HL4** only one nitrogen species was detected, in accordance with tertiary

Table 8.3 Comparison of catalytic parameters of immobilized and free zinc complexes

	k_{cat} [s^{-1}]	K_m [mM]	pK_a	k_{cat}/K_m [$M^{-1}s^{-1}$]
MR-[Zn_2(NH**L2**)($CH_3COO)_2]^+$	$8.82 \pm 0.26 \times 10^{-5}$	1.24 ± 0.36	6.40	0.07
MR-[Zn_2(CH_3**L4**)($CH_3COO)_2]^+$	$11.00 \pm 6.00 \times 10^{-6}$	1.57 ± 1.34	7.61	0.01
[Zn_2(CH_3**L4**)($CH_3COO)_2]^+$	$2.45 \pm 0.27 \times 10^{-3}$	9.48 ± 1.74	7.39	0.26
[Zn_2(CH_3**L5**)($CH_3COO)_2]^+$	$0.97 \pm 0.21 \times 10^{-3}$	7.01 ± 2.57	7.50	0.14
[Zn_2(CH_3**L2**)($CH_3COO)_2]^+$	$5.70 \pm 0.04 \times 10^{-3}$	20.8 ± 5.0	6.76	0.27

nitrogens (N_{tert} and N_{py}) being the only species present [34]. For NH_2H**L2** however an additional nitrogen environment is found in the XPS spectrum (N_{sec}), as expected when the NH_2 group is functionalized with the chloromethyl group of the resin [34]. Unsurprisingly, after addition of zinc(II) acetate both MR-[Zn_2(NH**L2**)($CH_3COO)_2]^+$ and MR-[Zn_2(CH_3**L4**)($CH_3COO)_2]^+$ spectra display the spin-orbit doublets peaks typical for zinc ($2p_{1/2}$ and $2p_{3/2}$) near 1021 and 1041 eV [34, 39].

The resins were also analyzed for their ability to degrade organophosphates. Since a homogeneous assay was not possible, a heterogeneous assay was developed (see Chap. 2). The data in Table 8.3 demonstrate that the immobilized complexes are in general less active biomimetics than the free complexes.

In Fig. 8.24 the proposed structures of MR-[Zn_2(CH_3**L4**)($CH_3COO)_2]^+$ and MR-[Zn_2(NH**L2**)($CH_3COO)_2]^+$ are shown along with the free complexes (Chaps. 4 and 7). While [Zn_2(CH_3**L4**)($CH_3COO)_2]^+$ is the precursor for the former complex, [Zn_2(CH_3**L5**)($CH_3COO)_2]^+$ is a more suitable 'monomer model' for the immobilized form. The most appropriate monomer counterpart for MR-[Zn_2(NH**L2**)($CH_3COO)_2]^+$ is suggested to be [Zn_2(CH_3**L2**)($CH_3COO)_2]^+$.

That the asymmetric MR-[Zn_2(CH_3**L4**)($CH_3COO)_2]^+$ is less active that the symmetric catalysts is unsurprising, as a similar result was found for the free asymmetric complexes (see Chap. 7). The general lower efficiency (k_{cat}/K_m) of the immobilized complexes (also including the symmetric MR-[Zn_2(NH**L2**)($CH_3COO)_2]^{+)}$ could be a result of the swelling of the resin [35] and the loss of zinc at pH values higher than 7 [42].

8.5.2 Immobilization of GpdQ and a Biomimetic on Magnetic Nanoparticles

PAMAM dendrimer modified magnetite nanoparticles were tested as supports for GpdQ and a biomimetic complex. Previous studies have shown that a G3-PAMAM dendrimer has a beneficial effect for protein immobilization opposed to unsubstituted MNP [1]. The dendrimer was build up stepwise by alternating additions of methanolic methyl acrylate and ethylenediamine solutions. The pendant amine functions were then further functionalized with glutaraldehyde.

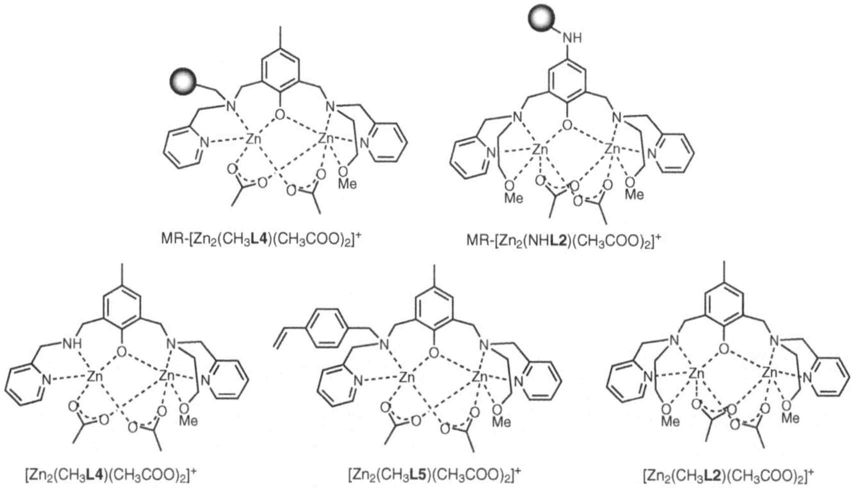

MR-$[Zn_2(CH_3\mathbf{L4})(CH_3COO)_2]^+$ MR-$[Zn_2(NH\mathbf{L2})(CH_3COO)_2]^+$

$[Zn_2(CH_3\mathbf{L4})(CH_3COO)_2]^+$ $[Zn_2(CH_3\mathbf{L5})(CH_3COO)_2]^+$ $[Zn_2(CH_3\mathbf{L2})(CH_3COO)_2]^+$

Fig. 8.24 Structures of the immobilized complexes and the 'free' counterparts

This yielded MNP with pendant aldehyde functions which were then combined with the lysine residues of the GpdQ surface and the amine residue of the ligands $NH_2H\mathbf{L2}$. The degree of immobilization varied depending on the glutaraldehyde concentration in the previous step. It is proposed that a higher glutaraldehyde concentration leads to partial crosslinking of the nanoparticles, thus decreasing the available sites for enzyme and complex immobilization. The amount of GpdQ that was immobilized with the optimized procedure was 1.488 μmol per g MNP. The specific activity directly after immobilization was 3.55 μmol mg^{-1} min^{-1}, after one week 3.39 μmol $mg^{-1}min^{-1}$ and after 120 days 3.36 μmol mg^{-1} min^{-1} (when using a 5.63 mM solution of BPNPP). This shows that after hydrolysis of organophosphates, the nanoparticles with the enzyme can be recovered from the reaction medium and reused multiple times with little loss of activity.

8.6 Conclusion

Two phosphoesterase model complexes (MR-$[Zn_2(CH_3\mathbf{L4})(CH_3COO)_2]^+$ and MR-$[Zn_2(NH\mathbf{L2})(CH_3COO)_2]^+$) were successfully immobilized on Merrifield resin and subsequently characterized with IR, microanalysis and XPS. After kinetic analysis it could be shown that the immobilized systems were active towards BDNPP but less efficient than the free Zn(II) complexes reported in Chaps. 4 and 7. This was attributed to resin swelling and loss of Zn(II) in alkaline medium. The Zn(II) complex based on $NH_2H\mathbf{L2}$ and in addition the Ser127Ala mutant of GpdQ were attached to PAMAM dendrimer modified magnetite nanoparticles using glutaraldehyde. The MNP were characterized with IR, microanalysis, XPS and TEM at different stages of functionalization. TEM confirmed the nanoparticle character and XPS affirmed that the nanoparticles were made of Fe_3O_4 and also provided insight into the elemental composition of the surface after PAMAM generation etc. Kinetic experiments with the immobilized GpdQ mutant showed that the enzyme was active on the nanoparticles and that it could be stored over a prolonged time in the immobilized form without loss of activity. This system has the potential with further development to be an effective method for bioremediation of organophosphate contaminated environments.

References

1. B.-F. Pan, F. Gao, H.-C. Gu, J. Colloid Interface Sci. **284**, 1–6 (2005)
2. F. Gao, B.-F. Pan, W.-M. Zheng, L.-M. Ao, H.-C. Gu, J. Magn. Magn. Mater. **293**, 48–54 (2005)
3. C. Grüttner, V. Böhmer, A. Casnati, J.-F. Dozol, D.N. Reinhoudt, M.M. Reinoso-Garcia, S. Rudershausen, J. Teller, R. Ungaro, W. Verboom, P. Wang, J. Magn. Magn. Mater. **293**, 559–566 (2005)
4. R.D. Ambashta, M. Sillanpää, J. Hazard. Mater. **180**, 38–49 (2010)
5. B. Yoza, A. Arakaki, T. Matsunaga, J. Biotechnol. **101**, 219–228 (2003)
6. Y. Zheng, C. Duanmu, Y. Gao, Org. Lett. **8**, 3215–3217 (2006)
7. A.H. Mansee, W. Chen, A. Mulchandani, Biotechnol. Bioprocess Eng. **5**, 436–440 (2000)
8. D.M. Munnecke, Biotechnol. Bioeng. **21**, 2247–2261 (1979)
9. S. Caldwell, F. Raushel, Appl. Biochem. Biotechnol. **31**, 59–73 (1991)
10. W. Chen, A. Mulchandani, A.H. Mansee, J. Ind. Microbiol. Biotechnol. **32**, 554–560 (2005)
11. A.H. Mansee, W. Chen, A. Mulchandani, J. Ind. Microbiol. Biotechnol. **32**, 554–560 (2005)
12. R. Schoevaart, M.W. Wolbers, M. Golubovic, M. Ottens, A.P.G. Kieboom, F. van Rantwijk, L.A.M. van der Wielen, R.A. Sheldon, Biotechnol. Bioeng. **87**, 754–762 (2004)
13. R.A. Sheldon, Biochem. Soc. Trans. **35**, 1583 (2007)
14. R.A. Sheldon, Appl. Microbiol. Biotechnol. **92**, 467–477 (2011)
15. P. Styring, C. Grindon, C.M. Fisher, Catal. Lett. **77**, 219–225 (2001)
16. K. Li, J. Zhang, Z.-W. Zhang, Y.-Z. Xiang, H.-H. Lin, X.-Q. Yu, J. Appl. Polym. Sci. **111**, 2485–2492 (2009)
17. J. Yang, P. Li, L. Wang, Synthesis **2011**(1295), 1301 (2011)
18. Q.H. Xia, H.Q. Ge, C.P. Ye, Z.M. Liu, K.X. Su, Chem. Rev. **105**, 1603–1662 (2005)
19. V. Sciannamea, A. Debuigne, Y. Piette, R. Jerome, C. Detrembleur, Chem. Commun. (40), 4180–4182 (2006)
20. C. Piovezan, R. Jovito, A.J. Bortoluzzi, Hn Terenzi, F.L. Fischer, P.C. Severino, C.T. Pich, G.G. Azzolini, R.A. Peralta, L.M. Rossi, A. Neves, Inorg. Chem. **49**, 2580–2582 (2010)
21. Y.L.M. Zee, L.R. Gahan, G. Schenk, Aust. J. Chem. **64**, 258–264 (2011)
22. G. Zaupa, L.J. Prins, P. Scrimin, Bioorg. Med. Chem. Lett. **19**, 3816–3820 (2009)
23. A. Mangalum, R.C. Smith, Tetrahedron **65**, 4298–4303 (2009)
24. S. Striegler, M. Dittel, Inorg. Chem. **44**, 2728–2733 (2005)
25. E. Díez-Barra, J.M. Fraile, J.I. García, E. García-Verdugo, C.I. Herrerías, S.V. Luis, J.A. Mayoral, P. Sánchez-Verdú, J. Tolosa, Tetrahedron Asymmetry **14**, 773–778 (2003)
26. S. Carloni, V. Borzatta, L. Moroni, G. Tanzi, G. Sartori, R. Maggi, Patent WO2005123254A1 Catalysts based on metal complexes for the synthesis of optically active chrysanthemic acid, p. 55, 2005
27. R.V. Jones, L. Godorhazy, N. Varga, D. Szalay, L. Urge, F. Darvas, J. Comb. Chem. **8**, 110–116 (2006)

28. K. Nakamoto, *Infrared and Raman Spectra of Inorganic and Coordination Compounds* (Wiley, New York, 1978)
29. C.S. Ltd, CasaXPS: Processing Software for XPS, AES, SIMS and More (2009)
30. L. Jie, X. Chao, J Non-Cryst Solids **119**, 37–40 (1990)
31. NIST, X-ray Photoelectron Spectroscopy Database, Version 4.1, http://srdata.nist.gov/xps/. Accessed 20 Feb 2013
32. G.L. Weissler, R.W. Carlson, *Vacuum Physics and Technology* (Academic, New York, 1979)
33. H.-L. Lee, N. Flynn, in *Handbook of Applied Solid State Spectroscopy*, ed. by D.R. Vij (Springer, US, 2006, ch. 11), pp. 485–507
34. C.D. Wanger, W.M. Riggs, L.E. Davis, J.F. Moulder, G.E. Muilenberg, *Handbook of X-ray Photoelectron Spectroscopy* (Perkin-Elmer Corp, Physical Electronics Division, Minnesota, USA, 1979)
35. R. Santini, M.C. Griffith, M. Qi, Tetrahedron Lett. **39**, 8951–8954 (1998)
36. B. Bauer-Siebenlist, S. Dechert, F. Meyer, Chem.Eur. J. **11**, 5343–5352 (2005)
37. A. Baykal, M.S. Toprak, Z. Durmus, M. Senel, H. Sozeri, A. Demir, J. Supercond. Novel Magn. **25**, 1541–1549 (2012)
38. B.J. Tan, K.J. Klabunde, P.M.A. Sherwood, Chem. Mater. **2**, 186–191 (1990)
39. J.C. Vickerman, *Surface Analysis: the Principal Techniques* (Wiley, New York, 1997)
40. J. Guild, *The Interference Systems of Crossed Diffraction Gratings: theory of Moire Fringes* (Clarendon, Oxford, 1956)
41. L.S. Wong, F. Khan, J. Micklefield, Chem. Rev. **109**, 4025–4053 (2009)
42. B. Bauer-Siebenlist, F. Meyer, E. Farkas, D. Vidovic, J.A. Cuesta-Seijo, R. Herbst-Irmer, H. Pritzkow, Inorg. Chem. **43**, 4189–4202 (2004)

Chapter 9
Conclusion and Outlook

The enzyme GpdQ has been investigated previously as a potential bioremediator. However, although the enzyme was known to degrade a range of organophosphate substrates and a mechanism has been proposed, little was known about the metal ion content in vivo. This thesis presents the first study of heterodinuclear derivatives of this enzyme and has shed light on the binding of different metals and the mechanistic implications in the presence of a range of metal ions. Moreover, in order to tune this enzyme towards organo-phosphate substrates, a mutant obtained via directed evolution was examined. This mutant showed breakdown of the hexameric structure with the different oligomers (monomer and dimer) having higher catalytic activities as the highest weight oligomer. Further studies of this mutant were disregarded due to its poor stability. In the search for more potent bioremediation tools, a stable, highly active mutant of GpdQ was found. This mutant (Ser127Ala) has improved activity and good stability. It was shown by MCD of the Co(II) derivative that it also has improved metal binding abilities. To create potentially an economical and recyclable bioremediation system, this mutant was immobilized on PAMAM dendrimer modified magnetite nanoparticles. This system showed activity towards OPs, reusability and was stable over time.

Future work in this area could include optimization and testing of (mutant and wt) GpdQ on a range of pesticides and nerve agents. Also co-immobilization should be considered to generate multifunctional bioremediation tools. For example OpdA or OPH could be envisioned as co-immobilization partners. They are specialized for organophosphate triesters while GpdQ is mainly a phosphodiesterase.

A number of dinucleating ligands for Zn(II) complexes were synthesized which allowed the investigation of the impacts of ligand modifications on the phosphoesterase mechanism (CH_3H**L2**, NO_2H**L2**, BrH**L2** and CH_3H**L3**). It was found that *para*-substituents with lower electron withdrawing properties yielded more active complexes in OP hydrolysis. This is an important finding since model complexes are in general less active than enzymes, and chemists are continuously searching to improve model complex function. Also the nature of the nucleophile was investigated. Previously, Chen et al. had proposed that alkoxide ligands arms (CH_3H**L1**), when present, are the active nucleophiles in OP hydrolysis [1]. This

L. J. Daumann, *Spectroscopic and Mechanistic Studies of Dinuclear Metallohydrolases and Their Biomimetic Complexes*, Springer Theses, DOI: 10.1007/978-3-319-06629-5_9,

assertion has been disproved by examination of the pK_a values of a range of Zn(II) complexes. Furthermore, an ^{18}O-labelling experiment was conducted with a complex based on Chen's ligand and a methyl ether arm ligand reported in this thesis. It was shown, based on the incorporation of the ^{18}O-label using ^{31}P NMR, that in fact a terminal water molecule acts a nucleophile.

Two Cd(II) complexes with the ligands CO_2EtH_3**L1** and CO_2EtH**L2** were introduced as competent mimics for phosphoesterases. The kinetic properties like pK_a values were similar to the Cd(II) substituted GpdQ phosphoesterase enzyme. While Cd(II) is not necessarily a biologically relevant metal, the high activity of the complexes is intriguing. Also, with ^{113}Cd NMR, a method giving insight about coordination geometry in solution is readily available. One of the complexes based on an alkoxide arm ligand showed activity towards β-lactam substrates and is the first lactamase biomimetic known to hydrolyze penicillin and nitrocefin using one of the ligand's alcohol arms. One of the most significant findings of this thesis is that while complex and pH remain the same, the nucleophile utilized changes upon variation of the substrate. Furthermore, a blue intermediate was isolated in nitrocefin hydrolysis which was stable over a prolonged time at ambient temperature. As opposed to the enzyme-bound intermediates which have often ms lifetimes, this biomimetic intermediate could be conveniently analyzed.

Co(II) complexes based on five different ligands (CO_2EtH_3**L1**, CO_2EtH**L2**, CH_3H**L2**, NO_2H**L2** and BrH**L2**) were described. They were fully magnetically and spectroscopically characterized. Fitting the magnetism of dinuclear Co(II) systems is a challenging task. However, using the software MagSaki, parameters such as ZFS, J, or g-tensors could be obtained. The results were compared with VTVH MCD fits. It was shown that some complexes were antiferromagnetically coupled while others displayed a ferromagnetic coupling behavior. This thesis describes the first full study of both the kinetic and spectroscopic characterization of phosphoesterase Co(II) biomimetics. Also, similar to the Cd(II) complex of CO_2EtH_3**L1**, the Co(II) complex acted as a metallo-β-lactamase mimic using one of the ligands' alkoxide arms to hydrolyze the substrate.

Two ligands, mimicking the asymmetric coordination environment of GpdQ more accurately were synthesized and complexed with Zn(II). While in the crystal structure a 5,6-coordination sphere was retained, this is not the case for the solution structure. MCD measurements of the Co(II) derivative and mass spectral studies of the di-Zn(II) derivative underline this. The complexes were shown to be less active than substances based on symmetric ligands.

The final chapter deals with the immobilization of the systems on solid supports. Two supports were chosen due to their availability and favorable properties. Merrifield resin was used to immobilize NH_2H**L2** and CH_3H**L4** and their Zn(II) complexes generated. Successful immobilization was supported with IR, XPS and microanalysis measurements. The complexes were shown to be active towards organophosphates and provide promising systems for the use in filter systems.

The challenge is now to enhance activity of GpdQ and biomimetics towards commonly used pesticides. Given the importance of stability, the model complexes could be good alternatives to the enzyme systems. However, their activity needs to be enhanced towards 'real life substrates'. One option could be the incorporation of functional groups, capable of forming hydrogen bonds to the substrate. For example, amine residues could be used to mimic active site pocket interactions. These interactions might aid in substrate activation and active site regeneration. To generate artificial enzymes, the model complexes could also be incorporated into proteins, to make use of the beneficial hydrogen bond interactions present in those systems.

Reference

1. J.W. Chen, X.Y. Wang, Y.G. Zhu, J. Lin, X.L. Yang, Y.Z. Li, Y. Lu, Z.J. Guo, Inorg. Chem. **44**, 3422–3430 (2005)

The manufacturer's authorised representative in the EU is Springer Nature Customer Service Centre GmbH, Europaplatz 3, 69115 Heidelberg, Germany. If you have any concerns regarding our products, please contact ProductSafety@springernature.com

Printed and bound by CPI Group (UK) Ltd, Croydon, CR0 4YY

15/07/2026

02167632-0004